NEIL CURTIS

LONGMAN BIOLOGY HANDBOOK

living organisms in all forms
explained and illustrated

LONGMAN   YORK PRESS

YORK PRESS
Immeuble Esseily, Place Riad Solh, Beirut.

LONGMAN GROUP UK LIMITED
Longman House, Burnt Mill
Harlow, Essex CM20 2JE, England
and Associated Companies throughout the world.

© Librairie du Liban 1985

First published 1992

ISBN 0 582 08764 3

Illustrations by Charlotte Kennedy and Jane Cheswright
with Philip Corke and Brian Ainsworth
Photocomposed in Britain by Prima Graphics, Camberley, Surrey, England.

Produced by Longman Group (FE) Ltd
Printed in Hong Kong

Contents

How to use the handbook

This handbook contains over 1800 words used in the biological sciences. These are arranged in groups under the main headings listed on pp. 3–4. The entries are grouped according to the meaning of the words to help the reader to obtain a broad understanding of the subject.

At the top of each page the subject is shown in bold type and the part of the subject in lighter type. For example, on pp. 18 and 19:

18 · **THE CELL**/CARBOHYDRATES

THE CELL/CARBOHYDRATES · **19**

In the definitions the words used have been limited so far as possible to about 1500 words in common use. These words are those listed in the 'defining vocabulary' in the *New Method English Dictionary* (fifth edition) by M. West and J. G. Endicott (Longman 1976). Words closely related to these words are also used: for example, *characteristics*, defined under *character* in West's *Dictionary*.

1. To find the meaning of a word

Look for the word in the alphabetical index at the end of the book, then turn to the page number listed.

In the index you may find words with a letter or number at the end. These only occur when the same word appears twice in the handbook. [a] indicates a word which is defined as it relates to animals and [p] a word defined as it relates to plants. For example, **cone**

cone[a] is part of the retina of the eye;

cone[p] is a reproductive structure in some plants.

The numbers also indicate a word which is defined twice in different contexts. For example, **translocation**

translocation[1] is the transport of materials in plants;

translocation[2] is a kind of chromosome mutation.

The description of the word may contain some words with arrows in brackets (parentheses) after them. This shows that the words with arrows are defined near by.

(↑) means that the related word appears above or on the facing page;

(↓) means that the related word appears below or on the facing page.

A word with a page number in brackets after it is defined elsewhere in the handbook on the page indicated. Looking up the words referred to may help in understanding the meaning of the word that is being defined.

The explanation of each word usually depends on knowing the meaning of a word or words above it. For example, on p. 178 the meaning of *spore mother cell*, *microsporangium*, and the words that follow depends on the meaning of the word *spore*, which appears above them. Once the earlier words are understood those that follow become easier to understand. The illustrations have been designed to help the reader understand the definitions but the definitions are not dependent on the illustrations.

2. To find related words

Look in the index for the word you are starting from and turn to the page number shown. Because this handbook is arranged by ideas, related words will be found in a set on that page or one near by. The illustrations will also help to show how words relate to one another.

For example, words relating to principles of classification are on pp. 40–41. On p. 40 *classification* is followed by words used to describe taxonomy and the binomial system and illustrations showing the different taxa involved in the classification of a species and the binomial system; p. 41 continues to explain and illustrate classification, explaining natural and artificial classifications and illustrating the relationships between the major groups of organisms.

3. As an aid to studying or revising

The handbook can be used for studying or revising a topic. For example, to revise your knowledge of gas exchange, you would look up *gas exchange* in the alphabetical index. Turning to the page indicated, p. 112, you would find *respiration*, *respiratory quotient*, *breathing*, *gas exchange*, and so on; on p. 113 you would find *air*, *gill*, *gill filament*, and so on. Turning over to p. 114 you would find *counter current exchange system* etc.

In this way, by starting with one word in a topic you can revise all the words that are important to this topic.

4. To find a word to fit a required meaning

It is almost impossible to find a word to fit a meaning in most dictionaries, but it is easy with this book. For example, if you had forgotten the word for the outer whorl of the perianth of a flower, all you would have to do would be to look up *perianth* in the alphabetical index and turn to the page indicated, p. 179. There you would find the word *calyx* with a diagram to illustrate its meaning.

5. Abbreviations used in the definitions

abbr.	abbreviated as	p.	page
adj	adjective	pl.	plural
e.g.	*exempli gratia* (for example)	pp.	pages
etc.	*et cetera* (and so on)	sing.	singular
i.e.	*id est* (that is to say)	v	verb
n	noun	=	the same as

THE
HANDBOOK

cell theory an idea, developed in 1839, by
Theodore Schwann, which states that all living
organisms are made up of individual cells and
that it is in these cells and by their division that
processes such as growth and reproduction
(p. 173) take place.

cell (*n*) the basic unit of a plant or animal. It is an
individual, usually microscopic (↓) mass of
living matter or protoplasm (p. 10). An animal
cell consists of a nucleus (p. 13), which
contains the chromosomes (p. 13), the
cytoplasm (p. 10) which is usually a viscous
fluid or gel surrounded by a very thin skin, the
plasma membrane (p. 13). A plant cell is similar
except that it is surrounded by a cellulose (p. 19)
cell wall (↓) and has a fluid-filled vacuole (p. 11).

cell wall the non-living external layer of a cell in
plants. It is comparatively rigid but slightly
elastic and provides support for the cell. There
may be a primary cell wall (p. 14) composed of
cellulose (p. 19) and calcium pectate and, in
older plants, a secondary cell wall (p. 14) made
of layers of cellulose containing other
substances, such as the woody lignin (p. 19).

organelle (*n*) any part of a cell, such as the
nucleus (p. 13) or flagellum (p. 12), that has a
particular and specialized function.

prokaryote (*n*) a cell in which the chromosomes
(p. 13) are free in the cytoplasm (p. 10) and not
enclosed in a membrane (p. 14): there is no
nucleus. Bacteria (p. 42) and blue-green algae
(p. 43) are prokaryotes.

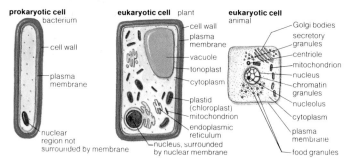

prokaryotic cell
bacterium
— cell wall
— plasma membrane
nuclear region not surrounded by membrane

eukaryotic cell plant
— cell wall
— plasma membrane
— vacuole
— tonoplast
— cytoplasm
— plastid (chloroplast)
— mitochondrion
— endoplasmic reticulum
— nucleus, surrounded by nuclear membrane

eukaryotic cell animal
— Golgi bodies
— secretory granules
— centriole
— mitochondrion
— nucleus
— chromatin granules
— nucleolus
— cytoplasm
— plasma membrane
— food granules

optical microscope

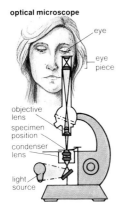

eye

eye piece

objective lens

specimen position

condenser lens

light source

electron microscope

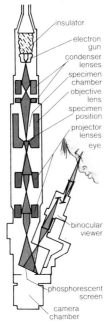

insulator

electron gun

condenser lenses

specimen chamber

objective lens

specimen position

projector lenses

eye

binocular viewer

phosphorescent screen

camera chamber

eukaryote (*n*) a cell in which the nucleus (p. 13) is separated from the cytoplasm (p. 10) by a nuclear membrane (p. 13). All organisms, except bacteria (p. 42) and blue-green algae (p. 43) are composed of eukaryotic cells.

unicellular (*adj*) of an organism consisting of one cell only.

multicellular (*adj*) of an organism composed of many cells.

cytology (*n*) the study or science of cells and their activities.

microscopy (*n*) the study, using a microscope (↓), of organisms too small to be seen with the naked eye.

microscope (*n*) an instrument used to give a magnified image of an object that is too small to be seen with the naked eye.

optical microscope a microscope (↑) in which light is passed through the object to be enlarged and passed to the eye through an objective lens system and an eyepiece. This instrument can magnify an object by a maximum of about 1500 times. For larger magnifications, an electron microscope (↓) must be used.

electron microscope a microscope (↑) which can be used to magnify objects by greater than 1500 times and to as much as 500 000 times by using electrons, which have a smaller wavelength than light, to examine the object.

ultrastructure (*n*) the structure of an object which can only be resolved using an electron microscope (↑).

sectioning (*n*) the cutting of an extremely thin slice of tissue (p. 83) which can then be examined using a microscope (↑). The tissue is first frozen or embedded in a material such as paraffin wax before it is cut. **section** (*n*).

microtome (*n*) an instrument used to cut very thin slices of a material.

staining (*n*) a method of examining particular structures inside cells by making parts of the cells opaque to light or electrons using chemicals. Certain kinds of staining materials will stain different structures e.g. iodine stains starch (p. 18).

centrifugation (*n*) a method of separating substances of different densities by accelerating them, usually in a rotating container (centrifuge), for quite long periods. Cells may be broken open and suspended in a liquid before centrifugation so that, after centrifugation, the solid particles, the sediment, will fall to the bottom of the container while a supernatant fluid will be left behind above the sediment

dialysis (*n*) a method of separating small molecules from larger molecules in a mixed solution (p. 118) by separating the solution from water by a membrane (p. 14) through which the small molecules will diffuse (p. 119) leaving behind the larger molecules which are too big to pass through the membrane.

chromatography (*n*) a method of separating mixtures of substances, such as amino acids (p. 21), by making a solution (p. 118) of the substances and allowing the substances to be absorbed (p. 81) and flow through a medium such as paper. The different substances will travel at different rates and so be separated.

chromatogram (*n*) the column or strip of solid on which substances have been separated by chromatography (↑).

electrophoresis (*n*) a method of separating mixtures of substances by suspending them in water and subjecting them to an electrical charge. Different substances will move in different directions and at different rates in response to the charge.

protoplasm (*n*) the contents of a cell.

cytoplasm (*n*) all the protoplasm (↑), or material, inside a cell other than the nucleus (p. 13), which can be thought of as alive. It is usually a viscous fluid or gel containing other organelles (p. 8), such as the Golgi body (↓). **cytoplasmic** (*adj*).

ribosome (*n*) a particle of protein (p. 21) and RNA (p. 24) which is contained in the cytoplasm (↑). Under the control of the DNA (p. 24) in the nucleus (p. 13), protein is produced on the ribosomes by linking together amino acids (p. 21). Ribosomes often occur in groups or chains.

dialysis

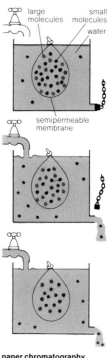

large molecules   small molecules

water

semipermeable membrane

paper chromatography

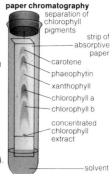

separation of chlorophyll pigments

strip of absorptive paper

carotene

phaeophytin

xanthophyll

chlorophyll a

chlorophyll b

concentrated chlorophyll extract

solvent

endoplasmic reticulum

rough endoplasmic reticulum
ribosomes
cell matrix

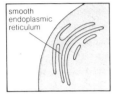

smooth endoplasmic reticulum

mitochondrion

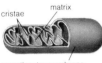

cristae
matrix
smooth outer membrane

vacuole

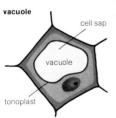

cell sap
vacuole
tonoplast

endoplasmic reticulum a meshwork of parallel, interconnected cavities within the matrix (p. 88) of a cell. These are bounded by unit membranes (p. 14) which are continuous with the nuclear membrane (p. 13). **ER** (*abbr*).

ER = endoplasmic reticulum (↑).

rough ER ER (↑) which is covered on the cytoplasmic (↑) side with ribosomes (↑).

smooth ER ER (↑) with no ribosomes (↑).

Golgi body a group or groups of flattened cavities within the cytoplasm (↑) of a cell bounded by membranes (p. 14) and connected with the ER (↑). It is similar to smooth ER (↑) but may be used for linking carbohydrates (p. 17) to proteins (p. 21), and it is associated with secretion (p. 106).

mitochondria (*n.pl.*) rod-shaped bodies in the cytoplasm (↑) of a cell. They are bounded by two unit membranes (p. 14) of which the inner is folded inwards into crests or cristae. Cell respiration (p. 30) and energy production take place in these bodies and there are more of them in cells that use a lot of energy.

lysosomes (*n.pl.*) spherical bodies that occur in the cytoplasm (↑) of cells. They are bounded by membranes (p. 14) and contain enzymes (p. 28) which may be released to destroy unwanted organelles (p. 8) or even whole cells.

microtubule (*n*) a fibrous (p. 143) structure made of protein (p. 21) found in the cytoplasm (↑). They may occur singly or in bundles. Their function may be cellular transport, e.g. the spindle (p. 37) fibres in nuclear division (p. 35).

microfilament (*n*) a very fine, thread-like structure made of protein (p. 21) which occurs in the cytoplasm (↑) of most cells.

fibril (*n*) a small fibre (p. 143) or thread-like structure.

vacuole (*n*) a droplet of fluid bounded by a membrane (p. 14) or tonoplast (↓) and contained within the cells of plants and animals except bacteria (p. 42) and blue-green algae (p. 43).

tonoplast (*n*) the inner plasma membrane (p. 13) of a cell in plants which separates the vacuole (↑) from the cytoplasm (↑).

protoplast (*n*) the protoplasmic (↑) material between the tonoplast (↓) and the plasma membrane (p. 13).

cell sap fluid contained in a plant vacuole (p. 11).
plastid (*n*) in plants, except bacteria (p. 42), blue-
green algae (p. 43), and fungi (p. 46), a
membrane (p. 14) bounded body in the
cytoplasm (p. 10) which contains DNA (p. 24),
pigments (p. 126), and food reserves.
chloroplast (*n*) in plants only, the plastid (↑)
containing chlorophyll (↓) and the site of
photosynthesis (p. 93). It is always bounded by
a double unit membrane (p. 14).

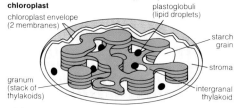

chloroplast
chloroplast envelope
(2 membranes)
plastoglobuli
(lipid droplets)
starch
grain
stroma
granum
(stack of
thylakoids)
intergranal
thylakoid

chlorophyll (*n*) a green pigment (p. 126) found in
the chloroplasts (↑) of plants which is important
in photosynthesis (p. 93). There are two forms
of chlorophyll; chlorophyll *a* and chlorophyll *b*.
leucoplast (*n*) a colourless plastid (↑) e.g. starch
(p. 18) grains.
stroma (*n*) the matrix (p. 88) within a chloroplast
(↑) containing starch (p. 18) grains and enzymes
(p. 28).
grana (*n.pl.*) disc-shaped, flattened vesicles (↓) in
the stroma (↑) of a chloroplast (↑) holding the
chorophyll (↑). **granum** (*sing.*).
vesicle (*n*) a thin-walled drop-like structure or a
cavity containing fluid.
lamella (*n*) a thin plate-like structure. **lamellae**
(*pl.*).
cilium (*n*) in animals and a few plants, a fine
thread which projects from the surface of a cell
and moves the fluid surrounding it by a beating
or rowing action. **cilia** (*pl.*).
flagellum (*n*) a fine, long thread which projects
from the surface of a cell and moves with an
undulating action. In bacteria, the flagellum
provides locomotion (p. 143) by a whip-like
action during part of their life history. It is longer
than a cilium (↑) **flagella** (*pl.*).

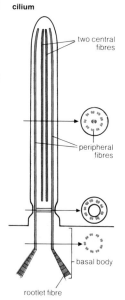

cilium
two central
fibres
peripheral
fibres
basal body
rootlet fibre

basal body a tiny, rod-shaped body situated at the base of a cilium (↑) or flagellum (↑) composed of nine fibrils (p. 11) arranged in a ring at the edge of the cilium. There are also two central fibrils which do not form part of the basal body.

microvilli (*n.pl.*) finger-like projections from the surface of the plasma membrane (↓) of a cell which improve the absorption (p. 81) powers of the cell by increasing its surface area. *See also* villi (p. 103).

nucleus

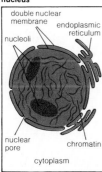

double nuclear membrane
endoplasmic reticulum
nucleoli
nuclear pore
chromatin
cytoplasm

nucleus (*n*) a body present within the cells of eukaryotic (p. 9) organisms which contains the chromosomes (↓) of the organism. **nuclear** (*adj*).

nuclear membrane the firm, double unit membrane (p. 14) surrounding the nucleus (↑) and separating it from the cytoplasm (p. 10) while allowing the exchange of materials between the nucleus and the cytoplasm through its pores (p. 120).

nucleolus (*n*) a small, round dense body, one or two of which may be present within the nucleus (↑). It is rich in RNA (p. 24) and protein (p. 21) but is not contained within a membrane (p. 14).

chromosome (*n*) a rod- or thread-shaped body occurring within the nucleus (↑) and which is readily stained by various dyes, hence its name. A chromosome is composed of DNA (p. 24) or RNA (p. 24), and protein (p. 21). Each chromosome is in the form of a long helix (p. 25) of DNA. Chromosomes mostly occur in pairs called homologous chromosomes (p. 39). They are composed of thousands of genes (p. 196) which give rise to and control particular characteristics and functions of the organism, such as eye colour, and which are passed on through the offspring by inheritance (p. 196). Each organism has a consistent number of chromosomes, e.g. in human cells, 23 pairs.

chromatin (*n*) a granular compound of nucleic acid (p. 22) and protein (p. 21) in the chromosome (↑) and which is strongly stained by certain dyes.

plasma membrane an extremely thin membrane (p. 14) separating the cell from its surroundings. It allows the transfer of substances between the cell and its surroundings.

plasmalemma (*n*) the plasma membrane (p. 13) or cell membrane (↓).

unit membrane the common structure, divided into three layers, of the plasma membrane (p. 13), or other membranes, such as the endoplasmic reticulum (p. 11). It comprises a monomolecular film (↓) and a bimolecular leaflet (↓).

monomolecular film a layer, one molecule thick, of protein (p. 21) which occurs either side of the bimolecular leaflet (↓) and forms part of the organization of the unit membrane (↑). Stained and under an electron microscope (p. 9), it appears as a dark stratum (↓).

bimolecular leaflet a layer, two molecules thick, of lipid (p. 20) which is found between two monomolecular films (↑) and forms part of the organization of the unit membrane (↑). Stained and under an electron microscope (p. 9) it appears as a light stratum (↓).

stratum (*n*) a layer. **strata** (*pl.*).

phagocytosis (*n*) the process in which a cell flows around particles in its surroundings and takes them into the cytoplasm (p. 10) to form a vacuole (p. 11).

pinocytosis (*n*) a process in which a cell folds back within itself and surrounds a tiny drop of fluid in its surroundings and takes it into the cytoplasm (p. 10) to form a vesicle (p. 12).

middle lamella in plants, the material which is laid down between adjacent cell walls (p. 8) and sticks the cells together. It is laid down as new cells form.

primary cell wall the first cell wall (p. 8) of a young cell which is laid down as the new cell forms. *See also* secondary cell wall (↓).

secondary cell wall a cell wall (p. 8) which is laid down inside the primary cell wall (↑). It surrounds some of the cells in older plants.

pit (*n*) a small area of the secondary cell wall (↑) which has remained almost unthickened or absent during the formation of the secondary wall. It allows substances to pass between the cells. The pits in one cell correspond in position with the pits in a neighbouring cell.

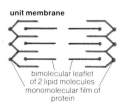

unit membrane

bimolecular leaflet of 2 lipid molecules
monomolecular film of protein

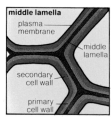

middle lamella

plasma membrane
middle lamella
secondary cell wall
primary cell wall

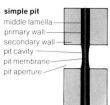

simple pit

middle lamella
primary wall
secondary wall
pit cavity
pit membrane
pit aperture

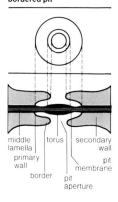

bordered pit

middle lamella
primary wall
border
torus
pit aperture
secondary wall
pit membrane

plasmodesmata

plasmodesma comprising
cytoplasm and tube of
endoplasmic reticulum

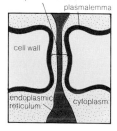

plasmalemma

cell wall

endoplasmic
reticulum

cytoplasm

**hydrogen bond between
water molecules**

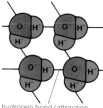

hydrogen bond (attraction
between positive hydrogen
atom and negative oxygen
atom)

plasmodesmata (*n.pl.*) fine threads of cytoplasm (p. 10) which connect the cytoplasm of neighbouring cells, and may be grouped through the membranes (↑) of pits (↑). Plasmodesmata run through narrow pores (p. 120) in the cellulose (p. 19) cell wall (p. 8). **plasmodesma** (*sing.*).

biochemistry (*n*) the study or science of the chemical substances and their reactions in animals and plants.

organic compound any substance which is a compound of carbon, except for the oxides and carbonates of carbon, and from which all living things are made. Oxygen and carbon are the main components of organic compounds.

inorganic compound a compound which, except for the oxides and carbonates, does not contain carbon, and which is not an organic compound (↑). Salt is an example of an inorganic compound.

hydrogen bond a bond which holds one molecule of water to another molecule making water more stable than it otherwise would be. A molecule of water consists of two hydrogen atoms bonded to one oxygen atom by sharing electrons. The resulting molecule is weakly polar with hydrogen atoms being positively charged and the oxygen negatively charged. These polar molecules are weakly attracted to one another.

acid (*n*) a substance that releases hydrogen (H^+) ions in a watery solution (p. 118) or accepts electrons in chemical reactions. An acid can be an inorganic compound (↑), such as hydrochloric acid, HCl or an organic compound (↑) such as ethanoic acid, CH_3COOH. The acidity of a solution can be measured on the pH scale ($-logH^+$ concentration). **acidic** (*adj*).

base[1] (*n*) a substance that releases hydroxyl (OH) ions in a watery solution (p. 118) or gives up electrons in chemical reactions, e.g. sodium hydroxide, NaOH. **basic** (*adj*).

pH *see* acid (↑).

buffer (*n*) a substance which helps a solution (p. 118) to resist a change in pH (↑) when an acid (↑) or base (↑) is added to the solution. Many biological fluids function as buffers.

condensation example of a condensation reaction

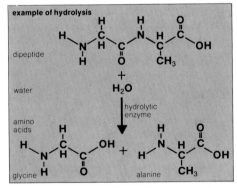

condensation (*n*) a reaction whereby two simple organic compounds (p. 15), such as glucose (↓) and fructose (↓), combine to form another compound, such as sucrose (p. 18) and a molecule of water.

hydrolysis (*n*) a reaction in which water combines with an organic compound (p. 15), such as sucrose (p. 18), to form two new organic compounds, such as glucose (↓) and fructose (↓). The reverse of condensation (↑).

molecular biology the study or science of the structure and activities of the molecules which make up animals and plants.

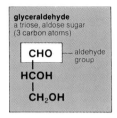

glyceraldehyde
a triose, aldose sugar
(3 carbon atoms)

CHO —— aldehyde group
HCOH
CH₂OH

ribose
a pentose sugar

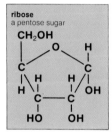

glucose

fructose

carbohydrate (*n*) an organic compound (p. 15) containing the elements carbon, hydrogen, and oxygen with the general formula $(CH_2O)_n$. Carbohydrates are essential in the metabolism (p. 26) of all living things.

monosaccharide (*n*) a carbohydrate (↑) composed of small molecules. Monosaccharides are the building blocks from which disaccharides (p. 18) and polysaccharides (p. 18) are built. Common monosaccharides found in cells contain from three to seven carbon atoms. A monosaccharide is the simplest sugar and, if further broken down, ceases to be a sugar.

sugar (*n*) the simplest carbohydrate (↑), a mono-, di- or polysaccharide (p. 18).

triose sugar a monosaccharide (↑) in which n for the general formula of the carbohydrate (↑) is 3. Glyceraldehyde is a triose sugar with the formula $C_3H_6O_3$.

pentose sugar a monosaccharide (↑) in which n for the general formula of the carbohydrate (↑) is 5. Ribose (p. 22) is a pentose sugar with the formula $C_5H_{10}O_5$.

hexose sugar a monosaccharide (↑) in which n for the general formula of the carbohydrate (↑) is 6. Glucose (↓) is a hexose sugar with the formula $C_6H_{12}O_6$. The atoms of hexose sugars may be arranged differently to give different types of sugars e.g. glucose and fructose (↓).

glucose (*n*) a hexose sugar (↑) which is widely found in animals and plants. Glucose provides a major source of energy in living things by being oxidized (p. 32) during respiration (p. 112) into carbon dioxide and water, releasing energy. In plants it is the product of photosynthesis (p. 93) and is stored as starch (p. 18) while in animals it is produced by the digestion (p. 98) of disaccharides (p. 18) and polysaccharides (p. 18) and is stored as glycogen (p. 19). Glucose combines with fructose (↓) to form sucrose (p. 18) by condensation (↑).

fructose (*n*) a hexose sugar (↑) which is widely found in plants. It combines with glucose (↑) to form sucrose (p. 18) by condensation (↑).

galactose (*n*) a hexose sugar (p. 17) which is a constituent of lactose (↓) and is found in many plant polysaccharides (↓) as well as in animal protein (p. 21)-polysaccharide combinations.

disaccharide (*n*) a carbohydrate (p. 17) which results from the combination of two monosaccharides (p. 17) by condensation (p. 16), e.g. maltose (↓) and sucrose (↓).

disaccharide
e.g. sucrose

CH₂OH

glucose + fructose

hydrolysis

maltose (*n*) a disaccharide (↑) which is formed from the condensation (p. 16) of two molecules of glucose (p. 17). It is a product of the breakdown of starch (↓) during germination (p. 168) in plants, and digestion (p. 98) in animals. Also known as **malt sugar.**

sucrose (*n*) a disaccharide (↑) which is a compound of one molecule of glucose (p. 17) and one molecule of fructose (p. 17). It is widespread in plants but not in animals. Also known as **cane sugar**.

lactose (*n*) a disaccharide (↑) which is a compound of one molecule of glucose (p. 17) and one molecule of galactose (↑). It occurs in the milk of mammals (p. 80). Also known as **milk sugar**.

polysaccharide (*n*) a carbohydrate (p. 17) which results from the combination of more than two monosaccharides (p. 17) by condensation (p. 16). A polysaccharide has the general formula $(C_6H_{10}O_5)_n$.

starch (*n*) a polysaccharide (↑) which forms one of the main food reserves of green plants. It is found in the leucoplasts (p. 12). It stains blue-black with iodine.

polysaccharide
e.g. starch (amylopectin)

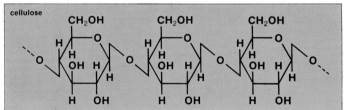

cellulose microfibrils in
surface view of plant cell
wall (×24,000)

glycogen (*n*) a polysaccharide (↑) stored by
animals and by fungi (p. 46). It is made up of
many glucose (p. 17) molecules. In vertebrates
(p. 74) it is present in large quantities in the liver
(p. 103) and muscles (p. 143).

cellulose (*n*) a long-chain polysaccharide (↑)
made up of units of glucose (p. 17). It is used
for structural support and is the main
component of the cell wall (p. 8) in plants.

cellulose

lignin (*n*) a complex organic compound (p. 15)
whose structure is not fully understood. With
cellulose (↑) it forms the chief components of
wood in trees. It is laid down in the cell walls
(p. 8) of sclerenchyma (p. 84), xylem (p. 84)
vessels, and tracheids (p. 84). It stains red with
acidified phloroglucinol. **lignified** (*adj*).

lipid (*n*) any of a number of organic compounds
(p. 15) found in plants and animals with very
different structures but which are all insoluble in
water and soluble in substances like ethoxyethane
(ether) and trichloromethane (chloroform). It
is formed by the condensation (p. 16) of glycerol
(↓) and fatty acids (↓). Lipids have a variety of
functions including storage, protection, insulation,
waterproofing, and as a source of energy.

fat (*n*) a lipid (↑) formed from the alcohol glycerol
(↓) and one or more fatty acids (↓). It is solid at
room temperature.

oil (*n*) a lipid (↑) formed from the alcohol glycerol
(↓) and one or more fatty acids (↓). It is liquid at
room temperature.

glycerol (*n*) an alcohol with the formula $C_3H_8O_3$
which is formed by the hydrolysis (p. 16) of a
fat. It is a sweet, sticky, odourless, colourless
liquid. Its modern name is propane-1,2,3,-triol.

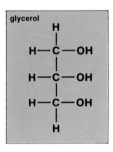

glycerol

fatty acid an organic acid (p. 15) with the general
formula $(R(CH_2)_nCOOH)$ which can be united
with glycerol (↑) by condensation (p. 16) to give
a lipid (↑). In living organisms, fatty acids
usually have unbranched chains and an even
number of carbon atoms.

triglyceride (*n*) the major component of animal
and plant lipids (↑). It is derived from glycerol (↑)
which has three reactive hydroxyl groups, by
condensation (p. 16) with three fatty acids (↑).

phospholipid (*n*) a lipid (↑) which contains a
phosphate group as an essential part of the
molecule. It is derived from glycerol (↑) attached
to two fatty acids (↑), a phosphate group, and a
nitrogenous base. Phospholipids are essential
components of cell membranes (p. 14).

saturated (*adj*) of a carbon chain, such as that in
a fatty acid (↑), in which each carbon atom is
attached by single bonds to carbons, hydrogen
atoms, or other groups. It is unreactive.

unsaturated (*adj*) of a carbon chain, such as that
in a fatty acid (↑), in which carbon atoms are
attached to other groups with at least one
double or triple bond. An unsaturated fatty acid
is reactive and may be essential to maintain a
vital structure or function in an organism.

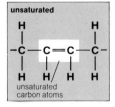

unsaturated

unsaturated
carbon atoms

primary, secondary, tertiary and quaternary structure of proteins

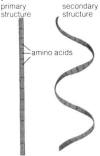

primary structure

secondary structure

amino acids

tertiary structure

quaternary structure

peptide bond between amino acids

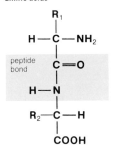

R_1 and R_2 are side groups

steroid (*n*) a complex, saturated (↑) hydrocarbon in which the carbon atoms are arranged in a system of rings. All steroids are chemically similar but may have very different functions in organisms. The most common steroid in animals is cholesterol.

protein (*n*) a very complex organic compound (p. 15) made up of large numbers of amino acids (↓). Proteins make up a large part of the dry weight of all living organisms.

amino acid an organic compound (p. 15) with an amino group of atoms (-NH_2) and acidic (p. 15) carboxyl (-COOH) groups of atoms on the molecule. The general formula is $RCHNH_2COOH$ with R representing a hydrogen or carbon chain. There are more than twenty naturally occurring amino acids with different R groups. Hundreds of thousands of amino acids are linked together to form a protein (↑). *See also* the diagram, amino acids and the genetic code on p. 204.

dipeptide (*n*) an organic compound (p. 15) which results from linking together two amino acids (↑) by condensation (p. 16).

polypeptide (*n*) an organic compound (p. 15) which results from linking together many amino acids (↑) by condensation (p. 16). In turn, polypeptides may be linked together to form proteins (↑).

peptide bond the link which joins one amino acid to the carboxyl (-COOH) group of another, resulting in the formation of a dipeptide (↑) or polypeptide (↑). A peptide bond can only be broken by the action of a hot acid (p. 15) or alkali.

conjugated protein a protein (↑) which occurs in combination with a non-protein or prosthetic group (p. 30). Haemoglobin (p. 126) is an example of a conjugated protein.

globular protein a protein (↑) which, because of the positive and negative charge on it, forms a complex three-dimensional structure as the opposite charges are attracted together and form weak bonds. A hormone (p. 130) is an example of a globular protein.

fibrous protein a protein (p. 21) which occurs as long parallel chains with cross links. Fibrous proteins are insoluble and are used for support and other structural purposes. Keratin in hair, hooves, feathers etc is an example of a fibrous protein.

colloid (*n*) a substance, such as starch (p. 18), that will not dissolve or be suspended in a liquid but which is dispersed in it.

nucleic acid a large, long-chain molecule composed of chains of nucleotides (↓) and found in all living organisms. The carrier of genetic (p. 196) information.

nucleotide (*n*) an organic compound (p. 15) formed from ribose (↓), phosphoric acid (↓), and a nitrogen base (↓).

ribose (*n*) a monosaccharide (p. 17) or pentose sugar (p. 17) which forms an essential part of a nucleotide (↑).

deoxyribose (*n*) a monosaccharide (p. 17) with one less oxygen than ribose (↑).

phosphoric acid an inorganic compound (p. 15) with the formula H_3PO_4 which forms an essential part of nucleotides (↑). The phosphate molecule from phosphoric acid forms a bridge between two pentose (p. 17) molecules.

base[2] (*n*) a substance, such as a purine (↓) or pyrimidine (↓), containing nitrogen, which is attached to the main sugar-phosphate chain in a nucleic acid (↑).

cytosine (*n*) a nitrogen base (↑) derived from pyrimidine (↓) and found in both ribonucleic acid (p. 24) and deoxyribonucleic acid (p. 24).

uracil (*n*) a nitrogen base (↑) derived from pyrimidine (↓) and found only in ribonucleic acid (p. 24).

adenine (*n*) a nitrogen base (↑) derived from purine (↓) and found in both ribonucleic acid (p. 24) and deoxyribonucleic acid (p. 24).

guanine (*n*) a nitrogen base (↑) derived from purine (↓) and found in both ribonucleic acid (p. 24) and deoxyribonucleic acid (p. 24).

thymine (*n*) a nitrogen base (↑) derived from pyrimidine (↓) and found only in deoxyribonucleic acid (p. 24).

nucleotide basic structure

the common bases in the nucleotides of DNA and RNA

	purines	pyrimidines
DNA only		O=C, HN, C—CH₃, C=CH, O=C, N, H *thymine*
DNA and RNA	NH₂, N=C, N, C, CH *adenine*; HC, N, C, NH; O=C, HN, C, N, CH *guanine*; H₂N, N, C, NH	NH₂, N=C, CH, C, CH *cytosine*; O=C, N, H
RNA only		O=C, HN, C, CH, C, CH *uracil*; O=C, N, H

pyrimidine (*n*) an organic compound (p. 15) with
the basic formula $C_4H_4N_2$ and with a cyclic
structure from which important nitrogen bases
(↑) are derived.

pyrimidine base any of the several compounds
related to pyrimidine (↑) and present in nucleic
acids (↑).

purine (*n*) an organic compound (p. 15) with the
basic formula $C_5H_4N_5$, with a double cyclic
structure, from which important nitrogen bases
(↑) are derived.

purine base any of several compounds related to
purine (↑) and present in nucleic acids (↑).

basic molecular shape of nitrogen base

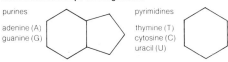

purines
adenine (A)
guanine (G)

pyrimidines
thymine (T)
cytosine (C)
uracil (U)

RNA ribonucleic acid. A nucleic acid (p. 22) consisting of a large number of nucleotides (p. 22) arranged to form a single strand. The base (p. 22) in each nucleotide is one of cytosine (p. 22), uracil (p. 22), adenine (p. 22), or guanine (p. 22). The sugar is ribose (p. 22). RNA is found in the nucleus (p. 13) of a cell and in the cytoplasm (p. 10). It usually occurs as ribosomes (p. 10) but also as *transfer RNA* and *messenger RNA*. Strands of RNA are produced in the nucleus from DNA (↓), passed to the cytoplasm, and then a ribosome is joined to the RNA. The ribosome moves along the strand of RNA and produces a polypeptide (p. 21) whose structure is controlled by the RNA. *See also* transcription and translation p. 205.

structure of portion of RNA molecule

phosphate

ribose — nitrogen base

phosphate

ribose — nitrogen base

phosphate

ribose — nitrogen base

DNA deoyxribonucleic acid. A nucleic acid (p. 22) consisting of a large number of nucleotides (p. 22) arranged to form a single strand. Usually, two strands are coiled round each other to form a double helix (↓). The base (p. 22) in each nucleotide consists of one of cytosine (p. 22), adenine (p. 22), guanine (p. 22), or thymine (p. 22). The sugar is deoxyribose (p. 22). DNA is found in the chromosomes (p. 13) of prokaryotes (p. 8) and eukaryotes (p. 9) and in the mitochondria (p. 11) of eukaryotes. It is the material of inheritance (p. 196) in almost all living organisms and is able to copy itself during nuclear divisions (p. 35).

structure of part of DNA molecule with helix unwound

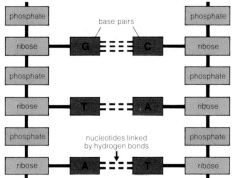

diagram of the DNA double helix

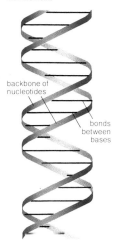

backbone of nucleotides

bonds between bases

polynucleotide chain a chain of linked nucleotides (p. 22) which makes up a nucleic acid (p. 22).

Watson-Crick hypothesis a hypothesis (p. 235) based on X-ray crystallography which suggests that DNA (↑) is a double helix (↓) of two coiled chains of alternating phosphate and sugar groups with the sugars linked by pairs of bases (p. 22).

double helix the arrangement of two helical (↓) polynucleotide chains (↑) in DNA (↑).

helix a helix is the curve that results from drawing a straight line on a plane which is then wrapped round a circular cylinder. The two helixes of DNA (↑) intertwine to form a double helix (↑) and are linked by nitrogen bases (p. 22). **helical** (adj).

base pairing the links holding the double helix (↑) of DNA (↑) together, each link consisting of a purine (p. 22) linked to a pyrimidine (p. 22) by hydrogen bonds (p. 15).

vitamin (n) the name given to a variety of organic compounds (p. 15) which are required by organisms for metabolism (p. 26) and which cannot usually be synthesized by the organism in sufficient quantities to replace that which is broken down during metabolism. See p. 238.

Benedict's test a method to determine the presence of monosaccharides (p. 17) and some disaccharides (p. 18) by adding a solution (p. 118) of copper sulphate, sodium citrate, and sodium carbonate to a solution of the sugar which produces a red precipitate (p. 26) when boiled because the sugar reduced the copper sulphate to copper (I) oxide. Sucrose (p. 18) and other non-reducing sugars do not reduce copper sulphate but it can be detected by hydrolysing (p. 16) it first into its component reducing sugars.

Fehling's test this is similar to Benedict's test (↑) but the reagent (p. 26) used is a solution (p. 118) containing copper sulphate, sodium potassium tartrate, and sodium hydroxide.

iodine test a method to determine the presence and distribution of starch (p. 18) in cells by cutting a thin section (p. 9) of the material and mounting it in iodine dissolved in potassium iodide. The starch grains turn blue-black.

emulsion test a method of testing for the presence of a lipid (p. 20) by dissolving the substance in alcohol (usually ethanol) and adding an equal volume of water. A cloudy white precipitate (↓) indicates a lipid.

alcohol/water test = emulsion test (↑).

Sudan III test a method of testing for a lipid (p. 20) which stains red with Sudan III solution.

greasemark test a method of testing for a lipid (p. 20) by taking a drop of the substance to be tested and placing it on a filter paper. When it is dry, only a lipid leaves a translucent mark when held up to the light.

translucent (*adj*) of a material that lets light pass through but through which objects cannot be seen clearly.

Millon's test a method of testing for protein (p. 21) by adding a few drops of Millon's reagent (↓) to a suspension of the protein and boiling it. The protein stains brick red.

Biuret test a method of testing for protein (p. 21) by adding an equal volume of 2 per cent sodium hydroxide solution (Biuret A) followed by 0.5 per cent copper sulphate solution (Buiret B). The protein stains purple.

emulsion (*n*) a colloidal (p. 22) suspension of one liquid in another.

suspension (*n*) a mixture in which the particles of one or more substances are distributed in a fluid.

fluid (*n*) a substance which flows, i.e. a liquid or a gas.

precipitate (*n*) an insoluble solid formed by a reaction which occurs in solution (p. 118).

reagent (*n*) a substance or solution (p. 118) used to produce a characteristic reaction in a chemical test.

metabolism (*n*) a general name for the chemical reactions which take place within the cells of all living organisms.

metabolite (*n*) any of the substances, inorganic (p. 15) or organic (p. 15) such as water or carbon dioxide, amino acids (p. 21) or vitamins (p. 25) which take part in metabolism (↑).

metabolic pathway a series of small steps in which metabolism (↑) proceeds.

metabolism chemical reations in a plant cell

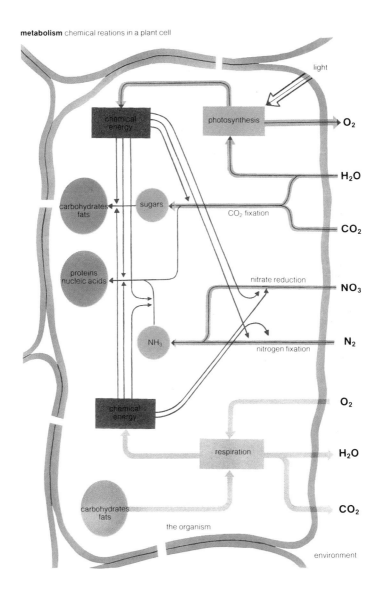

the function of enzymes in catalysis of reactions

synthesis

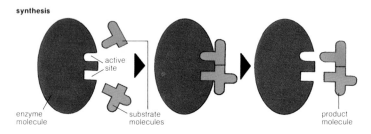

enzyme molecule · active site · substrate molecules · product molecule

enzyme (*n*) a protein (p. 21) which increases the
rate at which the chemical processes of
metabolism (p. 26) take place without being
used up by the reaction which it affects.
Enzymes are present in all living cells. They are
easily destroyed by high temperatures
(denatured) and require certain conditions
before they will act. The rate of an enzyme-
catalyzed (↓) reaction depends upon the
concentration of substrate (↓) and enzyme
temperature and pH (p. 15). Enzymes increase
the rate of reactions by lowering the activation
energy.

intracellular (*adj*) within a cell. For example, most
enzyme (↑) activity is intracellular i.e. takes
place within the cell that produces the enzymes.

extracellular (*adj*) outside a cell. For example,
digestive (p. 98) enzymes (↑), which are
extracellular in their activity may be secreted
(p. 106) into the gut (p. 98) of an animal from
other cells where they are produced.

in vivo 'in life' (*adj*) of all the processes which
take place within the living organism itself.

in vitro 'in glass' (*adj*) of processes, such as the
culture of cell tissues (p. 83) which are carried
out experimentally outside the living organism,
and originally derived from experiments carried
out on parts of an organism in a test tube.

catalyst (*n*) any substance, such as an enzyme
(↑), which increases the rate at which a
chemical reaction takes place but which is not
consumed by the reaction. **catalyze** (*v*).

breakdown

enzyme
molecule

active
site

substrate
molecule

product
molecules

enzymes control of reaction rate through substrate concentration

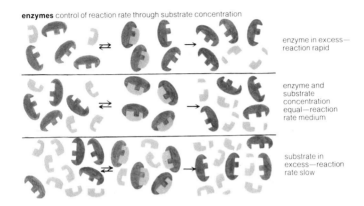

enzyme in excess—reaction rapid

enzyme and substrate concentration equal—reaction rate medium

substrate in excess—reaction rate slow

substrate (*n*) the substance on which something acts e.g. most enzymes (↑) only work on one substrate each and become attached to the substrate molecules.

active site that part of the enzyme (↑) molecule to which specific substrate (↑) molecules become attached.

enzyme-substrate complex the combination of the enzyme (↑) molecule with the substrate (↑) molecule.

lock and key hypothesis a hypothesis (p. 235) which explains the properties of enzymes (↑) by supposing that the particular shape of an enzyme protein (p. 21) corresponds with the shape of particular molecules like a lock and key so that one enzyme will only act as a catalyst (↑) for one specific kind of molecule.

inhibitor (*n*) a substance which slows down or stops a reaction which is controlled by an enzyme (↑). **inhibition** (*n*). **inhibit** (*v*).

competitive inhibition inhibition (↑) when the substrate (↑) and the inhibitor compete for the enzyme (↑). Also known as reversible (p. 30) inhibition.

non-competitive inhibition inhibition (↑) when the inhibitor combines permanently with the enzyme (↑) so that the substrate (↑) is excluded. Also known as non-reversible (p. 30) inhibition.

non-competitive inhibition

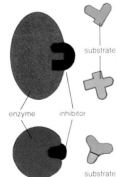

substrate

enzyme   inhibitor

substrate

reversible (*adj*) of a reaction or process that is not permanent i.e. it can work in the opposite direction. **reverse** (*v*).

non-reversible (*adj*) not reversible (↑).

cofactor (*n*) an additional inorganic compound (p. 15) which must be present in a reaction before the enzyme (p. 28) will catalyze (p. 28) it.

co-enzyme (*n*) an additional, non-protein (p. 21) organic compound (p. 15) which must be present in a reaction before the enzyme (p. 28) will catalyze (p. 28) it.

prosthetic group a non-protein (p. 21) organic compound (p. 15) which forms an essential part of the enzyme (p. 28) and which must be present in a reaction before the enzyme will catalyze (p. 28) it.

prosthetic group
role in enzyme reaction

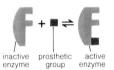

inactive   prosthetic   active
enzyme   group   enzyme

hydrolase (*n*) an enzyme (p. 28) which catalyzes (p. 28) hydrolysis (p. 16) reactions.

carbohydrase (*n*) an enzyme (p. 28) which catalyzes (p. 28) digestion (p. 98) reactions and aids in the breakdown of carbohydrates (p. 17).

oxidase (*n*) an enzyme (p. 28) group which catalyzes (p. 28) oxidation (p. 32) reactions.

dehydrogenase (*n*) an enzyme (p. 28) group which catalyzes (p. 28) reactions in which hydrogen atoms are removed from a sugar.

carboxylase (*n*) an enzyme (p. 28) group which catalyzes (p. 28) reactions in which carboxyl (COOH) groups are added to a substrate (p. 29).

transferase (*n*) an enzyme (p. 28) group which catalyzes (p. 28) reactions in which a group is transferred from one substrate (p. 29) to another.

isomerase (*n*) an enzyme (p. 28) group which catalyzes (p. 28) reactions in which the atoms of molecules are rearranged.

cell respiration the breakdown by oxidation (p. 32) of sugars yielding carbon dioxide, water and energy.

endergonic (*adj*) of a reaction which absorbs (p. 81) energy.

exergonic (*adj*) of a reaction which releases energy.

electron (*n*) a very small, negatively charged particle in an atom which may be raised to higher energy levels and then released during cell respiration (↑).

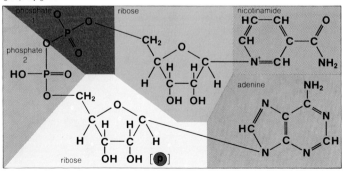

Krebs cycle
NADH + H⁺   FADH₂
NAD⁺   2H⁺ + 2e⁻   FAD   2H⁺ +

ADP→ATP   ADP→ATP

ADP→ATP

½O₂
O²⁻
H₂O
2H⁺

**electron carrier system
in respiration**

electron (hydrogen) carrier system a system
which operates during cell respiration (↑) in
which electrons (↑) (initially released as part of
a hydrogen atom which splits into an electron
and a proton) are collected by an electron
acceptor (↓) and passed to another electron
acceptor at lower energy levels. The energy
released in the process is used to convert ADP
(p. 33) to ATP (p. 33).

electron acceptor a molecule which functions as
a coenzyme (↑) with a dehydrogenase (↑) that
catalyzes (p. 28) the removal of hydrogen
during cell respiration (↑). It accepts electrons
(↑) and passes them on to electron acceptors
at lower energy levels.

NAD nicotinamide adenine dinucleotide. One of
the most important coenzymes (↑) or electron
acceptors (↑) concerned with cell respiration (↑).

NADP nicotinamide adenine dinucleotide
phosphate. An important coenzyme (↑) or
electron acceptor (↑) similar to NAD (↑).

**nicotinamide adenine
dinucleotide (NAD)**
adding a further phosphate
group at **p** gives NADP

phosphate
1
phosphate
2
HO—P=O
O—CH₂

ribose
CH₂
C   H   H   C
H   C—C   H
OH   OH

nicotinamide
HC—CH   O
CH   C—C
N⁺=CH   NH₂

adenine
NH₂
C
N   C   N
HC   C   CH
N   N

ribose   OH   OH   [**p**]

oxidation (*n*) a reaction in which a substance (1) loses electrons (p. 30); (2) has oxygen added to it; or (3) has hydrogen removed from it. **oxidize** (*v*).

reduction (*n*) a reaction in which a substance (1) gains electrons (p. 30); (2) has oxygen removed from it; or (3) has hydrogen added to it. **reduce** (*v*).

cytochrome (*n*) one of a system of coenzymes (p. 30) involved in cell respiration (p. 30) having prosthetic groups (p. 30) which contain iron. Cytochromes are involved in the production of ATP (↓) by oxidative phosphorylation (p. 34).

flavoprotein (*n*) FP. An important coenzyme (p. 30) involved in cell respiration (p. 30).

vitamin B the collective name for a group of vitamins (p. 25) which play an important role in cell respiration (p. 30) by functioning as coenzymes (p. 30).

aerobic (*adj*) of a reaction, for example, respiration (p. 112) which can only take place in the presence of free, gaseous oxygen. In aerobic respiration, organic compounds (p. 15) are converted to carbon dioxide and water with the release of energy. Organisms that use aerobic respiration are called aerobes.

anaerobic (*adj*) of a reaction, for example, respiration (p. 112) which takes place in the absence of free gaseous oxygen. In anaerobic respiration organic compounds (p. 15) such as sugars are broken down into other compounds such as carbon dioxide and ethanol with a lower release of energy. Organisms that use anaerobic respiration are called anaerobes.

basal metabolism the smallest (or minimum) amount of energy needed by the body to stay alive. It varies with the age, sex and health of the organism.

BMR basal metabolic rate = basal metabolism (↑).

metabolic rate in cell respiration (p. 30) the rate at which oxygen is used up and carbon dioxide is produced.

calorific value the amount of heat produced, measured in calories, when a given amount of food is completely burned. *See also* joule (p. 97).

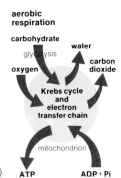

aerobic respiration

carbohydrate

glycolysis

oxygen

water

carbon dioxide

Krebs cycle and electron transfer chain

mitochondrion

ATP

ADP + Pi

ATP adenosine triphosphate. An organic compound (p. 15) composed of adenine (p. 22), ribose (p. 22), and three inorganic phosphate groups. It is a nucleotide (p. 22) and is responsible for storing energy temporarily during cell respiration (p. 30). It is formed by the addition of a third phosphate group to ADP (↓) which stores the energy that is released when required in other metabolic (p. 26) processes.

ADP adenosine diphosphate. The organic compound (p. 15) which accepts a phosphate group to form ATP (↑).

ADP, ATP and their reactions

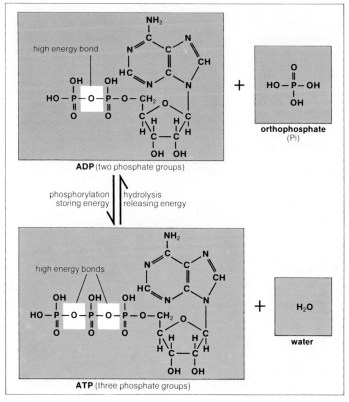

phosphate bond a bond which links the
phosphate groups in ATP (p. 33) and which is
often misleadingly referred to as a high energy
bond. Energy is stored throughout the ATP
molecule but is released as the phosphate
bonds are broken and other bonds are formed.

oxidative phosphorylation the process in which
ATP (p. 33) is produced from ADP (p. 33) in the
presence of oxygen during aerobic (p. 32) cell
respiration (p. 30).

glycolysis (*n*) the first part of cellular respiration
(p. 30) in which glucose (p. 17) is converted
into pyruvic acid (↓) in the cytoplasm (p. 10) of
all living organisms. It uses a complex system
of enzymes (p. 28) and coenzymes (p. 30). It
produces energy for short periods in the form
of ATP (p. 33) when there is a shortage of
oxygen. *See* endpaper.

pyruvic acid an organic compound (p. 15) which
is formed as the end product of glycolysis (↑).
For every molecule of glucose (p. 17) two
molecules of pyruvic acid are formed.

Kreb's cycle a part of cellular respiration (p. 30)
in which pyruvic acid (↑) in the presence of
oxygen and via a complex cycle of enzyme-
(p. 28) controlled reactions produces energy in
the form of ATP (p. 33) and intermediates which
give rise to other substances such as fatty acids
(p. 20) and amino acids (p. 21). It takes place in
the mitochondria (p. 11). *See* endpaper.

fermentation (*n*) a process in which pyruvic acid
(↑) in the absence of oxygen uses up hydrogen
atoms and so produces NAD (p. 31) allowing it
to be used again in glycolysis (↑).

lactic acid fermentation fermentation (↑) from
which lactic acid is produced. In higher animals
this takes place especially in the muscles
(p. 143) where there is an oxygen debt (p. 117).

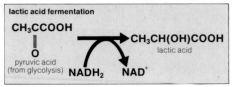

lactic acid fermentation

CH_3CCOOH → $CH_3CH(OH)COOH$
lactic acid

pyruvic acid
(from glycolysis) $NADH_2$ NAD^+

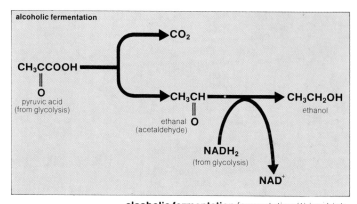

alcoholic fermentation

CH_3CCOOH
‖
O
pyruvic acid
(from glycolysis)

CO_2

CH_3CH
‖
O
ethanal
(acetaldehyde)

CH_3CH_2OH
ethanol

$NADH_2$
(from glycolysis)

NAD^+

alcoholic fermentation fermentation (↑) in which
ethanol (alcohol) and carbon dioxide are
produced. This process is made use of in the
brewing and wine-making industries in which
yeasts (p. 49) decompose sugars to provide
energy for their reproduction (p. 173) and growth.

nuclear division the process in which the nucleus
(p. 13) of a cell divides into two in the
development of new cells and new tissue
(p. 83) so that growth may occur or damaged
cells be replaced. There are two types: mitosis
(p. 37) and meiosis (p. 38).

centriole (*n*) a structure similar to a basal body
(p. 13). Centrioles are found outside the nuclear
membrane (p. 13) and divide at mitosis (p. 37)
forming the two ends of the spindle (p. 37).

chromatid (*n*) one of a pair of thread-like structures
which together appear as chromosomes (p. 13)
and which shorten and thicken during the
prophase (p. 37) of nuclear division (↑).

centromere (*n*) a region, somewhere along the
chromosome (p. 13), where force is exerted
during the separation of the chromatids (↑) in
mitosis (p. 37) and meiosis (p. 38).

chromomere (*n*) one of a number of granules of
chromatin (p. 13) which occur along a dividing
chromosome (p. 13) probably as a result of the
coiling and uncoiling within the chromatids (↑).
It appears as a 'bump' or constriction.

somatic cell any cell in a living organism other
than a germ cell (↓) and which contains the
characteristic number of chromosomes (p. 13),
normally diploid (↓), for the organism.

germ cell a cell that gives rise to a gamete
(p. 175). A cell in a living organism, other than a
somatic cell (↑), and which takes part in the
reproduction (p. 173) of the organism. It
contains only half of the characteristic number
of chromosomes (p. 13) of the organism i.e. it
is haploid (↓).

diploid and haploid stages in the life-cycle of a flowering plant

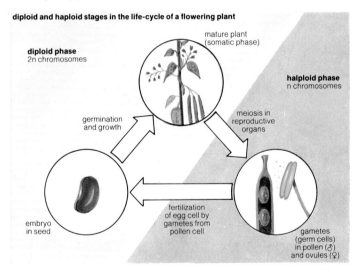

mature plant
(somatic phase)

diploid phase
2n chromosomes

halploid phase
n chromosomes

germination
and growth

meiosis in
reproductive
organs

fertilization
of egg cell by
gametes from
pollen cell

embryo
in seed

gametes
(germ cells)
in pollen (♂)
and ovules (♀)

haploid (*adj*) of a cell which has only unpaired
chromosomes (p. 13); half the diploid (↓)
number of chromosomes which are not paired
in the haploid state. Germ cells (↑) of most
animals and plants are haploid. *See also*
polyploidy etc p. 207.

diploid (*adj*) of a cell which has chromosomes
(p. 13) which occur in homologous (p. 39) pairs.
Somatic cells (↑) of most higher plants and
animals are described as diploid. Double the
haploid (↑) number.

mitosis
(only two pairs of homologous chromosomes shown for clarity)

chromosomes
nuclear membrane
nucleolus

prophase chromosomes become visible in the nucleus, each one duplicated into two chromatids, joined by a centromere

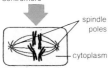

spindle poles
cytoplasm

metaphase nuclear membrane and nucleolus have disintegrated. Spindle fibres form. Chromosomes shorter and thicker, arranged midway between the spindle poles.

anaphase chromatids separate at centromeres. Sister chromatids drawn to opposite poles of the spindle.

telophase nuclear membrane and nucleoli reform. Chromosomes begin to lose their compact structure

interphase chromosomes no longer visible.

mitosis (*n*) the usual process of nuclear division (p. 35) into two daughter nuclei (p. 13) during vegetative growth. During mitosis each chromosome (p. 13) duplicates itself, each one of the duplicates going into separate daughter nuclei. The daughter cells are identical to each other and to the parent cell.

spindle (*n*) fibrous (p. 143) material which forms from the centrioles (p. 35) during mitosis (↑) and meiosis (p. 38). It takes part in the distribution of chromatids (p. 35) to the daughter cells. The chromosomes (p. 13) are arranged at its equator (↓) during metaphase (↓).

pole (*n*) one of the two points on the spindle (↑) which is the site of the formation of the spindle fibres (p. 143) from the centrioles (p. 35).

equator (*n*) part of the spindle (↑) midway between the poles (↑) to which the chromosomes (p. 13) become attached by the spindle attachment.

interphase (*n*) a stage in the cell cycle when the cell is preparing for nuclear division (p. 35). At this stage, the DNA (p. 24) is replicating to produce enough for the daughter cells.

prophase (*n*) the first main stage in nuclear division (p. 35) in which the chromosomes (p. 13) become visible and then the chromatids (p. 35) appear while the nucleolus (p. 13) and nuclear membrane (p. 13) begin to dissolve.

metaphase (*n*) a main stage in nuclear division (p. 35) at which the nuclear membrane (p. 13) has disappeared and the chromosomes (p. 13) lie on the equator (↑) of the spindle (↑). Then the chromatids (p. 35) start to move apart.

anaphase (*n*) a main stage in nuclear division (p. 35) in which the centromeres (p. 35) divide and the chromatids (p. 35) move to opposite poles (↑) by the contraction of the spindle (↑).

telophase (*n*) a main stage in nuclear division (p. 35) at which the chromatids (p. 35) arrive at the poles (↑) and the cytoplasm (p. 10) may divide to form two separate daughter cells in interphase (↑). The spindle (↑) fibres (p. 143) dissolve while the nucleolus (p. 13) and nuclear membrane (p. 13) in each daughter cell reform and the chromosomes (p.13) regain their thread-like form.

meiosis (*n*) nuclear division (p. 35) of a special
kind which begins in a diploid (p. 36) cell and
takes place in two stages. Each stage is similar
to mitosis (p. 37) but the chromosomes (p. 13)
are duplicated only once before the first division
so that each of the four resulting daughter cells
is haploid (p. 36). It occurs during the formation
of the gametes (p. 175).

meiosis
(cytoplasm and membrane
not shown)

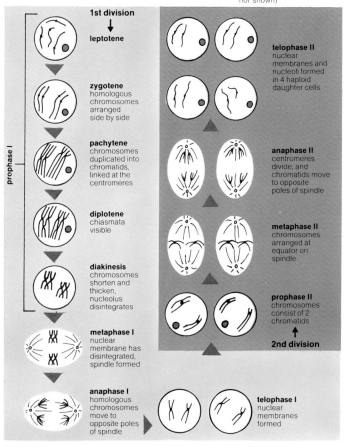

1st division
↓
leptotene

zygotene
homologous
chromosomes
arranged
side by side

pachytene
chromosomes
duplicated into
chromatids,
linked at the
centromeres

diplotene
chiasmata
visible

diakinesis
chromosomes
shorten and
thicken,
nucleolus
disintegrates

prophase I

metaphase I
nuclear
membrane has
disintegrated,
spindle formed

anaphase I
homologous
chromosomes
move to
opposite poles
of spindle

telophase I
nuclear
membranes
formed

telophase II
nuclear
membranes and
nucleoli formed
in 4 haploid
daughter cells

anaphase II
centromeres
divide, and
chromatids move
to opposite
poles of spindle

metaphase II
chromosomes
arranged at
equator on
spindle

prophase II
chromosomes
consist of 2
chromatids
↑
2nd division

mitosis	meiosis
occurs in somatic cells during growth and repair	occurs in the sex organs during gamete formation
no pairing or separation of homologous chromosomes	pairing and separation of homologous chromosomes
no chiasmata formed	chiasmata formed which may lead to crossing over and recombination
one separation of nuclear material i.e. separation of chromatids only	two separations of nuclear material i.e. separation of homologous chromosomes (1st division) and chromatids (2nd division)
2 daughter nuclei formed	4 daughter nuclei formed
daughter nuclei identical	daughter nuclei not identical
daughter nuclei diploid	daughter nuclei haploid

differences between mitosis and meiosis

pairs of homologous chromosomes

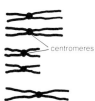

centromeres

bivalent (*n*) one of the pairs of homologous (↓) chromosomes (p. 13) which associate during the first prophase (p. 37) of meiosis (↑).

chiasmata (*n.pl.*) the points at which homologous (↓) chromosomes (p. 13) remain in contact as the chromatids (p. 35) move apart during the first prophase (p. 37) of meiosis (↑). There may be up to eight chiasmata in a bivalent (↑) pair of chromosomes. **chiasma** (*sing.*).

terminalization (*n*) the process in which the chiasmata (↑) move to the ends of the chromosomes (p. 13) during the prophase (p. 35) of meiosis (↑).

homologous chromosomes two chromosomes (p. 13) which form a pair in which the genes (p. 196) arranged along their length control identical characteristics of the organism, such as eye colour or height.

first meiotic division the first of two major stages of meiosis (↑) in which a nuclear division (p. 35) similar to mitosis (p. 37) takes place resulting in the separation of homologous (↑) chromosomes (p. 13).

second meiotic division the second of two major stages in meiosis (p. 37) in which a second nuclear division (p. 35) takes place and the two daughter cells formed from the first meiotic division (↑) each divide into two to result in four haploid (p. 36) daughter cells each containing one of the sister chromatids (p. 35).

classification (*n*) the arrangement of all living organisms into an ordered series of named and related groups. **classify** (*v*).

organisms (*n*) any living thing. Organisms can grow and reproduce (p. 175).

taxon (*n*) the general term for any group in a classification (↑) no matter what its rank (↓). **taxa** (*n.pl.*).

taxonomy (*n*) the science of classification (↑).

binomial system a system of naming every known living organism, first devised by the Swedish botanist, Carolus Linnaeus (1707–78), in which the organism is given a two-part scientific name which is usually Latinized. The first word indicates the genus (↓) while the second word indicates the species (↓). While the common names of organisms may only be understood in their place of origin, the scientific name is recognized internationally by scientists. For example, the bird with the English common name, peregrine falcon, is given the scientific name, *Falco peregrinus*.

species (*n*) a group of similar living organisms whose members can interbreed to produce fertile (p. 175) offspring but which cannot breed with other species groups. **specific** (*adj*).

genus (*n*) a group of organisms containing a number of similar species (↑). Of the scientific name, *Falco peregrinus*, *Falco* is the generic name referring to all birds that are classified (↑) as falcons.

classfication of the Peregrine falcon showing the series of ranks and their names		
rank	scientific name of taxonomic groups (taxa)	common name
kingdom	Animalia	animals
phylum	Chordata	vertebrates
class	Aves	birds
order	Falconiformes	birds of prey
family	Falconidae	falcons
genus	*Falco*	true falcons
species	*peregrinus*	Peregrine falcon

rank (*n*) one of a number of major groups into which living organisms are classified (↑). The largest group which contains organisms that have different body plans from those in any other large group is called a kingdom (↓). Each kingdom may be further divided, on the basis of diversity (p. 213), into a number of phyla, and so on. The principal rank names arranged in order from the largest groups to the most basic are kingdom, phylum, class, order, family, genus (↑) and species (↑).

kingdom (*n*) the highest rank (↑) or taxon (↑). Most simply, all life can be grouped into either the Plant or Animal kingdoms. This, however, is an oversimplification and in this book we divide living organisms into five kingdoms: Monera (p. 42), Protista (p. 44), Fungi (p. 46), Plants, Animals.

artificial classification a classification (↑) in which the organisms are arranged into groups on the basis of apparent analogous (p. 211) similarities which, in fact, have no common ancestry.

natural classification a classification (↑) in which the organisms are arranged into groups on the basis of homologous (p. 211) similarities which demonstrate a common ancestry.

evolution and relationship of main plant and animal groups

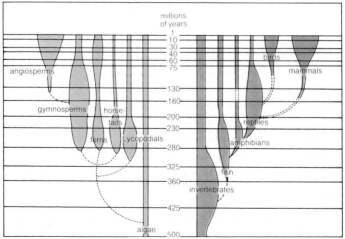

microbiology (n) the study or science of very
small (microscopic (p. 9)) or submicroscopic
living organisms. It includes bacteriology and
virology.

Monera (n) the kingdom (p. 41) of prokaryotic
(p. 8) organisms which includes the bacteria (↓)
and blue-green algae (↓).

bacteria (n.pl.) a group of microscopic (p. 9)
prokaryotic (p. 8) organisms that may be
unicellular (p. 9) or multicellular (p. 9). They lack
organelles (p. 8) bounded by membranes (p.14)
and contain no large vacuoles (p. 11). Most
bacteria are heterotrophic (p. 92) but some are
autotrophic (p. 92). Their respiration (p. 112)
may be either aerobic (p. 32) or anaerobic
(p. 32). Bacteria reproduce (p. 173) mainly by
asexual cell division. Heterotrophic bacteria
may cause disease. They are important in the
decay of plant and animal tissue (p. 83) to
release food materials for higher plants and in
sewage breakdown. **bacterium** (sing.).

bacillus (n) a rod-shaped bacterium (↑). **bacilli** (pl).

coccus (n) a spherical-shaped bacterium (↑).
cocci (pl)

streptococcus (n) a coccus (↑) which occurs in
chains.

staphylococcus (n) a coccus (↑) which occurs in
clusters.

spirillum (n) a spiral-shaped bacterium (↑).
spirilla (pl).

Gram's stain a stain used in the study of bacteria
(↑). Bacteria which take the violet stain are
gram-positive while others that do not are
gram-negative. Gram-positive bacteria are
more readily killed by antibiotics (p. 233).

myxobacterium (n) a bacillus (↑) that has a
delicate flexible cell wall (p. 8) and is able to
glide along solid surfaces.

spirochaete (n) a spirillum (↑) which is able to
move by flexing its body. Some are parasitic
(p. 92) and cause diseases such as syphilis.

rickettsia (n) any of the various bacilli (↑) which
live as parasites (p. 110) on some arthropods
(p. 67) and which can be transmitted to humans
causing diseases such as typhus.

structure of a generalized bacterium

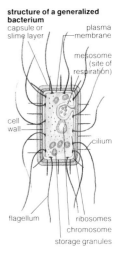

capsule or slime layer — plasma membrane — mesosome (site of respiration) — cell wall — cilium — flagellum — ribosomes — chromosome — storage granules

bacteria

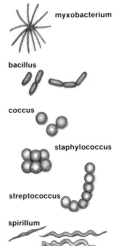

myxobacterium

bacillus

coccus

staphylococcus

streptococcus

spirillum

filamentous blue green algae

plant viruses
spherical

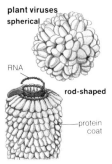

RNA

rod-shaped

protein
coat

actinomycete (*n*) a soil-dwelling gram-positive
(↑) bacterium (↑) with its cells arranged in
filaments (p. 181). It may be used to produce
antibiotics (p. 233) such as streptomycin.

pathogen (*n*) any parasitic (p. 92) bacterium (↑),
virus (↓), or fungus (p. 44) which produces
disease.

toxin (*n*) a poison produced by a living organism,
especially a bacterium (↑), and which may
cause the symptoms of disease which results
from the action of a pathogen (↑). **toxic** (*adj*).

blue-green algae a group of microscopic (p. 9)
prokaryotic (p. 8) organisms known as
Cyanophyta. They contain chlorophyll (p. 12)
and other pigments (p. 126), and are widely
distributed wherever water is present. Some
are able to take in (fix) atmospheric nitrogen
into organic compounds (p. 15).

virus (*n*) a pathogen (↑) which may or may not be
a living organism and which is so small that it
can only be observed with the aid of an electron
microscope (p. 9). It has no normal organelles
(p.8). A virus will only grow within its host (p.111)
but occurs as non-living chemicals outside.
Viruses are often named after their hosts to
which they are specific and the symptoms they
cause, for example, tobacco mosaic virus. A
virus consists of an outer protein (p. 21) coat
which surrounds a core of nucleic acids (p. 22).

bacteriophage (*n*) a virus (↑) which infects a
bacterium (↑). It consists of a head, which
contains its DNA (p. 24) or RNA (p. 24),
enclosed by a protein (p. 21) coat, and a tail
which ends in a plate that bears a number of
tail fibres (p. 143). **phage** (*abbr*).

life cycle of a bacteriophage

bacteriophage

head

tail

tail
fibres

1 bacteriophage attacking
a bacterium

bacterium
cell wall

nucleic
acid
injected

2 parts of new
bacteriophages
synthesized in
bacterial cell

3 bacterium destroyed,
new bacteriophages
released

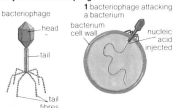

Protista (*n*) a kingdom (p. 41) of unicellular (p. 9) eukaryotic (p. 9) organisms, some of which have been previously allocated to the plant or animal kingdoms, or even both, e.g. Protozoa (↓) and unicellular algae (↓). **protistan** (*adj*).

binary fission asexual reproduction (p. 173) in which a single parent organism gives rise to two daughter organisms. The nucleus (p. 13) divides by mitosis (p. 37) followed by the division of the cytoplasm (p. 10).

algae (*n.pl.*) organisms, with a unicellular (p. 9) or simple multicellular (p. 9) body plan, that are able to manufacture their own food material by photosynthesis (p. 93). Unicellular types belong to the Protista (↑) while multicellular types, e.g. seaweeds, are regarded as plants. **alga** (*sing.*).

phycology (*n*) the science or study of algae (↑).

Protozoa (*n*) a division of the Protista (↑) in which the microscopic (p. 9) organisms are unicellular (p. 9), exist as a continuous mass of cytoplasm (p. 10), ingest (p. 98) their food and lack chloroplasts (p. 12) and cell walls (p. 8). Protozoans are widespread and important in natural communities. **protozoan** (*adj*).

Amoeba (*n*) a genus (p. 40) of Protozoa (↑). Its members consist of a single motile (p. 173) cell, able to take in food particles by engulfing them using pseudopodia (↓). **amoebae** (*pl.*).

pseudopodium (*n*) a temporary protuberance into which the cytoplasm (p. 10) of a protozoan (↑) flows and which enables it to move and feed. **pseudopodia** (*pl.*).

amoeboid movement the process of locomotion (p. 143) which results from the formation of pseudopodia (↑).

food vacuole a vacuole (p. 11) containing a food particle and a drop of water, engulfed by the pseudopodia (↑) of a protozoan (↑).

ectoplasm (*n*) the external plasma membrane (p. 13) of a protozoan (↑). A fibrous (p. 143) gel with a less granular structure than the endoplasm (↓) which it surrounds. It takes part in amoeboid movement (↑) and in cell division.

endoplasm (*n*) the cytoplasm (p. 10) of a protozoan (↑). It is more fluid and granular than ectoplasm (↑).

binary fission in a bacterium

pair of chromosomes

replicate chromosome

two pairs of identical chromosomes

two cells identical to parent cell

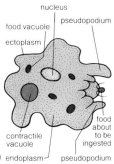

Amoeba

nucleus

pseudopodium

food vacuole

ectoplasm

food about to be ingested

contractile vacuole

endoplasm

pseudopodium

the movement of a cilium

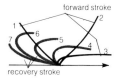

gel (n) a jelly-like material
granule (n) a small particle **granular** (adj)
Paramecium (n) a genus (p. 40) of Protozoa (↑)
Although it is unicellular (p. 9), its organization
is more complex than that of *Amoeba*. It moves
by means of cilia (p. 12), it possesses two kinds
of nuclei (p. 13), meganuclei (↓) and micronuclei
(↓), and it reproduces (p. 173) asexually by
transverse binary fission (↑)

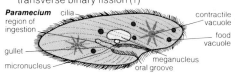

ciliate movement the process of locomotion
(p. 143) which involves the beating of stiffened
cilia (p. 12) against the water On the recovery
stroke the cilia relax so that they do not push
against the water in the reverse direction
eye spot an organelle (p. 8) which is sensitive to
light. It occurs in many Protozoa (↑)
oral groove a ciliated (p. 12) groove in *Paramecium*
(↑) into which food particles are drawn by the
beating of cilia. It leads to the gullet and the
region in which the food is ingested (p. 98)
micronucleus (n) the smaller of the two nuclei
(p. 13) of *Paramecium* which divides by mitosis
(p. 37) and supplies gametes (p. 175) during
conjugation (↓)
meganucleus (n) the larger of the two nuclei
(p. 13) of *Paramecium* (↑) which is concerned
with making protein (p. 21) for the organism
conjugation (n) a process of sexual reproduction
(p. 173) in *Paramecium* (↑) and other Protozoa
(↑) in which two cells temporarily come together
and exchange gametes (p. 175)
Euglena (n) a genus (p. 40) of Protista (↑). It
moves by means of a flagellum (p. 12) and
reproduces (p. 173) by transverse binary fission
(↑). It has no rigid cell wall (p. 8) but an elastic
transparent pellicle. It contains chloroplasts
(p. 12) by which it produces its own food
substances by photosynthesis (p. 93), but is
also able to ingest (p. 98) food through a gullet.

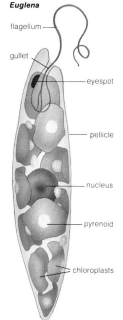

Euglena

mycology (*n*) the science or study of fungi (↓).

Fungi (*n*) a kingdom (p. 41) of eukaryotic (p. 9) organisms that are unable to make food material by photosynthesis (p. 93). Instead they take up all their nutrients (p. 92) from their surroundings. They may be microscopic (p. 9) or quite large. They may be unicellular (p. 9) or made up of hyphae (↓). They live either as saprophytes (p. 92) or parasites (p. 110) of plants and animals. Fungi may reproduce (p. 173) sexually and asexually. **Fungus** (*sing*).

hypha (*n*) a branched, haploid (p. 36) filament (p. 181) which is the basic unit of most fungi (↑). It is a tubular structure composed of a cell wall (p. 8) with a lining of cytoplasm (p. 10) and surrounding a vacuole (p. 11). In some fungi, the hyphae (*pl*) may be divided by cross walls or septa. The cell wall is composed mainly of the material chitin (p. 49).

mycelium (*n*) a mass of hyphae (↑) which make up the bulk of a fungus (↑). **mycelia** (*pl*).

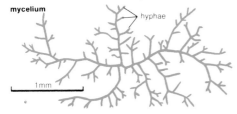

mycelium

hyphae

1 mm

coenocytic (*adj*) of hyphae (↑) which consist of tubular masses of protoplasm (p. 10) containing many nuclei (p. 13).

dikaryon (*adj*) of a hypha (↑) or mycelium (↑) made up of cells containing two haploid (p. 36) nuclei (p. 13) which divide simultaneously when a new cell is formed.

Phycomycetes (*n.pl.*) a group of Fungi (↑) which possess hyphae (↑) without septa (cross walls). Phycomycetes reproduce (p. 173) sexually by means of zygospores (↓) and asexually by means of zoospores (↓). This group includes the large genus (p. 40) of pin moulds, *Mucor*, and the related genus *Rhizopus*.

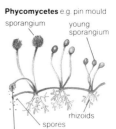

Phycomycetes e.g. pin mould

sporangium

young sporangium

rhizoids

spores

columella

zygospores
stages in formation

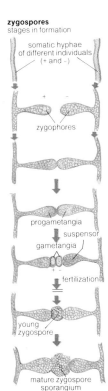

somatic hyphae
of different individuals
(+ and -)

zygophores

progametangia

suspensor

gametangia

fertilization

young
zygospore

mature zygospore
sporangium

**asci and ascospores
of Ascomycetes** ascospores
released
explosively
asci each from asci
with 8
ascospores

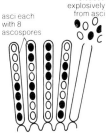

homothallic (*adj*) of the sexual reproduction
(p. 173) of certain fungi (↑) and algae (p. 44) in
which a single thallus (p. 52) produces the
opposite, differently sized gametes (p. 175) to
perform the sexual functions so that the species
is, in effect, hermaphrodite (p. 175).

heterothallic (*adj*) of the sexual reproduction
(p. 173) of certain fungi (↑) and algae (p. 44) in
which reproduction can only take place
between two genetically different thalli (p. 52)
which cannot reproduce independently. In
some fungi, the two thalli may be different in
form so that they are either male or female
while in others there may be no difference in
form but the gametes (p. 175) are different in
size between the two genetically different
strains of the same species (p. 40).

zygospore (*n*) a thick-walled resting spore
(p. 178) produced by a phycomycete (↑) during
sexual reproduction (p. 173) by the fusion of
two gametes (p. 175) called gametangia.

zoospore (*n*) the naked, flagellate (p. 12) spore
(p. 178) produced in a sporangium (p. 178)
during asexual reproduction (p. 173).

Ascomycetes (*n.pl.*) a group of Fungi (↑) which
possess hyphae (↑) with septa. Ascomycetes
reproduce (p. 173) sexually by means of
ascospores (↓) and asexually by means of
conidia (↓). *Penicillium* is an important genus
(p. 40) of Ascomycetes from which antibiotics
(p. 233) are manufactured.

ascus (*n*) a near-cylindrical or spherical cell in
which ascospores (↓) are formed. A number of
asci (*pl.*) may be grouped together into a fruit
body which is visible to the naked eye.

ascospore (*n*) the spore (p. 178) which forms in
the ascus (↑) as a result of the fusion of haploid
(p. 36) nuclei (p. 13) followed by meiosis (p. 38)
to restore the haploid state. Normally, each
ascus contains eight ascospores.

septum (*n*) a wall across a hypha (↑). **septa** (*pl.*).

conidium (*n*) a spore (p. 178) or bud which is
produced during asexual reproduction (p. 173)
from the tips of particular hyphae (↑). **conidia**
(*pl.*). See diagram on p. 48.

Basidiomycetes (*n.pl.*) a group of Fungi (p.46)
which possess hyphae (p.46) with septa. Hyphae
are often massed into substantial fruit bodies
such as mushrooms (↓) or toadstools (↓). They
reproduce (p. 173) sexually by basidiospores
(↓). *Agaricus*, including the field mushroom, is a
genus (p.40) of this group.

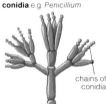

conidia e.g. *Penicillium*

chains of
conidia

mushrooms
and toadstools

mushroom (*n*) the common name for the fruit
body of Basidiomycetes (↑) belonging to the
order Agaricales. The name is usually used for
those species (p. 40) that are good to eat.

toadstool (*n*) the common name for the fruit body
of Basidiomycetes (↑) belonging to the order
Agaricales and which are not referred to as
mushrooms (↑). It is not necessarily a poisonous
species (p. 46).

basidiospore (*n*) a haploid (p. 36) spore (p. 178)
produced following sexual reproduction (p. 173)
and meiosis (p. 38) and borne externally on the
fruit bodies of Basidiomycetes (↑).

basidium (*n*) a club-shaped or cylindrical cell on
which the basidiospores (↑) are borne on short
stalks, usually four at a time. **basidial** (*adj*).

sterigmata (*n.pl.*) the stalks on a basidium (↑) on
which the basidiospores (↑) are borne. Each
basidial cell usually bears four sterigmata.

cap (*n*) the umbrella-shaped structure which
crowns the central stem of the larger fungi
(p. 46) forming the fruit body and in which the
spores (p. 178) are produced.

pileus (*n*) = cap (↑).

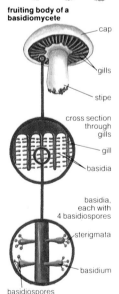

fruiting body of a
basidiomycete

cap

gills

stipe

cross section
through gills

gill

basidia

basidia,
each with
4 basidiospores

sterigmata

basidium

basidiospores

gills (*n.pl.*) the fin-like structures which occur on the underside of the cap (↑) of the fruit body of a fungus (p. 46). They bear the spore- (p. 178) producing cells or basidia (↑).

yeast (*n*) unicellular (p. 9) fungi (p. 46) which are very important in brewing and baking and as sources of proteins (p. 21) and minerals. Most yeasts are Ascomycetes (p. 47).

Fungi Imperfecti a loose grouping of fungi (p. 46) which only reproduce (p. 173) asexually.

rust (*n*) a parasitic (p. 92) basidiomycete (↑) fungus (p. 46). Rusts are serious pests of crops and may cause huge losses.

blight (*n*) a disease of plants, such as potatoes, which results from the rapid spread of the hyphae (p. 46) of fungi (p. 46), such as *Phytophthora*, through the leaves of the host (p. 110).

chitin (*n*) a horny material found in the cell walls (p. 8) of many fungi (p. 46) and composed of polysaccharides (p. 18). It is similar to the material that protects the bodies of insects (p. 69).

mycorrhiza (*n*) the symbiotic (p. 228) association (p. 227) which may occur between a fungus (p. 46) and the roots of certain higher plants, especially trees.

lichen (*n*) a symbiotic (p. 228) association (p. 227) of an alga (p. 44) and a fungus (p. 46) to form a slow-growing plant which colonizes (p. 221) such inhospitable environments (p. 218) as rocks in mountainous areas or the trunks of trees.

slime mould widely distributed fungi (p. 46) consisting of masses of protoplasm (p. 10) containing many nuclei (p. 13) and occurring in damp conditions. They reproduce by means of spores (p. 178) and are often classified (p. 40) with fungi. During part of their life history, slime moulds are able to undertake amoeboid movement (p. 44).

cellular slime mould

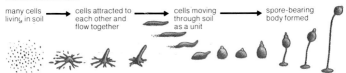

many cells living in soil → cells attracted to each other and flow together → cells moving through soil as a unit → spore-bearing body formed

botany (*n*) the study or science of plants or
plant life.

Plantae (*n*) one of the five kingdoms (p. 41) of
living organisms containing all plants that are
capable of making their own food by
photosynthesis (p. 93). It includes the
multicellular (p. 9) algae (p. 44), Musci (p. 52),
Filicales (p. 56), gymnosperms (p. 57) and
angiosperms (p. 57).

Thallophyta (*n*) in the two-kingdom (p. 41)
classification (p. 40) of living organisms, a
division made up of all those non-animal
organisms in which the body is not differentiated
into stem, roots, and leaves etc. Reproduction
(p. 173) takes place sexually by fusion of
gametes (p. 175) and asexually by spores
(p. 178). It includes bacteria (p. 42), blue-green
algae (p. 43), fungi (p. 46) and lichens (p. 49).

vascular plant a plant which possesses a
vascular system (p. 127) to transport water and
food materials through the plant and which also
provides support for the plant.

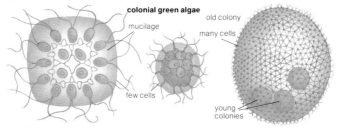

colonial green algae

old colony

mucilage

many cells

few cells

young
colonies

Chlorophyta (*n*) a division of mainly
multicellular (p. 9) algae (p. 44) in which the
plants are mainly freshwater although there are
some marine forms. These are the green algae
and contain chlorophyll (p. 12) for
photosynthesis (p. 93). They store food as
starch (p. 18) and fats.

Chlamydomonas (*n*) a unicellular (p. 9) genus
(p.40) of Chlorophyta (↑) which is found widely
in freshwater ponds. They possess two flagella
(p. 12) and a cup-shaped chloroplast (p. 12)
contained within the cell wall (p. 8).

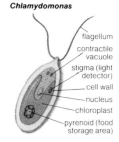

Chlamydomonas

flagellum

contractile
vacuole

stigma (light
detector)

cell wall

nucleus

chloroplast

pyrenoid (food
storage area)

Spirogyra

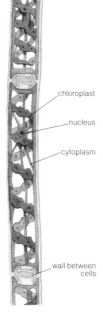

chloroplast

nucleus

cytoplasm

wall between cells

Spirogyra (*n*) a genus (p. 40) of typical
filamentous (p. 181) algae of the Chlorophyta
(↑) They are found in freshwater and consist of
a simple chain of identical cells each containing
the characteristic spiral chloroplast (p. 12).

Phaeophyta (*n*) a division of the algae (p. 44) in
which the plants are largely marine, such as
the large seaweeds of the shore. They contain
chlorophyll (p. 12) and the brown pigment
(p. 126), fucoxanthin, and they are referred to
as the brown algae. They are multicellular (p. 9)
and store food as sugars.

Fucus (*n*) a typical genus (p. 40) of the
Phaeophyta (↑). They are the common seaweeds
of the intertidal zone. Each plant is differentiated
into a holdfast for clinging to the rocks, a tough
stalk called a stipe, and flat fronds. They are
commonly described as wracks.

bladder[p] (*n*) an air-filled sac which occurs in
some members of the Phaeophyta (↑). Plants
containing these bladders are commonly called
bladderwracks.

conceptacle (*n*) one of the many cavities which
occur at the tips of the fronds of some members
of the Phaeophyta (↑) such as bladderwracks.
They open by a pore (p. 120) called an ostiole,
and, as well as sex organs, contain masses of
sterile hairs called paraphyses.

Phaeophyta brown algae

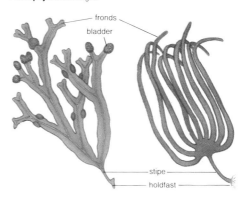

fronds

bladder

stipe

holdfast

Bryophyta (*n.pl.*) a division of the Plantae (p. 50) including the Hepaticae (↓) Anthocerotae (↓) and the Musci (↓) Bryophytes lack vascular tissues (p. 83) although the stems of some mosses have a central strand of conducting tissue. They are mostly plants of warm places but some are aquatic others live in desert habitats (p. 217) or cold places and may sometimes be the dominant form of plant life All bryophytes show a clear alternation of generations (p. 176) with a conspicuous, food-independent gametophyte (p. 177) generation and a short-lived sporophyte (p.177) generation dependent on the gametophyte. They are small flattened plants with leaves and a stem but no roots and are attached by a rhizoid (↓)

Hepaticae (*n.pl.*) liverworts A family of the Bryophyta (↑). These are the simplest bryophytes and may be either a flattened, leafless gametophyte (p. 177), a thallose (↓) liverwort, or a creeping, leafy gametophyte known as a leafy liverwort. A typical thallose liverwort is ribbon-like and has Y-shaped branches They are usually aquatic, living in damp soil or as epiphytes (p. 228).

Anthocerotae (*n.pl.*) hornworts. A family of the Bryophyta (↑). The plant consists of a lobed, green thallus (↓) anchored by a rhizoid (↓) to the substrate which is usually moist soil or mud:

Musci (*n.pl.*) mosses. A family of the Bryophyta (↑). These are the most advanced bryophytes and they either grow in erect, virtually unbranched cushions or in feathery, creeping, branched mats They are widely distributed throughout the world living in damp conditions, such as in woodland, or they may be aquatic, may even survive in drier conditions, such as on walls or roofs of houses

thallus (*n*) a general term for the plant body which is not differentiated into root, stem or leaf e.g. liverworts. **thalloid** (*adj*)

rhizoid (*n*) an elongate, single cell, such as in a liverwort, or a multicellular (p. 9) thread, such as in a moss, that anchors the gametophyte (p. 177) to the substrate. It is not a true root.

liverwort sporophyte

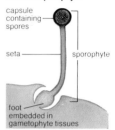

capsule containing spores

seta

sporophyte

foot embedded in gametophyte tissues

thalloid liverwort

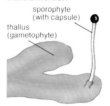

sporophyte (with capsule)

thallus (gametophyte)

rhizoids

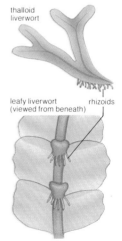

thalloid liverwort

leafy liverwort (viewed from beneath)

rhizoids

protonema

cells

acrocarpous moss

pleurocarpous moss

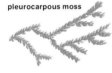

foot[p](*n*) the lower part of the sporophyte (p. 171) generation of a bryophyte (↑) which remains embedded in the archegonium (p. 177).

capsule (*n*) (1) the end part of the sporophyte (p. 177) generation of liverworts or mosses which, at maturity, contains the spores (p. 178). (2) a dry fruit, such as that of the poppy, formed from two or more carpels (p. 179) which, during dehiscence (p. 185), open by a variety of slits or pores (p. 120) to release the seeds.

seta[p] (*n*) the stalk of the capsule (↑).

columella (*n*) the central, sterile tissue (p. 83) within the capsule (↑) of liverworts and mosses.

calyptra (*n*) a hood-like structure which covers the capsule (↑) of mosses until mature. It is the remains of the archegonium (p. 177).

operculum[p] (*n*) the lid of a moss capsule (↑) which is shed to reveal the peristome teeth (↓).

peristome teeth a ring of teeth at the tip of the capsule (↑) of mosses which open and close in response to varying levels of moisture – they open and close when dry and close when moist.

elater (*n*) a spindle-shaped body contained in the capsule (↑) of liverworts. Spiral thickenings change shape with varying moisture levels causing the elaters to flick spores (p. 178) from the capsule.

protonema (*n*) the branching filament (p. 181) that grows from the germinating spore (p. 178) of mosses. It develops buds which grow into the leafy gametophyte (p. 177) generation. Also known as **first thread protonemata** (*pl.*).

paraphyses

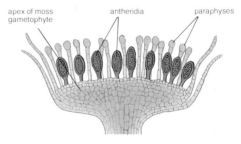

apex of moss gametophyte · antheridia · paraphyses

gemmae in thalloid liverworts: vegetative reproduction

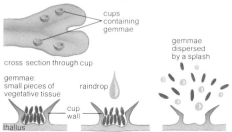

cups containing gemmae

gemmae dispersed by a splash

cross section through cup

gemmae: small pieces of vegetative tissue

raindrop

cup wall

thallus

liverwort sporophyte discharging spores

capsule walls

spores

elaters

elater with helical thickenings in cell wall

gemmae (*n.pl.*) minute, lens-shaped bodies produced by liverworts as a means of asexual reproduction (p. 173). **gemma** (*sing.*)

gemmae cup the receptacle or cup-shaped body on the upper surface of the gametophyte (p. 177) generation in liverworts which contains the gemmae (↑).

Pteridophyta (*n.pl.*) a division of the Plantae (p. 50) including the Lycopodiales (↓), Equisetales (↓) and the Filicales (↓) Pteridophytes have a well-developed vascular system (p. 127). They are widely distributed, especially in the tropics, and live mainly on land. There is alternation of generations (p.176) between gametophyte (p.177) and sporophyte (p. 177) phases in which the latter is the most prominent when the plant is differentiated into roots, stems, leaves and rhizomes (p. 174).

vascular cryptogams an alternative name for the Pteridophyta (↑), so called because there is a clear vascular system (p. 127) but no prominent organs of reproduction (p. 173), such as in the angiosperms (p. 57).

homosporous (*adj*) of plants with only a single kind of spore (p. 178) which gives rise to a hermaphrodite (p. 175) generation of gametophytes (p. 177). It occurs in some of the Pteridophyta (↑).

heterosporous (*adj*) of plants with two distinct kinds of spores (p. 178) which give rise to male and female gametophyte (p. 177) generations respectively. It occurs in some Pteridophytes (↑) and is thought to represent an evolutionary (p. 208) step towards the production of seeds.

homospory and heterospory in vascular plants

homosporous
bryophytes
some pteridophytes (e.g. ferns)

heterosporous
some pteridophytes (e.g. club mosses)
gymnosperms
angiosperms

heterosporous plants the production of microspores

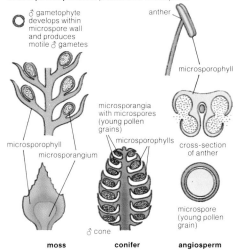

♂ gametophyte develops within microspore wall and produces motile ♂ gametes

anther

microsporophyll

microsporangia with microspores (young pollen grains)

microsporophyll

microsporangium

microsporophylls

cross-section of anther

microspore (young pollen grain)

♂ cone

moss   **conifer**   **angiosperm**

horsetail
Equisetales

strobilus

10 cm

1 cm

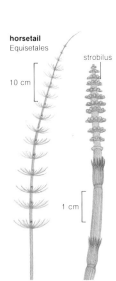

club moss Lycopodiales

spore-bearing shoots

strobilus (*n*) the reproductive (p. 173) structure of certain members of the Pteridophyta (↑). It consists of sporophylls (↓) on an axis. **strobili** (*pl.*).

cone[p] = strobilus (↑).

Lycopodiales (*n.pl.*) club mosses. A division of the Pteridophyta (↑). They are an ancient group and even attained tree-like forms. They may be heterosporous (↑) or homosporous (↑) and bear densely packed small leaves on branched stems. They are evergreen.

sporophyll (*n*) a modified leaf that bears a sporangium (p. 178).

Equisetales (*n.pl.*) horsetails. A division of the Pteridophyta (↑). They are an ancient group and even attained tree-like forms. They are characterized by having whorls (p. 83) of small leaves on upright stems with strobili (↑) at the tips. They are homosporous (↑).

microphyll (*n*) a foliage leaf typical of the Lycopodiales (↑) and the Equisetales (↑) which may be very small and has a simple vascular system (p. 127) comprising a single vein running from the base to the apex.

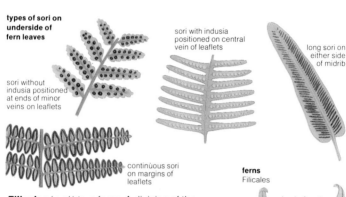

types of sori on underside of fern leaves

sori with indusia positioned on central vein of leaflets

long sori on either side of midrib

sori without indusia positioned at ends of minor veins on leaflets

continuous sori on margins of leaflets

ferns
Filicales

Filicales (*n.pl.*) true ferns. A division of the Pteridophyta (p. 54). The plants are characterized by obvious frond-like leaves, often with sporangia (p. 178) on the undersides, and rhizomes (p. 174) underground. They are homosporous (p. 54).

megaphyll (*n*) a large frond-like foliage leaf with a branched system of veins. It is typical of the Filicales (↑).

frond (*n*) a large well-divided leaf typical of the Filicales (↑).

sorus (*n*) a reproductive (p. 173) organ made up of a group of sporangia (p. 178) which occur on the undersides of leaves in Filicales (↑).

indusium (*n*) a flap of tissue (p. 83) covering the sorus (↑). **indusia** (*pl.*).

annulus (*n*) an arc or ring of cells in the sporangia (p. 178) of Filicales (↑) which are involved in opening the sporangium on drying to release the spores (p. 178).

circinate vernation the way in which the young fronds (↑) of the Filicales (↑) occur rolled-up.

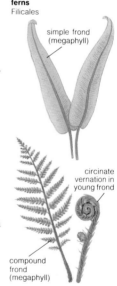

simple frond (megaphyll)

circinate vernation in young frond

compound frond (megaphyll)

sporangia of ferns
fern sorus

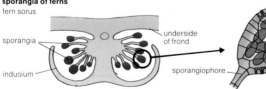

sporangia

underside of frond

wall of sporangium

spores

indusium

sporangiophore

annulus

Spermatophyta (*n.pl.*) seed plants. A division of the Plantae (p. 54) including the Gymnospermae (↓) and the Angiospermae (↓). They are widely distributed and are the dominant plants on land today. The body is highly organized and differentiated into root, stem and leaf, and there is a well-developed vascular system (p. 127). They are heterosporous (p. 54) with a dominant sporophyte (p. 177) generation which is the plant itself. The male gametophyte (p. 177) is the pollen (p. 181) grain while the female gametophyte is the egg which becomes the seed after fertilization (p. 175).

conifer an example of a gymnosperm

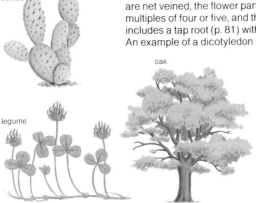

Gymnospermae (*n.pl.*) a division of the Spermatophyta (↑) which includes trees and shrubs in which the seeds are naked and not enclosed in a fruit. Most have cones (p. 55).

Angiospermae (*n.pl.*) flowering plants. A division of the Spermatophyta (↑) which includes the dominant plants on land. They are highly differentiated and have microsporophylls (p. 178) and megasporophylls (p. 179) combined into true flowers as stamens (p. 181) ˑand carpels (p. 179).

Dicotyledonae (*n*) dicotyledons. A class of the Angiospermae (↑) in which the seeds have two cotyledons (p. 168) or seed leaves, the leaves are net veined, the flower parts are usually in multiples of four or five, and the root system includes a tap root (p. 81) with lateral branches. An example of a dicotyledon is the buttercup.

dictotyledons some examples

cactus

oak

composite

legume

Monocotyledonae (*n*) monocotyledons. A class of the Angiospermae (p. 57) in which the seeds have one cotyledon (p. 168) or seed leaf, the leaves are usually parallel veined, and the flower parts are usually in multiples of three. Typical monocotyledons are the grasses.

ephemeral (*adj*) of a plant in which the complete life cycle from germination (p. 168) to the production of seed and death is very short so that many generations of the plant may be completed within a single year.

annual (*adj*) of a plant that completes its whole life cycle from the germination (p. 168) of seeds to the production of the next crop of seeds, followed by the death of the plant, within one year.

biennial (*adj*) of a plant that completes its whole life cycle from the germination (p. 168) of seeds to the production of the next crop of seeds, followed by the death of the plant, within two years. During the first year, the plant produces foliage and photosynthesizes (p. 93) to provide an energy store for the reproductive (p. 173) activities of the second year.

perennial (*adj*) of a plant which survives for a number of years and may or may not reproduce (p. 173) within the first year.

monocotyledons
some examples

grass

palm

orchid

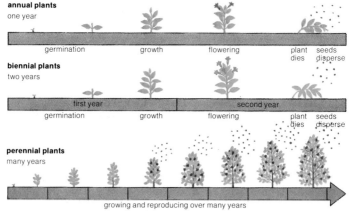

annual plants
one year

germination   growth   flowering   plant dies   seeds disperse

biennial plants
two years

first year   second year

germination   growth   flowering   plant dies   seeds disperse

perennial plants
many years

growing and reproducing over many years

two types of tree

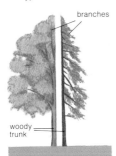

branches

woody trunk

deciduous with sympodial branching

evergreen with monopodial branching

deciduous (*adj*) (1) of a plant that sheds its leaves periodically, in accordance with the season, so that water loss by transpiration (p. 120) is reduced during periods of very dry or cold weather when water is in short supply. (2) in animals, of dentition (p. 104), for example, milk teeth (p. 104) which are shed and replaced by adult teeth.

evergreen (*adj*) of a plant that bears leaves throughout the year and which has adaptations, such as a leathery cuticle (p. 83) or needle-like leaves as in the gymnosperms (p. 57), conifers, to reduce water losses.

herbaceous (*adj*) of a perennial (↑) plant in which the foliage dies back each year while the plant survives as, for example, a bulb (p. 174), corm (p. 174), or tuber (p. 174). Herbaceous plants have no wood in their stems or roots.

tree (*n*) a woody, perennial (↑) plant which usually reaches a height of greater than 4 to 6 metres (13 to 20 feet) and which has a single stem from which branches grow at some distance from the ground level.

sapling (*n*) a young tree.

shrub (*n*) a woody, perennial (↑) plant which is smaller than a tree and from which branches grow quite close to ground level.

climber (*n*) a plant which though rooted in the ground uses other plants to support itself. Climbers use long coiled threadlike tendrils, suckers or adventitious roots (p. 81) to hold on to other plants and sometimes they twist around their stems.

foliage (*n*) all the leaves of a plant together. **foliar** (*adj*).

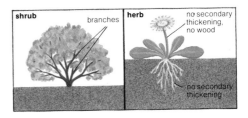

shrub
branches

herb
no secondary thickening, no wood

no secondary thickening

zoology (*n*) the study or science of animals or animal life.

Metazoa (*n*) a term used to describe all those truly multicellular (p. 9) animals as opposed to those animals which belong to the Protozoa (p. 44).

Coelenterata (*n*) a phylum of multicellular (p. 9) invertebrate (p. 75), aquatic and usually marine animals which includes the corals (↓) and jellyfishes. The body is radially symmetrical (↓) and consists of a simple body cavity which opens to the exterior by a mouth which is surrounded by a ring of tentacles (p. 71) that may have stinging cells or nematoblasts and are used for trapping prey and for defence. The body wall consists of an endoderm (p. 166) and an ectoderm (p. 166) separated by a jelly-like mesogloea. Reproduction (p. 173) takes place sexually, and asexually by budding (p. 173).

tissue grade the state of organization of animal cells into different types of tissue (p. 83) for different functions, such as muscular (p. 143) tissue and nervous tissue (p. 91) leading to greater co-ordination of activities such as response and locomotion (p. 143).

symmetrical (*adj*) of structures whose parts are arranged equally and regularly on either side of a line or plane (bilateral symmetry (p. 62)) or round a central point (radial symmetry (↓)).

asymmetrical (*adj*) not symmetrical (↑).

radial symmetry the condition in which the form of an organism is such that its structures radiate from a central point so that, if a cross section is made through any diameter, one half will be a mirror image of the other.

diploblastic (*adj*) of an animal whose body wall is composed of two layers, an endoderm (p. 166) and an ectoderm (p. 166), separated by a jelly-like mesogloea.

enteron (*n*) a sac-like body cavity which functions as a digestive (p. 98) tract or gut (p. 98).

planula larva the small, ciliated (p. 12) larva (p. 165) of a member of the Coelenterata (↑) which results from sexual reproduction (p. 173) and which swims to a suitable site before settling and growing into a polyp (↓).

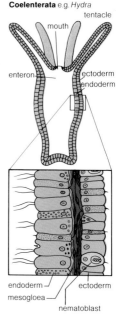

Coelenterata e.g. *Hydra*

tentacle

mouth

enteron

ectoderm
endoderm

endoderm

ectoderm

mesogloea

nematoblast

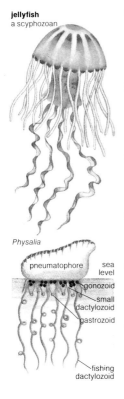

jellyfish
a scyphozoan

Physalia

pneumatophore — sea level

gonozoid

small dactylozoid

gastrozoid

fishing dactylozoid

Hydrozoa (*n*) a class of colonial and mainly marine Coelenterata (↑) in which alternation of generations (p. 176) is typical to give free-swimming medusae (↓) that reproduce (p. 173) sexually giving rise to sedentary polyps (↓) which reproduce asexually by budding (p. 173).

polyp (*n*) the sedentary stage in the life cycle of Coelenterata (↑) in which the body is tubular and surrounded by the tentacles (p. 71) at one end while attached to the substrate at the other. It reproduces (p. 173) asexually by budding (p. 173).

medusa (*n*) the free-swimming stage in the life cycle of Coelenterata (↑) in which the body is usually bell-shaped and surrounded by tentacles (p. 71) at one end. It reproduces (p. 173) sexually. **medusae** (*pl*.).

Scyphozoa (*n*) a class of the Coelenterata (↑) which comprises the jellyfishes and which may have no polyp (↑) form. The tentacles (p. 71) surround the mouth and bear stinging hairs.

Anthozoa (*n*) a class of marine Coelenterata (↑), including the sea anemones and corals (↓), in which the medusa (↑) stage is absent. The enteron (↑) is divided by vertical walls or septa and the animals may be colonial or solitary.

coral (*n*) any of the members of the Anthozoa (↑) which are today all colonial and in which the polyp (↑) is contained by a jelly-like, horny, or calcareous (containing $CaCO_3$) matrix (p. 88).

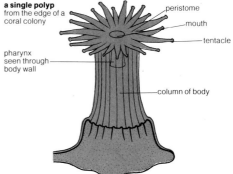

a single polyp
from the edge of a coral colony

peristome

mouth

tentacle

pharynx seen through body wall

column of body

Platyhelminthes (*n*) a phylum of multicellular (p. 9) invertebrate (p. 75) animals which includes the flatworms (↓). The body is bilaterally symmetrical (↓) and worm-like, and consists of a single opening to the gut which is often branched. There is no coelom (p. 167) or vascular system (p. 127). The body wall consists of an ectoderm (p. 166), mesoderm (p. 167), and endoderm (p. 166). They are usually hermaphrodite (p. 175).

flatworm (*n*) any of the members of the Platyhelminthes (↑) which have a flattened shape from above downwards that allows the oxygen used in respiration (p. 112) to diffuse into all parts of the body. There are three groups which include the mainly marine flatworms proper, the parasitic (p. 92) tapeworms, and the parasitic flukes.

Platyhelminthes e.g. *Planaria*

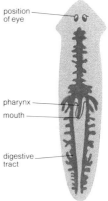

position of eye

pharynx

mouth

digestive tract

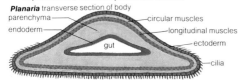

Planaria transverse section of body

parenchyma

endoderm

gut

circular muscles

longitudinal muscles

ectoderm

cilia

triploblastic (*adj*) of those animals, such as the Platyhelminthes (↑), in which the body wall consists of three layers, the ectoderm (p. 166), the mesoderm (p. 167), which is formed from cells which have moved from the surface layer, and the endoderm (p. 166).

bilateral symmetry the condition in which one half of the organism, from a section taken down its long axis, is a mirror image of the other half.

acoelomate (*adj*) of those animals without a coelom (p. 167) e.g. Platyhelminthes (↑).

flame cell one of a number of cup-shaped cells that occur in animals, such as the Platyhelminthes (↑), which, by the beating of their cilia (p. 12), draw fluid waste products into their cavity, and then to the exterior.

sucker (*n*) an organ of attachment, for example, in parasitic (p. 92) Platyhelminthes (↑) an adaptation of the pharynx (p. 99) is used to attach the organism to the host (p. 110).

Turbellaria (*n*) a class of the Platyhelminthes (↑) which includes free-living, mainly aquatic flatworms (↑) with a ciliated (p. 12) ectoderm (p. 166).

Planaria (*n*) a genus (p. 40) of the Turbellaria (↑) which includes freshwater forms that have numerous cilia (p. 12) on the underside that assist in locomotion (p. 143), respiration (p. 112) and direct food particles into the mouth.

Trematoda (*n*) a class of the Platyhelminthes (↑) which includes internal parasites (p. 110), such as the flukes, that have a complex life cycle including more than one host (p. 110), usually a vertebrate (p. 74) and an invertebrate (p. 75). They have suckers (↑), a branched gut (p. 98), and a thickened cuticle (p. 145) to resist digestion (p. 98) by the host.

bilharzia (*n*) a disease of humans living especially in Africa which is caused by a liver fluke that spends part of its life in freshwater snails, which are eaten by fish and then by humans. It enters the liver (p. 103) from the gut (p. 98) along the bile duct (p. 101).

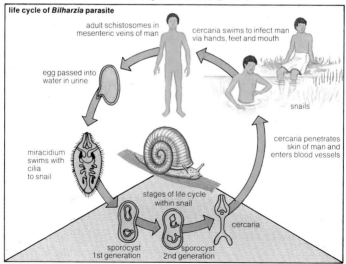

life cycle of *Bilharzia* parasite

adult schistosomes in mesenteric veins of man

cercaria swims to infect man via hands, feet and mouth

egg passed into water in urine

snails

miracidium swims with cilia to snail

cercaria penetrates skin of man and enters blood vessels

stages of life cycle within snail

cercaria

sporocyst 1st generation

sporocyst 2nd generation

Cestoda (*n*) a class of the Platyhelminthes (p. 62) which includes the internal parasites (p. 110), the tapeworm, that have a complex life cycle including more than one host (p. 110), both usually vertebrates (p. 74). They are armed with suckers as well as powerful grappling hooks on the head for attachment to the wall of the host's gut (p. 98). The body is divided into sections and has a tough cuticle (p. 145) to prevent digestion (p. 98) by the host.

Nematoda (*n*) a phylum of multicellular (p. 9) invertebrate (p. 75) animals which includes the roundworms (↓). The phylum includes terrestrial (p. 219), aquatic, and parasitic (p. 92) forms which have no cilia (p. 12), and an alimentary canal (p. 98) with a mouth and an anus (p. 103).

roundworm (*n*) any of the members of the Nematoda (↑) which have a characteristic rounded, unsegmented body that tapers at each end. They move by lashing the whole body into s shapes. The sexes are usually separate and the females lay large numbers of eggs. They are able to withstand adverse conditions by secreting (p. 106) a protective coat around the body.

threadworm (*n*) = roundworm (↑).

pseudocoel (*n*) a fluid-filled body cavity between the digestive (p. 98) tract and the other organs of roundworms (↑).

Annelida (*n*) a phylum of multicellular (p. 9) invertebrate (p. 75), mainly free-living, and typically marine aquatic animals which includes the 'true' segmented worms (↓). They have a central nervous system (p. 149), a thin cuticle (p. 145), and bristle-like chaetae (↓) on the body.

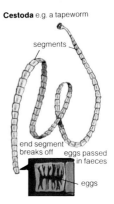

Cestoda e.g. a tapeworm

segments

end segment breaks off — eggs passed in faeces

eggs

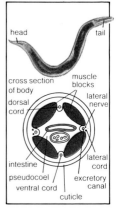

an adult roundworm

head | tail

cross section of body | muscle blocks

dorsal cord | lateral nerve

intestine | lateral cord

pseudocoel | excretory canal

ventral cord | cuticle

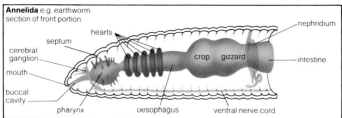

Annelida e.g. earthworm
section of front portion

nephridium

hearts

septum

cerebral ganglion

mouth

buccal cavity

crop | gizzard

intestine

pharynx | oesophagus | ventral nerve cord

segmented worm any of the members of the
Annelida (↑). They have a body which is divided
into obvious ring-like segments. Digestion
(p. 98) takes place in a simple, tube-like gut
(p. 98) which runs from the mouth at the front to
the anus (p. 103) at the rear. Between the body
wall and the gut is a fluid-filled coelom (p. 167).
They are hermaphrodite (p. 175).

nephridium (n) an organ which is used for excretion
(p. 134) in some invertebrates (p. 75), e.g. Annelida
(↑). It consists of a tube which opens to the exterior
at one end and, at the other, to flame cells (p. 62)
or to the coelom (p. 167). **nephridia** (pl.).

chaeta (n) one of a number of bristle-like
structures, composed of chitin (p. 49) which
are present, and arranged segmentally, along
the outside of the bodies of Annelida (↑). They
may assist in locomotion (p. 143) and, for
example, help earthworms (p. 66) to grip the
soil in which they live. **chaetae** (pl.).

chaeta transverse section
of earthworm body wall

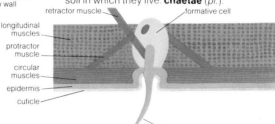

retractor muscle — formative cell
longitudinal muscles
protractor muscle
circular muscles
epidermis
cuticle
chaeta

seta[a] (n) = chaeta (↑).

cerebral ganglion one of the pair of solid strands
of nervous tissue (p. 91) which runs ventrally
and forms part of the central nervous system
(p. 149) in Annelida (↑) and other invertebrates
(p. 75) and to which the ganglia (p. 155) are
connected segmentally.

nerve cord = cerebral ganglion (↑).

Polychaeta (n) a class of marine Annelida (↑)
which includes the bristleworms, ragworms,
lugworms etc, that have many chaetae (↑). The
sexes are usually separate.

parapodium (n) in members of the Polychaetae (↑),
one of many extensions of the body wall on which
the chaetae (↑) are found. **parapodia** (pl.).

trochosphere larva the larva (p. 165) of Annelida
(p. 64) and some other groups of invertebrates
(p. 75) which may be related through evolution
(p. 208). It is free-swimming, planktonic (p. 227),
and covered with cilia (p. 12), especially around
the mouth which leads to the digestive (p. 98)
tract and anus (p. 103).

Hirudinea (*n*) a class of ectoparasitic (p. 110),
freshwater Annelida (p. 64) which includes the
leeches (↓). They have no chaetae (p. 65) or
parapodia (p. 65) and are hermaphrodite (p. 125)

leech (*n*) any of the Hirudinea (↑) which are
flattened and have a small sucker (p. 62) at
the front end and a larger, more obvious one at
the hind end. Some are carnivorous (p. 109)
but most are parasitic (p. 92), feeding on the
blood of their host (p. 110).

Oligochaeta (*n*) a class of mainly terrestrial
(p. 219) and freshwater Annelida (p. 64) which
includes the earthworms (↓). They have few
chaetae (p. 65), no parapodia (p. 65) and are
hermaphrodite (p. 175).

earthworm (*n*) any of the members of the
Oligochaeta (↑) which comprise the genus (p. 40)
Lumbricus. They live by burrowing in the soil
and digesting (p. 98) any organic matter in it
and are important in improving the structure of
the soil. They have a few chaetae (p. 65) and
secrete (p. 106) mucus (p. 99) from their skin.

clitellum (*n*) a saddle-like swelling of the
epidermis (p. 131) in earthworms (↑) which
binds the worms together during copulation
(p. 191) and then secretes (p. 106) the cocoon (↓).

cocoon (*n*) the protective covering , e.g. for the
eggs of an earthworm (↑) which is secreted
(p. 106) by the clitellum (↑).

leech   sucker

clitellum
copulation between earthworms

clitellum

parts of an insect

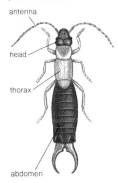

antenna

head

thorax

abdomen

Arthropoda (*n*) a phylum of multicellular (p. 9),
invertebrate (p. 75) animals that occupy aerial,
terrestrial (p. 219), freshwater and marine
environments (p. 218) and make up some 80
per cent of known animal life. Their bodies are
highly organized: a *head*, segments at the
forward end, containing the organs for feeding
and sensation as well as the brain (p. 155); a
thorax, segments between the head and the
abdomen which bear the jointed appendages
(↓), and when present, the wings; and an
abdomen, segments at the rear end. They are
bilaterally symmetrical (p.62), and protected by
a tough exoskeleton (p.145) which is segmented
for mobility. They often have compound eyes.
Growth takes place by ecdysis (p. 165). Each
segment usually bears a pair of jointed
appendages. The coelom (p. 167) is small and
the main body cavity is a haemocoel (p. 68)
containing a tube which is able to contract and
function as a heart (p. 124). There is a nerve
cord (p. 65) lying below the gut (p. 98)
connected to paired ganglia (p. 155) for each
segment. The best-known members of this
phylum are the insects (p. 69) and spiders (p.70).

metameric segmentation the condition in which
the body of an animal, especially certain
invertebrates (p. 75) such as Annelida (p. 64) and
Arthropoda (↑), is divided into a series of clearly
definable units which are essentially similar to
one another and repeat their patterns of blood
vessels (p. 127), organs of excretion (p. 134)
and respiration (p. 112), nerves (p. 149) etc. In
the Arthropoda, the similarity between the units
is reduced especially at the head end.

appendage (*n*) any relatively large protuberance
or projection from the main body of an organism.

jointed appendage any one of the projections from
the body of an arthropod (↑) which is divided into
a number of segments, seven in insects (p. 69),
and which are hinged between the segments to
allow for articulation (bending) in different planes.
The appendages are modified for different
functions e.g. locomotion (p. 143), feeding,
reproduction (p. 173), and respiration (p. 112).

antenna (*n*) one of the pair of highly mobile,
thread-like, jointed appendages (p. 67) which
occur on the head of an arthropod (p. 67)
which are used mainly for touch and smell
although, in some members of the group, they
may assist in locomotion (p.143). **antennae** (*pl.*).

haemocoel (*n*) a blood-filled (p. 90) cavity which
forms the main body cavity in arthropods
(p. 67). The coelom (p. 167) is reduced to
cavities surrounding the gonads (p. 187) etc
while the haemocoel is essentially an expanded
part of the blood system.

Crustacea (*n*) a class of the Arthropoda (p. 67),
which includes the aquatic shrimps and crabs
and the terrestrial (p. 219) woodlice. Typically,
the body is divided into a head with two pairs of
antennae (↑) and compound eyes, a thorax (p.115),
and an abdomen (p. 116). The exoskeleton
(p. 145) may be hardened by calcite ($CaCO_3$).

copepod (*n*) any of the group of small, aquatic
crustaceans (↑) which form an important part of
the marine plankton (p. 227). They have no
carapace (↓) or compound eyes and the first pair
of appendages (p. 67) on the head are modified
for filter feeding (p. 108) while the six pairs
on the thorax (p. 115) are used for swimming. The
abdomen (p. 116) has no appendages.

isopod (*n*) any of the flattened, terrestrial (p. 219),
freshwater, marine and often parasitic (p. 92)
members of the Crustacea (↑) which have no
carapace (↓) e.g. woodlice and shore slaters.

decapod (*n*) any of the terrestrial (p. 219),
freshwater and mainly marine members of the
Crustacea (↑) e.g. the highly specialized crabs,
lobsters and prawns. They often have an
elongated abdomen (p. 116) which ends in a
tail that enables them to escape predation
(p. 220) by swimming rapidly backwards. The
head and thorax (p. 116) may be fused and
protected by a carapace (↓). There are five
pairs of jointed appendages (p. 67) on the
thorax which are used in locomotion (p. 143)
and three pairs used in feeding. One or two of
the pairs of the legs may bear pincers which
are used in courtship and for defence.

Crustacea main groups

copepod

isopod

decapod

Myriapoda

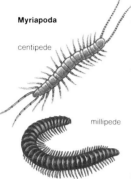

centipede

millipede

internal structure of an insect

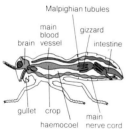

Malpighian tubules

main blood vessel
gizzard
brain
intestine

gullet crop
main
haemocoel nerve cord

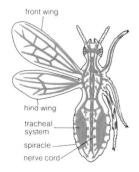

front wing

hind wing

tracheal system

spiracle

nerve cord

carapace (n) a toughened, shield-like part of the exoskeleton (p. 145) which protects the back and sides of the head and thorax (p. 115) in some arthropods (p. 67), such as the crabs.

crayfish (n) any of the relatively small, freshwater decapods (↑), which resemble and are related to the marine lobsters. It has an elongated carapace (↑) and a flexible abdomen (p. 116). The first of the five pairs of jointed appendages (p. 67) on the thorax (p. 115) are modified to form large pincers that are used for feeding and defence. The remaining four pairs are used for locomotion (p. 143).

Myriapoda (n) a class of terrestrial (p. 219) arthropods (p. 67), including centipedes (↓) and millipedes (↓), which have elongated bodies with many segments, each bearing one or more pairs of jointed appendages (p. 67). They have a definite head bearing antennae (↑) and mouth parts.

chilopod (n) any of the flattened, carnivorous (p. 109) myriapods (↑), including the centipedes (↓), which have one pair of legs on each segment of which the first pair contains poison glands (p. 87).

centipede (n) any of the chilopods (↑), but especially members of the genus (p. 40) *Lithobius*, the members of which live beneath stones.

millipede (n) any of the rounded, herbivorous (p. 109) myriapods (↑) which have four single segments at the front end of the body and numerous double segments each with two pairs of legs.

Insecta (n) the largest, most diverse (p. 213) and important class of the Arthropoda (p. 67), among which the majority are terrestrial (p. 219) or aerial. Insects include the majority of all the known animals with more than a million species (p. 40) described and perhaps another thirty million awaiting identification. The body is characteristically divided into a head with a pair of antennae (↑), compound and simple eyes and mouth parts adapted for different feeding methods; a thorax (p. 115) with three pairs of legs and, usually, two pairs of wings for flight; and a limbless abdomen (p. 116). They have a waterproof exoskeleton (p. 145) and a very efficient means of respiration (p. 112) by means of tracheae (p. 115).

metamorphosis (*n*) the process by which, under the control of hormones (p. 130) some animals, e.g. insects (p. 69) change rapidly in form from the larva (p. 165) into the adult with considerable destruction of the larval tissue (p. 83).

proboscis (*n*) an extension of the mouth parts of an insect (p. 69) which is used for feeding.

commissure (*n*) a nerve cord (p. 65) which connects the segmental ganglia (p. 155) in the arthropods (p. 67).

Arachnida (*n*) the class of mainly terrestrial (p. 219) arthropods (p. 67), which includes the spiders (↓), scorpions and pseudoscorpions. The body is divided into two main regions, the prosoma (↓) and opisthosoma (↓), and there are four pairs of walking legs, a pair of chelicerae and a pair of pedipalps (↓). The head has simple eyes and no antennae (p. 68). Respiration (p. 112) is achieved by book lungs (↓).

prosoma (*n*) the front region of the body of an arachnid (↑) which is made up of the head and thorax (p. 115) fused together.

opisthosoma (*n*) the hind region or abdomen (p. 116) of an arachnid (↑).

spider (*n*) an arachnid (↑) in which the prosoma (↑) and opisthosoma (↑) are separated by a narrow waist and which has spinnerets (↓).

pedipalps (*n.pl.*) the second pair of jointed appendages (p. 67) on the prosoma (↑) of arachnids (↑). They may be used for seizing prey, adapted as antennae (p. 68), or used for purposes of copulation (p. 191).

book lung one of the organs of respiration (p. 112) in arachnids (↑). They are composed of many fine layers of tissue (p. 83), resembling a book, through which blood (p. 90) flows and absorbs (p. 81) oxygen.

web (*n*) the thin, silken material which is spun by spiders (↑) in a variety of forms and used to capture prey such as flying insects, as in a net.

spinneret (*n*) one of the pair of appendages (p. 67) which is situated on the opisthosoma (↑) of spiders (↑) and secretes (p. 106) a liquid that hardens into silk for making webs (↑), wrapping the cocoon (p. 66), or binding the prey.

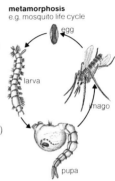
metamorphosis
e.g. mosquito life cycle

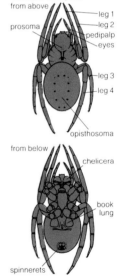
structure of a spider

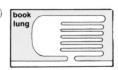
book lung

Mollusca

Gastropoda e.g. a snail

Cephalopoda
e.g. a squid

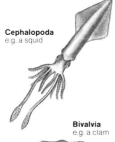

Bivalvia
e.g. a clam

**adult snail and larva
(trochophore)**

trochophore

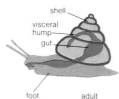

shell

visceral
hump

gut

foot   adult

Mollusca (*n*) a phylum of bilaterally symmetrical
(p. 62), invertebrate (p. 75), multicellular (p. 9)
animals that occupy terrestrial (p. 219),
freshwater, and marine environments (p. 218)
and include cockles, slugs, snails etc. They
have soft, unsegmented bodies which are
divided into a head region, a visceral hump (↓),
and a foot (↓). In some groups of molluscs, the
mantle (↓) secretes (p. 106) a hard shell. The
coelom (p. 167) is reduced and there is a
haemocoel (p. 68).

visceral hump the soft mass of tissue (p. 83)
which makes up the bulk of a mollusc (↑) and
which contains the main digestive (p. 98)
system.

foot[a] (*n*) a soft, muscular (p. 143) development of
the underside of the body of a mollusc (↑) which
is used for locomotion (p. 143).

mantle (*n*) a fold of the body wall which covers
the visceral hump (↑). In some molluscs (↑) it
secretes (p. 106) a shell composed of calcium
carbonate while, in others, it is folded to form a
cavity which encloses the organs of respiration
(p. 112).

radula (*n*) a tongue-like strip in most molluscs (↑)
which is covered with horny teeth and is used
to grind away food particles. As it is worn away,
it is continuously replaced.

tentacle (*n*) a flexible appendage (p. 67). In
cephalopods (p. 72), there are normally eight or
ten extending from the foot (↑) which is
incorporated into the head. Each tentacle bears
many suckers and they are used for sense
organs, for defence, and for grasping prey.

trochophore larva the free-swimming, ciliated
(p. 12) larva (p. 165) of aquatic molluscs (↑).

Gastropoda (*n*) a class of terrestrial (p. 219),
freshwater and marine molluscs (↑), including
winkles, slugs, and snails, in which the visceral
hump (↑) is coiled. This torsion (↓) of the
visceral hump is reflected in the coiling of the
shell. There is a muscular foot (↑) which is used
in locomotion (p. 143), the eyes are on tentacles
(↑), and gastropods feed using a radula (↑).

torsion (*n*) the act or condition of being twisted.

snail (*n*) any of the terrestrial (p. 219) members of
the Gastropoda (p. 71), which have no gills
(p. 113) but the mantle (p. 71) cavity functions
as a lung (p. 115). This includes the slugs
which lack the shell found in true snails.

Bivalvia (*n*) a class of flattened freshwater and
marine molluscs (p. 71) in which the mantle
(p. 71) occurs in two parts and secretes (p. 106)
a shell consisting of two hinged valves which
may be pulled together by powerful muscles
(p. 143). They have a poorly developed head
and are filter feeders (p. 108). Some bivalves
burrow into sand, mud, rock, or wood, some are
attached to the substrate by strong threads,
and others may be free swimming, propelling
themselves backwards by forcibly opening and
closing the valves.

mussel (*n*) any of a group of typical members of
the Bivalvia (↑) which include freshwater and
marine forms that have powerful muscles
(p. 143) to clamp their valves tightly closed for
protection. They are attached firmly to the
substrate, such as rocks, by strong threads.

siphon (*n*) a tube, e.g. one of two tubes which
protrude at the posterior end between the open
valves of a bivalve (↑) mollusc (p. 71) and
which form part of the system that circulates
water through the mantle (p. 71) cavity for
feeding and respiration (p. 112).

Cephalopoda (*n*) a class of marine molluscs
(p. 71) with a well-developed head containing a
complex brain (p. 155) and eyes. The head is
surrounded by a ring of sucker-covered
tentacles (p. 71) which is a modification of the
foot (p. 71). There is a muscular siphon (↑) for
respiration (p. 112) and the shell is much
reduced and usually internal.

octopus (*n*) any of the members of the
Cephalopoda (↑) with eight arm-like tentacles
(p. 71) and a soft, ovally-shaped body.

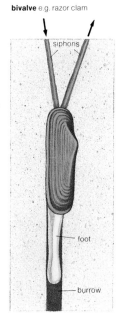
bivalve e.g. razor clam
siphons · foot · burrow

octopus

tentacle · sucker

Echinodermata

a starfish

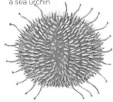

a sea urchin

a sea lily

Echinodermata (*n*) a phylum of radially symmetrical (p. 60) and usually five-rayed (↓), multicellular (p. 9), invertebrate (p. 75) animals that occupy marine environments (p. 218) and include the starfishes (↓) and sea urchins (p. 74). They have no head and a simple nervous system (p. 149). Part of the coelom (p. 167) is adapted to become a water vascular system (↓) which is unique to the group and connects with the tube feet (↓) which are used in locomotion (p. 143) and feeding. They have an internal skeleton (p. 143) of plates composed of calcite ($CaCO_3$) and most of them have spines.

spiny-skinned animal any of the members of the Echinodermata (↑) in which the ectoderm (p. 166) is covered with sharp, moveable, calcareous ($CaCO_3$) spines which connect with the calcareous ossicles (↓).

five-rayed radial symmetry radial symmetry (p. 60) which is typical of the Echinodermata (↑) in which there are five axes of symmetry.

tube foot any of the mobile, hollow, tube-like appendages (p. 67) which connect with the water vascular system (↓) and may end in suckers. They are used for locomotion (p. 143), feeding, and, in the sedentary sea-lilies, have cilia (p. 12) and are used for collecting food particles.

water vascular system a vascular system (p. 127) which is unique to the Echinodermata (↑) and consists of a series of canals containing sea water which, under pressure, operate the tube feet (↑).

madreporite (*n*) a sieve plate on the upper surface of echinoderms (↑) which is the opening of the water vascular system (↑) to the exterior.

calcareous ossicle any of the bone-like plates, made of calcium carbonate which make up the internal skeleton (p. 143) of the Echinodermata (↑).

starfish (*n*) any of the group of flattened, star-shaped Echinodermata (↑) which, typically, have five flexible arms radiating from the central disc which contains the main organs and the mouth on the underside. The arms have tube feet (↑) on the underside which are used for locomotion (p. 143) and for gripping prey. They usually live in the littoral (p. 219) zone.

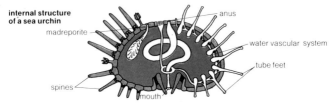

internal structure of a sea urchin — madreporite — anus — water vascular system — tube feet — spines — mouth

sea urchin any of the group of usually, globular, heart-shaped, or disc-shaped Echinodermata (p. 73) which have no arms and in which the calcareous ossicles (p. 73) are fused together to form a rigid, shell-like skeleton (p. 143) to which are attached spines that can be moved by the water vascular system (p. 73). They usually live on or buried in the sea bed feeding on plants and other debris through the mouth on the underside.

Chordata (*n*) a phylum of bilaterally symmetrical (p. 60), invertebrate (↓) and vertebrate (↓) multicellular (p. 9) animals, that includes humans and other mammals (p. 80) and is characterized by possessing a stiff, rod-like notochord (p. 167) during some stage of their life cycle.

cranium (*n*) the part of the skeleton (p. 143), composed of bone, of a vertebrate (↓) member of the Chordata (↑) which is also referred to as the skull and which contains the brain (p. 155).

vertebral column the part of the skeleton (p. 143) of a vertebrate (↓) member of the Chordata (↑) which is situated along the dorsal length of the body, from the cranium (↑) to the tail (↓), and is made of a linked chain of small bones or cartilages (p. 90), the vertebrae. It is flexible and allows for movement and locomotion (p. 143). It replaces the notochord (p. 167) and is a hollow column containing the spinal cord (p. 154). Also known as **spine** or **backbone**.

visceral cleft one of the paired openings in the pharynx (p. 99) which occur at some stage in the life cycle of members of the Chordata (↑) and persist in the aquatic species (p. 40). They lead from the exterior to the gills (p. 113) and are involved with filter feeding (p. 108) and gas exchange (p. 112) as water is pumped through them.

vertebrate (*n*) an animal with a vertebral column (↑).

the three main groups of fish

Agnatha
e.g. lamprey

cartilaginous fish
e.g. shark

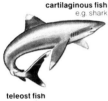

teleost fish
e.g. perch

invertebrate (*n*) an animal without a vertebral column (↑).

tail (*n*) an extension of the vertebral column (↑) which continues beyond the anus (p. 103) in most vertebrate (↑) members of the Chordata (↑.) It may be used for locomotion (p. 143), for balance and manoeuvrability (↓), or as a fifth limb.

manoeuvrability (*n*) the ability to make controlled changes of movement and direction.

Gnathostomata (*n*) a subphylum or superclass of the vertebrate (↑) Chordata (↑) which are characterized by the possession of a jaw (p. 105). The notochord (p. 167) is not retained throughout the life history.

Agnatha (*n*) a subphylum or superclass of the vertebrate (↑) Chordata (↑) which are characterized by having no jaw (p. 105). They are aquatic and primitive (p. 212).

Pisces the class of the Chordata (↑) which contains the fish. Fish are freshwater and marine animals with streamlined bodies that are usually covered with scales (p. 76). They have a powerful, finned (↓) tail (↑) which is used to propel them through the water, while their pairs of pelvic (↓) and pectoral (↓) fins are used for stability and manoeuvrability (↑). Gas exchange (p. 112) takes place in the gills (p. 113) and fish are exothermic (p. 130).

fin (*n*) a flattened, membraneous external organ on the body of a fish which usually occurs in pairs. It is used for steering, stability, and propulsion.

pectoral (*adj*) of the chest, e.g. the pectoral fins (↑) of a fish are attached to the shoulder and are used for steering up or down in the water and for counteracting pitching and rolling.

pelvic (*adj*) of the pelvic girdle (p. 147), e.g. the pelvic fins (↑) of a fish are attached to the pelvic girdle and are used for steering up or down and for counteracting pitching and rolling.

dorsal (*adj*) at, near or towards the back of an animal i.e. that part which is normally directed upwards (or backwards in humans).

ventral (*adj*) at, near or towards the part of an animal that is normally directed downwards (or forwards in humans).

scale (*n*) one of the many bony or horny plates which are made in the skin of fish and which may be above or beneath the skin. They overlap to form a protective and streamlined covering for the fish. Under the microscope (p. 9), it can be seen that they have a ring-like structure which represents the growth rate of the fish and can be used for aging purposes.

Chondrichthyes (*n*) a subclass of the Pisces (p. 75) which are entirely marine and include the sharks and rays. They are characterized by having an internal skeleton (p. 143) made of cartilage (p. 90) and are, therefore, also referred to as the cartilaginous fish. They have no swim bladder (↓) so that they sink if they cease moving.

cartilaginous fish = Chondrichthyes (↑).

Osteichthyes (*n*) a subclass of the Pisces (p. 75) which includes both freshwater and marine forms. They are characterized by having an internal skeleton (p. 143) and scales (↑) made from bone and are, therefore, referred to as bony fish. They possess a swim bladder (↓).

bony fish = Osteichthyes (↑).

teleost fish any of the main group of Osteichthyes (↑) in which the body tends to be laterally flattened and which have a swim bladder (↓) to adjust their buoyancy (↓). Their fins (p. 75) are composed of a thin, membraneous skin supported on bony rays. Their jaws (p. 105) are shortened so that the mouth can open widely and the visceral clefts (p. 74) are protected by a covering operculum (p. 113). The scales (↑) are thin, bony, and rounded. There is a wide variety of types of teleost fish and they occupy most aquatic environments (p. 218).

external features of a teleost fish

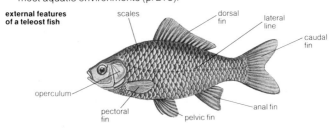

scales · dorsal fin · lateral line · caudal fin · anal fin · pelvic fin · pectoral fin · operculum

mermaid's purse

homocercal
tail fin

heterocercal
tail fin

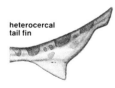

Amphibia
e.g. salamander

swim bladder a sac situated within the abdominal (p. 116) cavity of bony fish (↑). It contains a mixture of oxygen and nitrogen and oxygen can be pumped into it from the blood (p. 90) to increase the fish's buoyancy (↓) or vice versa so that the fish's depth can be controlled. It also functions as a sound detector and producer and, in lung fish, enables respiration (p. 112) out of water.

buoyancy (*n*) the ability to float in a liquid.

mermaid's purse the protective egg case which encloses the small number of eggs produced by cartilaginous fish (↑).

homocercal (*adj*) of a tail (p. 75), such as that of the teleost fish (↑), which is symmetrical (p. 60) in shape.

heterocercal (*adj*) of a tail (p. 75), such as that of the cartilaginous fish (↑), which is asymmetrically (p. 60) shaped such that the lower fin (p. 75) is larger than the upper fin to give the fish additional lift thereby compensating for the lack of a swim bladder (↑).

tetrapod (*n*) any of the vertebrate (p. 74) members of the Chordata (p. 74), such as a mammal (p. 80), which have two pairs of limbs for support, locomotion (p. 143), etc.

pentadactyl (*adj*) of the limb of a tetrapod (↑) which terminates in five digits, although the digits may be reduced or fused together as adaptations to various modes of life.

Amphibia (*n*) a class of primitive (p. 212) tetrapod (↑) chordates (p. 75), such as the frogs and toads, among which fertilization (p. 175) is external so that they must return to water to breed. Their larval (p. 165) forms are all aquatic and have gills (p. 113) but the majority of the adults are able to survive in damp conditions on land because they have a lung (p. 115) and are able to breathe air, respiring (p. 112) mainly through the thin, porous skin. Because of the thin skin, body fluids are easily lost. Like fish, they are exothermic (p. 130).

salamanders (*n.pl.*) members of the order Urodela of the Amphibia (↑) which includes tailed amphibia. The order Urodela also includes newts.

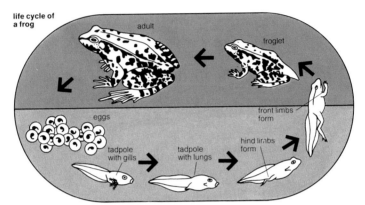

life cycle of a frog

adult

froglet

front limbs form

eggs

tadpole with gills

tadpole with lungs

hind limbs form

Anura (*n*) the order of the Amphibia (↑) which includes the aquatic, tree-dwelling, or damp-loving frogs and the warty skinned toads which can survive in drier conditions. Their hind limbs are elongated and powerful for jumping and have webbed feet for swimming.

Reptilia (*n*) a class of the Chordata (p. 74) which includes the most primitive (p. 212) tetrapods (p. 77), such as the snakes and lizards, that are wholly adapted to terrestrial (p. 219) environments (p. 218) although some, such as the turtles, have returned to an aquatic existence. They are air breathing and possess a true lung (p. 115). Their skin is scaly so that it is resistant to loss of body fluids. Since fertilization (p. 175) is internal, there is no need to return to the water to breed, and they lay amniote (p. 191) eggs with a leathery skin from which the young develop without passing through a larval (p. 165) stage. Like the fish and amphibians (p. 77), they are exothermic (p. 130).

cleidoic (*adj*) of an egg, such as that of a reptile (↑), which has a waterproof covering or shell that is permeable to air.

Chelonia (*n*) an order of the Reptilia (↑), which contains the turtles and tortoises, that are characterized by the plates of bone, overlaid with further horny plates, that enclose the body.

the four main groups of Reptilia

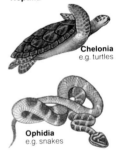

Chelonia e.g. turtles

Ophidia e.g. snakes

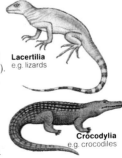

Lacertilia e.g. lizards

Crocodylia e.g. crocodiles

Squamata (*n*) an order of the Reptilia (↑), which contains the scaly skinned lizards and snakes.

Lacertilia (*n*) a suborder of the Squamata (↑) which contains the lizards. Most are truly tetrapod (p. 77) with a long tail (p. 75), have opening and closing eyelids, an eardrum (p. 158), and normal articulation of the jaws (p. 105).

Ophidia (*n*) a suborder of the Squamata (↑) which contains the snakes. They have elongated bodies with no limbs, no eardrum (p. 158) and no moveable eyelid. The jaws (p. 105) can be dislocated to allow a very wide gape so that large prey can be swallowed whole.

cloaca (*n*) the chamber which terminates the gut (p. 98) in all vertebrates (p. 74), other than the placental (p. 192) mammals (p. 80), and into which the contents of the alimentary canal (p. 98), the kidneys (p. 136), and the reproductive (p. 173) organs are discharged. There is a single opening leading to the exterior.

poison gland one of the modified salivary glands (p. 87) which may be present in some species (p. 40) of, for example, the Ophidia (↑), and which secrete (p. 106) toxic substances that may be, for example, injected into prey through the fangs.

Aves (*n*) a class of the Chordata (p. 74), which contains the birds. They are characterized by the possession of feathers (p. 147) for insulation and flight (p. 147) and other adaptations for flying. Although they are similar in many ways to the Reptilia (↑) from which they evolved (p. 208), they are endothermic (p. 130). There are some flightless species (p. 40). They lay amniote (p. 191) eggs with a calcareous ($CaCO_3$) shell.

internal organs of a bird

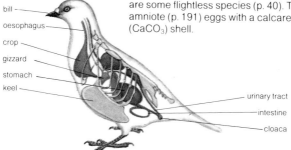

bill

oesophagus

crop

gizzard

stomach

keel

urinary tract

intestine

cloaca

bill (*n*) the horny structure which encloses the
jaws (p. 105) of birds. It lacks teeth and may
take a variety of forms adapted to different
methods of feeding. Also known as **beak**.

keel (*n*) a bony projection of the sternum (p. 149)
of birds to which the powerful pectoral (p. 75)
muscles (p. 148) are attached for flight (p. 147).

air sac one of a number of thin-walled, bladder-like
sacs in birds, which are connected to the lungs
and which are present in the abdominal (p. 116)
and thoracic (p. 115) cavities. They even
penetrate into some of the bones of the skeleton
(p. 143) to lighten the body of the bird without
reducing its strength. The tracheae (p. 115) of
some insects (p. 69) contain air sacs.

Mammalia (*n*) a class of the Chordata (p. 74)
which contains all the mammals, e.g. dogs,
cats and apes. They are endothermic (p. 130),
have a glandular (p. 87) skin, and are covered
with hair for insulation. They are characterized
by possessing mammary glands which secrete
(p.106) milk to feed the young. They possess
heterodont dentition (p. 104), a secondary
palate which enables them to eat and breathe
at the same time and relatively large brains
(p. 155).

Monotremata (*n*) a subclass of the Mammalia (↑)
which includes the primitive (p. 212) spiny
anteater and duck-billed platypus. They
possess a cloaca (p. 79) and lay eggs. The
young are transferred to a pouch and fed from
milk which is secreted (p. 106) on to a groove
in the abdomen (p. 116). They are covered with
hair but have a relatively low body temperature.
They have a poorly developed brain (p. 155).

Metatheria (*n*) a subclass of the Mammalia (↑)
which includes the marsupial or pouched forms,
such as the kangaroo. They are viviparous
(p. 192) but the poorly developed live young
are born after only a brief period of gestation
(p. 192) and then transferred to a pouch where
they are suckled and complete their growth.

Eutheria (*n*) a subclass of the Mammalia (↑)
which contains the 'true' viviparous (p. 192),
placental (p. 192) mammals.

**the three main
subclasses of mammals**

Monotremata (monotremes)
e.g. duck-billed platypus

Metatheria (marsupials)
e.g. kangaroo

Eutheria (placental mammals)
e.g. elephant

tap root

fibrous roots

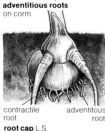

adventitious roots
on corm

contractile   adventitous
root   root

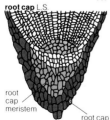

root cap L.S.

root
cap
meristem
        root cap

anatomy (n) the study or science of the internal structure of animals and plants.

histology (n) the study or science of tissues (p. 83).

morphology (n) the study or science of the external structure and form of animals and plants without particular regard to their function and internal structure or anatomy (↑).

physiology (n) the study or science of the processes which take place in animals and plants.

root (n) the structure of a plant which anchors it firmly to the soil and which is responsible for the uptake of water containing mineral salts from the soil and passing them into the stem. A root may also function as a food store. Unlike underground stems, a root does not contain chlorophyll (p. 12) and cannot bear leaves or buds.

tap root a main, usually central root which may be clearly distinguished from the other roots in a root system.

adventitious root one of a number of roots which grow directly from the stem of the plant as in bulbs (p. 174), corms (p. 174), and rhizomes (p. 174) and which do not grow from a main root.

fibrous root one of a number of roots which grow at the same time as the germination (p. 168) of a plant, such as a grass, and from which other lateral roots grow.

root cap a layer of cells at the tip of a root which protects the growing point from abrasion and wear by soil particles etc.

root hair a fine, thin-walled, tube-shaped structure which grows out from the epidermis (p. 131) just behind the root tip and which is in intimate contact with the soil surrounding a root. It greatly increases the root's surface area for the uptake of water. Water is drawn into the root hair by osmosis (p. 118) because the root hairs and the piliferous layer (p. 82) contain fluid with a lower osmotic potential (p. 118) than the water in the soil.

absorb (v) to take in liquid through the surface. **absorption** (n).

piliferous layer a single layer of cells which surrounds the root tip and part of the root of a plant and from which the root hairs (p. 81) grow. The cells contain fluid with a lower osmotic potential (p. 118) than that of soil water so that it is the main region of absorption (p. 81) of the root.

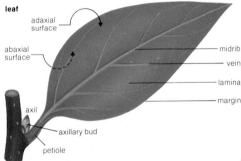

leaf

adaxial surface

abaxial surface

midrib

vein

lamina

margin

axil

axillary bud

petiole

leaf (n) a usually flattened, green structure which may or may not be joined to the stem of a plant by a stalk or petiole (↓). Its function is to make food for the plant in the form of carbohydrates (p. 17) by photosynthesis (p. 93).

vein[D] (n) one of a network of structures found in a leaf which provide support for the leaf and also transport the water, which is used during photosynthesis (p. 93), and organic (p. 15) solutes, into and out of the leaf tissue (↓).

lamina (n) the flat, thin, blade-shaped structure which comprises the major part of the foliage leaf.

petiole (n) the stalk or stem which may join the lamina (↑) to the stem of a plant.

midrib (n) the central or middle rib of a leaf which is an extension of the petiole (↑) into the leaf blade.

stem (n) that part of a plant which is usually erect and above ground and which bears the leaves, buds and flowers of the plant. Its function is to transport water and food throughout the plant, space out the leaves, and hold any flowers in a suitable position for pollination (p. 183).

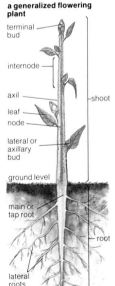

a generalized flowering plant

terminal bud

internode

axil

leaf

node

lateral or axillary bud

ground level

main or tap root

lateral roots

shoot

root

whorl

whorl (*n*) a group of three or more of the same organs arranged in a circle at the same level on a stem.

node (*n*) that part of the stem from which the leaves grow.

internode (*n*) the region of the stem between the nodes (↑).

bud (*n*) an undeveloped shoot which may develop into a flower or a new shoot. It consists of a short stem around which the immature leaves are folded and overlap. Buds may be at the tip of a shoot, when they are called terminal, or in the axils (↓) when they are called axillary.

shoot (*n*) the whole part of the plant which occurs above ground and which usually consists of a stem, leaves, buds and flowers.

axil (*n*) the angle between the upper side of a leaf and the stem on which the leaf is growing.

lenticel (*n*) a small raised gap or pore (p. 128) in the bark (p. 172) of a woody stem through which oxygen and carbon dioxide may pass.

tissue (*n*) a group of cells that perform a particular function in an organism.

vascular tissue a tissue (↑) which is specialized mainly for the transport of food and water throughout a plant, and is composed mainly of xylem (p. 84) and phloem (p. 84) together with sclerenchyma (p. 84) and parenchyma (↓).

ground tissue a tissue (↑), such as pith (p. 86) and cortex (p. 86), usually composed of parenchyma (↓), and which occupies all parts of the plant which do not contain the specialized tissue.

packing tissue = ground tissue (↑).

epidermal tissue a dermal (p. 131) tissue (↑) which forms a continuous outer skin over the surface of a plant. There are no spaces between the cells but it is penetrated by stomata (p. 120).

cuticle (*n*) the waterproof, waxy, or resinous outer surface of the epidermal tissue (↑) which occurs on the aerial (p. 219) parts of the plant.

parenchyma (*n*) a tissue (↑) which consists of rounded cells enclosed in a cellulose (p. 19) cell wall (p. 8) and containing air-filled intercellular (p. 110) spaces. Parenchyma supports the non-woody parts of a plant and also functions as storage tissue in the roots, stem and leaves.

parenchyma cells in cross section

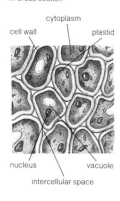

cytoplasm

cell wall plastid

nucleus vacuole

intercellular space

collenchyma (*n*) a tissue (p. 83) composed of elongated cells in which the primary cell wall (p. 14) is unevenly thickened with cellulose (p. 19). Collenchyma tissue is specialized to provide support to actively growing parts of the plant which may also need to be flexible.

sclerenchyma (*n*) a tissue (p. 83) which has a secondary cell wall (p. 14) of lignin (p. 19) and which is composed of sclereids (↓) and fibres (↓). Its function is to provide support.

sclereid (*n*) one of the two types of cells which comprise the sclerenchyma (↑). It is not always easy to differentiate between a sclereid and a fibre (↓) although they are generally very little longer than they are broad and are the stone cells of the shells of nuts and the stones of fruits.

fibre[p] (*n*) one of the two types of cells which comprise the sclerenchyma (↑). They are elongated lignified (p. 19) cells with no living contents and provide great support.

xylem (*n*) a vascular tissue (p. 83) consisting of hollow cells with no living contents and additional supporting tissue (p. 83) including fibres (↑), sclereids (↑) and some parenchyma (p. 83). The cell walls (p. 8) are lignified (p. 19), the thickness varying in shape and extent. The two main cell types found in xylem are vessels (↓) and tracheids (↓).

tracheid (*n*) one of the two types of cells found in xylem (↑). A tracheid is elongated and has tapering ends and cross walls. Tracheids run parallel to the length of the organ which contains them. Each tracheid is connected to its neighbour by pairs of pits (p. 14) through which water can easily pass.

phloem (*n*) a vascular tissue (p. 83) which transports food throughout the plant (translocation (p. 122)). It contains sieve tubes (↓) and companion cells (↓), and, in some plants, may also contain other cells, such as parenchyma (p. 83) and fibres (↑).

sieve tube a column of thin-walled, elongated cells which are specialized to transport food materials through the plant.

collenchyma cells
in cross section

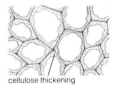

cellulose thickening

cell types in xylem

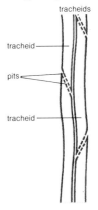

tracheids

tracheid

pits

tracheid

vessels

no end walls

spiral
thickening
of cell wall

phloem
(L.S.)

sieve plate
with pores

wall of sieve
element

**position of xylem and phloem
in young and old roots**

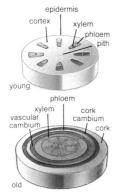

pericycle and
endodermis
cortex

phloem

xylem

young

wood
xylem | phloem
vascular
cambium

cork
cork
cambium
cork

old

**position of xylem and phloem
in young and old stems**

epidermis
cortex
xylem
phloem
pith

young

phloem
xylem
vascular
cambium

cork
cambium
cork

old

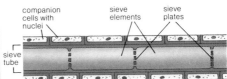

companion
cells with
nuclei

sieve
elements

sieve
plates

sieve
tube

sieve plate the perforated end wall of a sieve tube
(↑) through which strands of cytoplasm (p. 10)
pass to connect the neighbouring cells.

companion cell a small, thin-walled cell
containing dense cytoplasm (p. 10) and a well-
defined nucleus (p. 13) situated alongside the
sieve tube (↑) and which may aid the
metabolism (p. 26) of the sieve tube.

vessel[p] (*n*) one of the two types of cells found in
xylem (↑). Each vessel consists of a series of
cells arranged into a tube-like form with no
cross walls. It runs parallel to the length of the
organ containing it and is found mainly in the
angiosperms (p. 57). When mature, the vessel
has no living contents, and has thick lignified
(p. 19), walls for strength. There are several
types of thickening: *annular* which has rings of
lignin along the length of the cell; *spiral* which
has a spiral or coil of lignin round the inner
surface of the cell wall (p. 8); *scalariform* which
has a ladder-like series of bars of lignin on the
inner surface of the cell wall; *reticulate* which
has a network of lignin over the inner surface of
the cell wall; and *pitted* which has lignin over
the whole inner surface of the cell wall except
for many small pits (p. 14) or pores (p. 120).

vessels types of thickening
annular   spiral   scalariform   reticulate   pitted

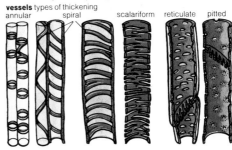

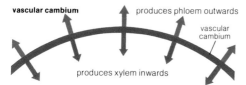

vascular cambium   produces phloem outwards

vascular
cambium

produces xylem inwards

cambium (*n*) the layer of narrow, thin-walled cells
which are situated between the xylem (p. 84)
and phloem (p. 84) and give rise by division to
secondary xylem (p. 172) and secondary
phloem (p. 172). The cambium does not lose its
ability to make new cells and is responsible for
lateral growth in plants.

secondary tissue the additional tissue (p. 83)
formed by the cambium (↑) leading to an
increase in the lateral dimensions of the stem
or root of a plant.

stele (*n*) the core or bundle of vascular tissue (p. 83)
in the centre of the roots and stems of plants.

exodermis (*n*) the outer layer of thickened cells
which may replace the epidermis (p. 131) in the
older parts of roots.

endodermis (*n*) the layer of cells surrounding the
stele (↑) on the innermost part of the cortex (↓)
of a root.

cortex (*n*) the tissue (p. 83) usually of
parenchyma (p. 83), which occurs in the stems
and roots of plants between the stele (↑) and
the epidermis (p. 131). It tends to make the
stem more rigid.

pith (*n*) the central core of the stem composed of
parenchyma (p. 83) tissue (p. 83) and found
within the stele (↑).

medullary ray one of a number of plates of
parenchyma (p. 83) cells which are arranged
radially and pass from the pith (↑) to the cortex
(↑) or terminate in secondary xylem (p. 84) and
phloem (p. 84).

pericycle (*n*) the outermost layer of the stele (↑)
with the endodermis (↑) and composed of
parenchyma tissue (p. 83).

mesophyll (*n*) the tissue (p. 83) which lies between
the epidermal (p. 131) layers of a leaf lamina
(p. 82) and is involved in photosynthesis (p. 93).

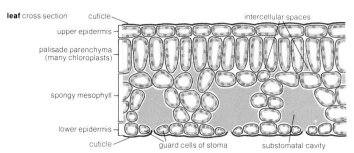

leaf cross section — cuticle — intercellular spaces — upper epidermis — palisade parenchyma (many chloroplasts) — spongy mesophyll — lower epidermis — cuticle — guard cells of stoma — substomatal cavity

types of epithelium

pavement (squamous) — nucleus — cytoplasm — basement membrane

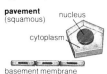

columnar

ciliated

glandular — goblet cell secreting mucus

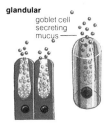

palisade mesophyll the mesophyll (↑) composed of cylindrical cells at right-angles to the leaf surface and situated just below the upper endodermis (↑). It contains numerous chloroplasts (p.12) and is concerned with photosynthesis (p. 93).

spongy mesophyll the mesophyll (↑) composed of loosely and randomly arranged cells with few chloroplasts (p. 12) and large air spaces which are connected with the atmosphere through the stomata (p. 120).

epithelium (n) an animal tissue (p. 83) composed of a sheet of cells which are densely packed and which covers a surface or lines a cavity.

endothelium (n) the epithelium (↑) which lines the heart (p. 124) and blood vessels (p. 127).

basement membrane a membrane (p. 14) composed of a thin layer of cement to which one of the cells of the epithelium (↑) is fixed.

ciliated epithelium an epithelium (↑) bearing cilia (p.12) found in the trachea (p.115) and bronchi (p.116).

glandular epithelium an epithelium (↑) which is specialized to form secretory (p. 106) glands (↓).

goblet cell a wine-glass-shaped cell which secretes (p. 106) mucus (p. 99) on to the outside of columnar epithelium (↑) to protect it.

gland (n) an organ which secretes (p. 106) chemicals to the outside. **glandular** (adj).

compound epithelium an epithelium (↑) which is made up from more than one layer of cells with columnar cells attached to the basal membrane (p. 14) and squamous (flattened) cells furthest from it. It is found at areas of stress such as the epidermis (p. 131) of skin.

stratified epithelium = compound epithelium (p.87).

transitional epithelium a stratified epithelium (↑) that is also capable of stretching, and found in areas such as the bladder (p. 135).

connective tissue the tissue (p. 83) that functions for support or packing purposes in animals. It has a few quite small cells with greater amounts of intercellular (p. 110) or matrix (↓) material.

matrix (*n*) the intercellular (p. 110) ground substance in which cells are contained.

areolar tissue a connective tissue (↑) which surrounds and connects organs. It is composed of collagen (↓) and elastic fibres (↓) in an amorphous matrix (↑).

fibroblast (*n*) an irregularly shaped but often elongated and flattened cell which functions in the production of collagen (↓).

mast cell a cell present in the matrix (↑) of areolar tissue (↑) which produces anticoagulant (p. 128) substances and is also found in the endothelium (p. 87) of blood vessels (p. 127).

macrophage (*n*) a large cell found widely in animals but particularly in the connective tissue (↑). Macrophages wander freely through the tissue and in the lymph nodes (p. 128) by amoeboid movement (p. 44) and destroy harmful bacteria (p. 42) by engulfing them as well as helping to repair any damage to tissue (p. 83).

collagen fibre a non-elastic fibre with high tensile strength found in connective tissue (↑), particularly in tendons (p. 146), skin and skeletal (p. 145) material. Also known as a **white fibre**.

elastic fibre a highly elastic fibre (p. 143) found in connective tissue (↑), particularly in ligaments (p. 146) and organs, such as lungs (p. 115). Also known as a **yellow fibre**.

adipose tissue a connective tissue (↑) similar to areolar tissue (↑) but containing closely packed fat cells and found under the skin and associated with certain organs to provide insulation, protection, and to store energy.

bone (*n*) a hard connective tissue (↑) composed of osteoblasts (p. 90) in a matrix (↑) made up of collagen fibres (p. 88) and calcium phosphate. It makes up the majority of the skeleton (p. 145).

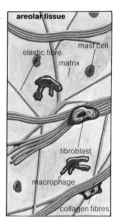

areolar tissue

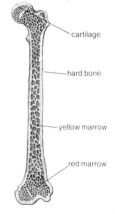

structure of a long bone

compact bone bone in which the Haversian canals (↓) are densely packed.

periosteum (*n*) connective tissue (↑) surrounding the bone and containing osteoblasts (p. 90) as well as many collagen fibres (↑) making it tough. The muscles (p. 143) and ligaments (p. 146) are attached to the periosteum.

Haversian canal a canal running along the length of bone and containing the nerves (p. 149) and blood vessels (p. 127) as well as the lymph vessels (p. 128) which secrete (p. 106) the osteocytes (p. 90).

Haversian system the system of Haversian canals (↑) surrounded by rings of bone and which connect with the surface of the bone and with its marrow (p. 90).

canaliculus (*n*) one of the fine canals linking the lacunae (↓) and containing the branches of the osteocytes (p. 90).

endosteum (*n*) a thin layer of connective tissue (↑) within a bone next to the cavity containing the marrow (p. 90).

lacuna (*n*) one of the spaces between the bone lamellae (↓) in which the osteoblasts (p. 90) are found. **lacunae** (*pl.*).

bone lamellae ring-like layers of calcified matrix (↑) in bone and surrounding the Haversian canals (↑).

compact bone
transverse section

periosteum

Haversian system

Haversian canal
bone lamellae
osteoblast { canaliculus
lacunae

endosteum

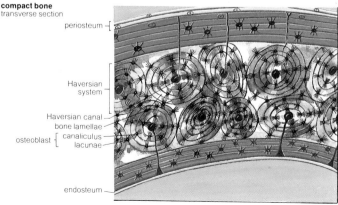

chondroblast (*n*) a cell which occurs in cartilage
(p. 90) and secretes (p. 106) the matrix (p. 88)
of clear chondrin (↓).

cartilage (*n*) a skeletal (p. 145) tissue (p. 83)
composed of chondroblasts (↑) in a matrix
(p. 88) of clear chondrin (↓). There are also
many collagen fibres (p. 88) contained within it.

chondrin (*n*) a bluish-white clear gelatinous
material which forms the ground substance of
cartilage (↑). Chondrin is elastic.

hyaline cartilage cartilage (↑) which contains
collagen fibres (p. 88) and which forms the
embryonic (p. 166) skeleton (p. 145).

osteoblast (*n*) a cell present in the hyaline
cartilage (↑) which is responsible for the laying
down of the calcified matrix (p. 88) of bone.

osteocyte (*n*) an osteoblast (↑) which has become
incorporated in the bone during its formation
and has stopped dividing.

spongy bone bone which contains a network of
bone lamellae (p. 89) surrounding irregularly
placed lacunae (p. 89) containing red marrow (↓).

epiphysis (*n*) the end of the limb (p. 147) bone in
mammals (p. 80) which enters and takes part in
the joint (p. 146).

marrow (*n*) the soft, fatty tissue (p. 83) which is
present in some bones and which produces
the white blood cells (↓).

blood (*n*) the specialized fluid in animals which is
found in vessels (p. 127) contained within
endothelial (p. 87) walls and which may contain
a pigment (p. 126) used in the transport of
respiratory (p. 112) gases as well as
transporting food and other materials
throughout the body.

plasma (*n*) the clear, almost colourless fluid part
of the blood (↑) which carries the white blood
cells (↓), the red blood cells (↓) and the platelets
(p. 128). It consists of 90 per cent water and 10
per cent other organic (p. 15) and inorganic
(p. 15) compounds.

serum (*n*) the clear, pale-yellow fluid which
remains after blood (↑) has clotted (p. 129) and
consists essentially of plasma (↑) without the
clotting agents.

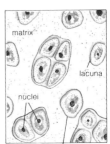

cartilage   chondroblast
                 (cartilage cell)

blood cells

**red blood cell
or erythrocyte**

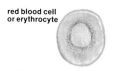

white blood cells

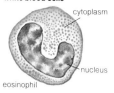

cytoplasm

nucleus

eosinophil

basophil

lymphocyte

nucleus   cytoplasm

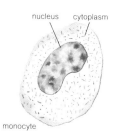

monocyte

red blood cell a blood cell which contains the
respiratory (p. 112) pigment (p. 126), such as
haemoglobin (p. 126).
erythrocyte (*n*) = red blood cell (↑).
white blood cell a blood cell which contains no
respiratory (p. 112) pigment (p. 126). White blood
cells are important in defending the body against
disease because they are able to engulf bacteria
(p. 42) as well as producing antibodies (p. 233).
leucocyte (*n*) = white blood cell (↑).
polymorphonuclear leucocyte a white blood
cell (↑) with a dark staining, lobed nucleus
(p. 13) and granular cytoplasm (p. 10). They
are produced in the bone marrow (↑).
granulocyte (*n*) = polymorphonuclear leucocyte (↑).
eosinophil (*n*) a polymorphonuclear leucocyte
(↑) which can be stained with acid (p. 15) dyes
such as eosin. Their numbers are normally
quite low in the blood (↑) but increase in
number if the body becomes infected with
parasitic (p. 92) or allergic (p. 234) disease.
basophil (*n*) a polymorphonuclear leucocyte (↑)
which can be stained with basic (p. 15) dyes.
Their numbers are normally very low in the blood
(↑) but they are able to engulf bacteria (p. 42).
neutrophil (*n*) the commonest type of leucocytes
(↑) which are able to migrate out of the blood (↑)
stream into the tissues (p. 83) of the body to
engulf bacteria (p. 42) wherever they invade.
On their death, they give rise to pus.
lymphocyte (*n*) a white blood cell (↑) which is
produced in the lymphatic system (p. 128) and
is important in defending the body against
disease. It has a large nucleus (p. 13) and clear
cytoplasm (p. 10).
monocyte (*n*) the largest type of white blood cell (↑)
and is produced in the lymphatic system (p. 128). It
has a spherical nucleus (p. 13) and clear cytoplasm
(p. 10). It actively engulfs and devours any
invading foreign bodies such as bacteria (p. 42).
nervous tissue tissue (p. 83) containing the nerve
cells (p. 149), which are specialized for the
transmission of nervous impulses (p. 150),
together with the supporting connective tissue
(p. 88).

nutrition (*n*) the means by which an organism provides its energy by using nutrients (↓).

nutrient (*n*) any material which is taken in by a living organism and which enables it to grow and remain healthy, replace lost or damaged tissue (p. 83), and provide energy for these and other functions.

holophytic (*adj*) of nutrition (↑), such as that of plants, in which simple inorganic compounds (p. 15) can be taken in and built up into complex organic compounds (p. 15) using the energy of light, either to provide energy for metabolism (p. 26) or growth or to make living protoplasm (p. 10).

chemosynthetic (*adj*) of nutrition (↑) in which energy is obtained by a simple inorganic (p. 15) chemical reaction such as the oxidation (p. 32) of ammonia to a nitrite by a bacterium (p. 42).

autotrophic (*adj*) of nutrition (↑) in which simple inorganic compounds (p. 15) are taken in and built up into complex organic compounds (p. 15).

heterotrophic (*adj*) of nutrition (↑), such as that in animals and some fungi (p. 46), in which the organic compounds (p. 15) can only be made from other complex organic compounds which have to be first taken into the body.

saprozoic (*adj*) of nutrition (↑) in which the organism takes in organic compounds (p. 15) only in solution (p. 118) rather than in solid form.

holozoic (*adj*) of nutrition (↑), as found in animals, in which complex organic compounds (p. 15) are broken down into simpler substances which are then used to make body structures or oxidized (p. 32) to supply the organism's energy needs.

saprophytic (*adj*) of nutrition (↑) in which the organism obtains complex organic compounds (p. 15) in solution (p. 118) from dead and/or decaying plant or animal material.

parasitic (*adj*) of nutrition (↑) in which the organism derives its food directly from another living organism at the expense of the host (p. 111) but without necessarily killing it.

types of nutrition

holophytic/autotrophic
green plant

heterotrophic/saprophytic
ink cap fungus

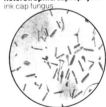

heterotrophic/parasitic
pathogenic bacterium

heterotrophic/holozoic
bird

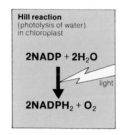

Hill reaction
(photolysis of water)
in chloroplast

$$2NADP + 2H_2O$$

light

$$2NADPH_2 + O_2$$

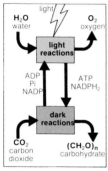

the links between the light
reactions and dark reactions
of photosynthesis

macronutrient (*n*) any nutrient (↑) which is required by an organism in substantial amounts. *See* p. 240.

major elements = macronutrients (↑).

micronutrient (*n*) a nutrient (↑) which is required in only minute or trace amounts. *See* p. 241.

chlorosis (*n*) the yellowing of the leaves of green plants caused by the loss of chlorophyll (p. 12).

active mineral uptake the uptake and transport through a plant, across a cell membrane (p. 14), of mineral ions from regions of low concentration to regions of high concentration. This process requires energy both to take up minerals and to retain them.

passive mineral uptake the uptake and transport of mineral ions through a plant, usually across a cell membrane (p. 14), from regions of high concentration to regions of low concentration by diffusion (p. 119) without using energy.

photosynthesis (*n*) the process that takes place in green plants in which organic compounds (p. 15) are made from inorganic compounds (p. 15) using the energy of light. It takes place in two main stages: in the light-dependent or photochemical stage, light is absorbed by chlorophyll (p. 12) in the chloroplasts (p. 12), located mainly on the leaves of plants, and used to produce ATP (p. 33) and to supply hydrogen atoms by oxidizing (p. 32) water. These are then used in the reduction (p. 32) of carbon dioxide. In the dark or chemical stage, carbon dioxide is reduced and carbohydrates (p. 17) are made. Photosynthesis will only take place at suitable temperatures and in the presence of chlorophyll, carbon dioxide, water, and light.

limiting factor one of a number of factors which controls the rate at which a chemical reaction, such as photosynthesis (↑), takes place. The rate is limited by that factor which is closest to its minimum or smallest value.

photosynthetic pigment one of the pigments (p. 126) which make up chlorophyll (p. 12) and absorb (p. 81) light. The following substances are photosynthetic pigments: chlorophyll a; chlorophyll b; carotene; and xanthophyll.

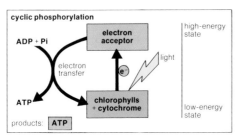

cyclic photophosphorylation a step in the light dependent stage of photosynthesis (p. 93) in which light is involved in the formation of ATP (p. 33) from ADP (p. 33) by the addition of phosphate.

non-cyclic photophosphorylation a step in the light-dependent stage of photosynthesis (p. 93) in which light is involved in the formation of ATP (p. 33) from ADP (p. 33) by the addition of phosphate and in which the water is split to provide hydrogen ions.

absorption spectrum a diagrammatic representation of the way in which a substance, such as chlorophyll (p. 12), absorbs (p. 81) radiation of different wavelengths by different amounts. Chlorophyll absorbs blue and red light readily so that it appears green.

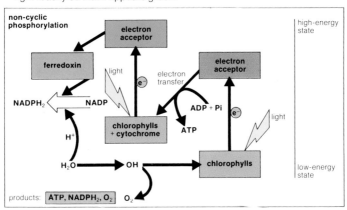

action spectra and absorption spectra in photosynthesis

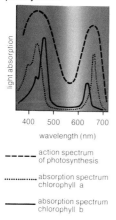

400   500   600   700

wavelength (nm)

– – – – action spectrum of photosynthesis

· · · · · · · · · absorption spectrum chlorophyll a

———— absorption spectrum chlorophyll b

action spectrum a diagrammatic representation of the way in which radiation of different wavelengths affects a process, such as photosynthesis (p. 93). In this case, it shows that red and blue light are the most effective in the action of photosynthesis.

photosystem I one of the two systems of pigments (p. 126) which each contains chlorophyll a (p. 12), accessory pigments, and electron carriers (p. 31) and which are involved in electron transfer reactions coupled with phosphorylation (↑). Also known as **pigment system I** or **PSI**.

photosystem II see photosystem I (↑). Also known as **pigment system II** or **PSII**.

ferredoxin (n) any of a number of red-brown (iron-containing) pigments (p. 126) which function as electron carriers (p. 31) in photosynthesis (p. 93).

plastoquinone (n) an electron carrier (p. 31) used in photosynthesis (p. 93).

C3 plant a plant in which PGA (p. 97) containing three carbon atoms is produced in the early stage of photosynthesis (p. 93). Photosynthesis in these plants is less efficient than in C4 plants (↓).

C4 plant a plant in which a dicarboxylic acid containing four carbon atoms is produced in the early stage of photosynthesis (p. 93). The method of fixing carbon dioxide has evolved from that of C3 plants (↑) and operates more efficiently.

C_4 pathway of CO_2 fixation

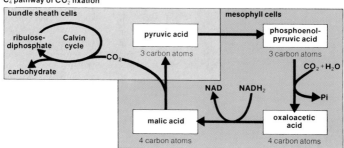

Calvin cycle the steps in the dark stage of
photosynthesis (p. 93) in which carbon dioxide
is reduced (p. 32) using the hydrogen produced
in the light-dependent stage and synthesized
into carbohydrates (p. 17) using the energy of
ATP (p. 33) also formed during the
light-dependent stage.

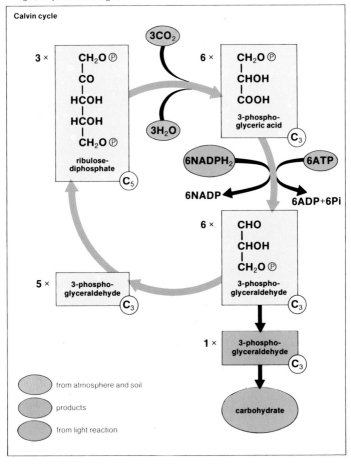

Calvin cycle

$3CO_2$

$3 \times$ CH$_2$O $\circled{P}$
      |
      CO
      |
      HCOH
      |
      HCOH
      |
      CH$_2$O $\circled{P}$

ribulose-
diphosphate C_5

$3H_2O$

$6 \times$ CH$_2$O $\circled{P}$
      |
      CHOH
      |
      COOH

3-phospho-
glyceric acid C_3

6NADPH$_2$ **6ATP**

6NADP   **6ADP+6Pi**

$6 \times$ CHO
      |
      CHOH
      |
      CH$_2$O $\circled{P}$

3-phospho-
glyceraldehyde C_3

$5 \times$ 3-phospho-
glyceraldehyde C_3

$1 \times$ 3-phospho-
glyceraldehyde C_3

from atmosphere and soil

products

from light reaction

carbohydrate

ribulose diphosphate RUDP. A pentose (p. 17) with which carbon dioxide is combined at the beginning of the Calvin cycle (↑).

phosphoglyceric acid PGA. A complex organic acid (p. 15) which is formed as the result of the combination of carbon dioxide with RUDP (↑) in the fixing of carbon dioxide at the beginning of the Calvin cycle (↑).

phosphoglyceraldehyde (*n*) a compound formed as the result of the reduction (p. 32) of PGA (↑) during the Calvin cycle (↑). This is then synthesized into starch (p. 18) which is the most important product of photosynthesis (p. 93). Also known as **triose phosphate**.

phosphoenol pyruvic acid PEP. An organic compound (p. 15) which is used by C4 plants (p. 96) in the fixation of carbon dioxide instead of RUDP (↑). Using this compound, carbon dioxide can be stored in chemical form and used later. This is very useful in areas, e.g. the tropics, where carbon dioxide may be in short supply.

compensation point the point at which the intensity of light is such that the amount of carbon dioxide produced by respiration (p. 112) and photorespiration (↓) exactly balances the amount consumed by photosynthesis (p. 93).

photorespiration (*n*) a light-dependent process in which carbon dioxide is produced and oxygen used up, wasting carbon and energy.

animal nutrition heterotrophic nutrition (p. 92) in which carbohydrates (p. 17) and fats are needed for structural materials and for energy, amino acids (p. 21) are needed to supply nitrogen, and to stimulate growth etc, minerals are required to ensure that the body functions healthily, and vitamins (p. 25) are required to promote and maintain growth.

joule (*n*) the work done when the point of application of a force of 1 newton is displaced through a distance of 1 metre in the direction of the force. 1 calorie (↓) is equivalent to 4.18 joules. The joule can be used as a measure of the energy value of nutrients (p. 92).

kilojoule (*n*) = 1000 joules (↑).

calorie (*n*) see joule (↑).

compensation point

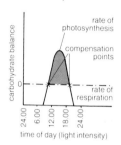

rate of photosynthesis

compensation points

rate of respiration

carbohydrate balance

time of day (light intensity)

photorespiration

O_2

CO_2

high O_2 concentration in plant tissue

low CO_2 concentration in plant tissue

gut or alimentary canal

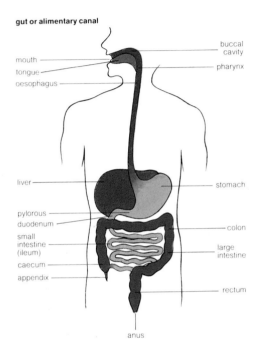

mouth
tongue
oesophagus
liver
pylorous
duodenum
small
intestine
(ileum)
caecum
appendix
anus
buccal
cavity
pharynx
stomach
colon
large
intestine
rectum

gut (*n*) a tube, the gastro-intestinal tract, usually leading from the mouth to the anus (p. 103), in animals, which in humans may be as much as 9 metres long and in which food is conveyed, digested (↓) and absorbed (p. 81).

alimentary canal = gut (↑).

ingestion (*n*) the process of taking nutrients (p. 92) into the body for digestion (↓).

digestion (*n*) the breakdown of complex organic compounds (p. 15) or nutrients (p. 92) into simpler, soluble materials which can then be used in the metabolism (p. 26) of the animal.

egestion (*n*) the process of eliminating or discharging food or waste products from the body.

faeces (*n.pl.*) the substances that remain after digestion (↑) and absorption (p. 81) of food in the alimentary canal (↑).

defaecation (*n*) the process of egesting (↑) unwanted food from the body. Material which is defaecated has not taken part in the metabolism (p. 26) of the organism and is therefore not an example of excretion (p. 134).

assimilation (*n*) after digestion (↑), the process of taking into the cells the simple, soluble organic compounds (p. 15) which can then be converted into the complex organic compounds from which the organism is made. **assimilate** (*v*).

buccal cavity the mouth cavity. In mammals (p. 80), that part of the alimentary canal (↑) into which the mouth opens and in which food particles are masticated (p. 104) before they are swallowed.

mucus (*n*) any slimy fluid produced by the mucous membranes (p. 14) of animals and used for protection and lubrication.

saliva (*n*) a fluid secreted into the buccal cavity (↑) by the salivary gland (p. 87) in response to the presence of food. It consists mainly of mucus (↑) and lubricates the food before swallowing. It contains an enzyme (p. 28) in some animals to aid the digestion (↑) of starch (p. 18).

pharynx (*n*) the part of the alimentary canal (↑) between the buccal cavity (↑) and the oesophagus (↓) into which food that has been masticated (p. 104) is pushed by the tongue. The pharynx then contracts by muscular (p. 143) action to force the food into the oesophagus. The gill slits (p. 113) open into the pharynx in fish.

oesophagus (*n*) the part of the alimentary canal (↑) between the pharynx (↑) and the stomach (p.100). It is lined with a folded mucous membrane (p.14) and has layers of smooth muscle fibres (p.144) which contract to force food into the stomach.

epiglottis (*n*) a flap which closes the trachea (p. 175) during swallowing so that food passes into the oesophagus (↑) and not into the trachea.

bolus (*n*) a rounded mass consisting of masticated (p. 104) food particles and saliva (↑) into which food is formed in the buccal cavity (↑) before swallowing.

stomach (*n*) the part of the alimentary canal
(p. 98) between the oesophagus (p. 99) and the
duodenum (↓) into which food is passed and
can be stored in quite large amounts for long
periods so that it is not necessary for the
animal to be eating continuously. Food is mixed
with gastric juices and, although little of it is
absorbed (p. 81), materials, such as minerals or
vitamins (p. 25), may be taken into the blood
(p. 90) stream. The stomach is muscular (p. 143)
and is lined with a mucous membrane (p. 14).

peristalsis (*n*) the waves of rhythmical
contractions which take place in the alimentary
canal (p. 98) by muscular (p. 143) action and
which force food through the canal.

chief cell one of a number of cells found in the
gastric glands (p. 87) which secrete (p. 106) the
enzymes (p. 28) pepsin (p. 107) and rennin
(p. 106) which digest (p. 98) proteins (p. 21)
and milk protein (in young mammals (p. 80))
respectively. Also known as **peptic cell**.

fundis gland a gland (p. 87) in the stomach (↑)
which secretes (p. 106) mucus (p. 99) to protect
and lubricate the wall of the stomach.

peristalsis

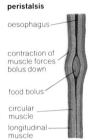

oesophagus

contraction of
muscle forces
bolus down

food bolus

circular
muscle

longitudinal
muscle

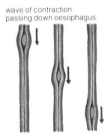

wave of contraction
passing down oesophagus

section of stomach wall

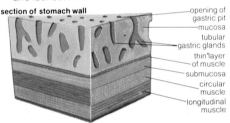

opening of
gastric pit
mucosa
tubular
gastric glands
thin layer
of muscle
submucosa
circular
muscle
longitudinal
muscle

detail of gastric gland

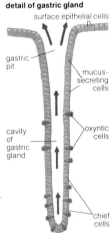

surface epithelial cells

gastric
pit

mucus-
secreting
cells

cavity
of
gastric
gland

oxyntic
cells

chief
cells

oxyntic cell one of a number of cells found in the
gastric glands (p. 87) which secrete (p. 106)
hydrochloric acid (HCl) which in turn kills
harmful bacteria (p. 42), makes available
calcium and iron salts and provides a suitably
low pH (p. 15) for the formation of pepsin (p. 107).

chyme (*n*) a partially broken-down, semi-fluid
mixture of food particles and gastric juices
which is then released in small quantities into
the duodenum (↓).

section through lining of duodenum

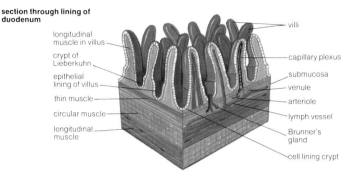

longitudinal muscle in villus
crypt of Lieberkuhn
epithelial lining of villus
thin muscle
circular muscle
longitudinal muscle

villi
capillary plexus
submucosa
venule
arteriole
lymph vessel
Brunner's gland
cell lining crypt

duodenum (*n*) the part of the alimentary canal
(p. 98) between the stomach (↑) and the ileum
(p. 102) which forms the first part of the small
intestine (p. 102) into which the chyme (↑)
passes from the stomach. Digestion (p. 98)
continues in the duodenum with the aid of
intestinal juices (p. 102) and, in addition, it
receives secretions (p. 106) from the pancreas
(p. 102) and the liver (p. 103). Its lining is
covered with villi (p. 103) and the glands (p. 87)
that secrete the intestinal juices.

chyle (*n*) lymph (p. 128) containing the results of
digestion (p. 98). The liquid looks milky because
it contains emulsified (p. 26) fats and oils.

bile (*n*) a secretion (p. 106) from the liver (p. 103)
which contains some waste material from the
liver and bile salts which emulsify fats, increase
the activity of certain enzymes (p. 28), aid in the
absorption (p. 81) of some vitamins (p. 25) and
is rich in sodium bicarbonate which neutralizes
stomach (↑) acid.

bile duct the tube through which bile (↑) is passed
from the liver (p. 103) to the duodenum (↑).

gall bladder a sac-like bladder extending from
the bile duct (↑) and situated within or near the
liver (p. 103). It functions as a store for bile (↑)
when it is not required for digestive (p. 98)
purposes and then, by muscular (p. 143)
contractions, empties into the duodenum (↑)
through the bile duct.

pancreatic juice a solution (p. 118) in water of alkaline salts, which neutralize the acid (p. 15) from the stomach (p. 100), and enzymes (p. 28) to aid in digestion (p. 98).

pancreas (*n*) a gland (p. 87) which is connected to the duodenum (p. 101) by a duct and which produces pancreatic juice (↑) and insulin (↓).

islets of Langerhans cells contained within the pancreas (↑) which produce insulin (↓).

insulin a hormone (p. 130) which controls the sugar level in the blood (p. 90) and, if it is deficient, the sugar level rises while, if it is in excess, the level falls leading to a coma.

jejunum (*n*) the part of the small intestine (↓) between the duodenum (p. 101) and the ileum (↓).

small intestine the narrow tube which forms part of the alimentary canal (p. 98) between the stomach (p. 100) and the colon (↓). It is the main region of digestion (p. 98) and absorption (p. 81) and includes the duodenum (p. 101).

ileum (*n*) the longest, usually coiled part of the small intestine (↑) between the jejunum (↑) and the colon (↓). It is muscular (p. 143) and causes food particles to pass along it by peristalsis (p. 100). Its lining is folded and covered with large numbers of villi (↓) which increase the surface area for absorption (p. 81).

intestinal juice a secretion (p. 106) produced in the glands (p. 87) of the intestine (↑) which contains a mixture of enzymes (p. 28) of digestion (p. 98), such as amylase (p. 106) and sucrase (p. 107). Also known as **succus entericus**.

Brunner's glands the deep-lying glands (p. 87) in the walls of the duodenum (p. 101) which secrete (p. 106) intestinal juices.

crypts of Lieberkuhn the glands (p. 87) found within the walls of the small intestine (↑) which secrete (p. 106) the intestinal juices.

appendix (*n*) in humans, a blind-ended tube at the end of the caecum (↓).

caecum (*n*) a blind-ended branch of the gut (p. 98) at the junction of the small and large intestines (↑). It is very large and important in the digestions (p. 98) of some mammals (p. 80), not humans.

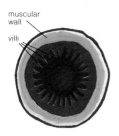

section across
small intestine (ileum)

muscular
wall

villi

villus

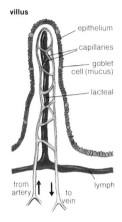

epithelium

capillaries

goblet cell (mucus)

lacteal

from artery

to vein

lymph

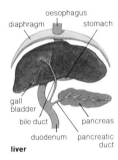

oesophagus

diaphragm

stomach

gall bladder

bile duct

pancreas

duodenum

pancreatic duct

liver

colon (*n*) the first part of the large intestine. It secretes (p. 106) mucus (p. 99) and contains the remains of food materials which cannot be digested (p. 98) as well as the digestive juices. From this material, water and vitamins (p. 25) are reabsorbed into the blood (p. 90) leaving the faeces (p. 99) which are moved on to the rectum (↓).

rectum (*n*) the part of the large intestine in which the faeces (p. 99) are stored before release through the anus (↓).

anus (*n*) the posterior opening to the alimentary canal (p. 98) through which the faeces (p. 99) may pass at intervals and which is closed by a ring of muscles (p. 143) called the anal sphincter (p. 127).

villi (*n.pl.*) the rod-like projections which cover the lining of the small intestine (↑) to increase the surface area for absorption (p. 81). **villus** (*sing.*).

liver (*n*) a gland (p. 87) which lies close to the stomach (p. 100) and is connected with the small intestine (↑) by the bile duct (p. 101) through which it secretes (p. 106) bile (p. 101) for digestive (p. 98) purposes. The liver also removes damaged red cells (p. 91) from the blood (p. 90), stores iron, synthesizes vitamin (p. 25) A, and stores vitamins A, D, and B, synthesizes blood proteins (p. 21), removes blood poisons, and synthesizes agents which help blood clot (p. 129), breaks down alcohol, stores excess carbohydrates (p. 17) and metabolizes (p. 26) fats.

hepatic portal vein one of a system of veins (p. 127) which carry blood (p. 90) rich in absorbed (p. 81) food materials, such as glucose (p. 17), direct from the intestine (↑) to the liver (↑).

liver cell one of the cells making up the liver (↑). Each cell is in direct contact with the blood (p. 90) so that material diffuses between blood and liver very rapidly. Liver cells are cube shaped with granular cytoplasm (p. 10).

reticulo-endothelial system a system of macrophage (p. 88) cells which is present in the liver (↑) and other parts of the body and which is in contact with the blood (p. 90) and other fluids. These macrophage cells are able to engulf foreign bodies and thus protect the body from infection, damage and disease.

tooth (*n*) one of a number of hard, resistant structures found growing on the jaws (↓) of vertebrate (p. 74) animals and which are used to break down food materials mechanically. They may be specialized for different functions among different animals and even within the same animal. **teeth** (*pl.*).

dentition (*n*) the kind, arrangement and number of the teeth (↑) of an animal.

heterodont dentition the condition in which the teeth (↑) of an animal, typically mammals (p. 80), are differentiated into different forms, such as molars (↓) and canines (↓), to perform different functions, such as grinding up the food or killing prey.

homodont dentition the condition in which the teeth (↑) of an animal are all identical.

mastication (*n*) the process which takes place in the buccal cavity (p. 99) whereby food is mechanically broken down by the action of teeth (↑), tongue, and cheeks into a bolus (p. 99) for swallowing.

dental formula a formula which indicates by letters and numbers the types and numbers of teeth (↑) in the upper and lower jaws (↓) of a mammal (p. 80). For example, the dental formula of a human would be
     i2/2, c1/1, p2/2, m3/3
which indicates that both the upper and lower jaws have two incisors (↓), one canine (↓), two premolars (↓), and three molars (↓) on each side of the jaw.

incisor (*n*) one of the chisel-shaped teeth (↑), very prominent in rodents, that occur at the very front of a mammal's (p. 80) jaw (↓) and which have a single root and a sharp edge with which to sever portions of the food.

canine (*n*) one of the 'dog teeth' or pointed conical teeth (↑) which occur between the incisors (↑) and the premolars (↓) and which are used to kill prey in carnivorous (p. 109) animals such as dogs and cats.

carnassial (*n*) one of the large flesh-cutting teeth (↑) found in terrestrial (p. 219) carnivores (p. 109).

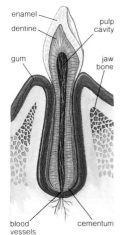

heterodont dentition
e.g. carnivore (a dog)

incisors
molars
upper jaw (maxilla)
canines
carnassials
premolars
lower jaw (mandible)

incisor

enamel
pulp cavity
dentine
gum
jaw bone
blood vessels and nerve fibres
cementum

molar

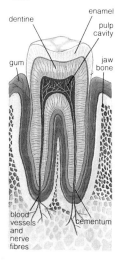

dentine

enamel

pulp cavity

gum

jaw bone

blood vessels and nerve fibres

cementum

premolar (*n*) one of the crushing and grinding cheek teeth (↑) that occur between the canines (↑) and molars (↓) in mammals (p. 80). They are usually ridged and furrowed with more than one root. They are represented in the first 'milk' or deciduous (p. 59) dentition (↑).

molar (*n*) one of the large crushing and grinding cheek teeth (↑) that occur at the back of the mouth of a mammal (p. 80). They are ridged and furrowed with more than one root. They are not represented in the first 'milk' or deciduous (p. 59) dentition (↑) and, in humans, there are four which do not erupt until later in life and are referred to as 'wisdom teeth'.

enamel (*n*) the hard outer layer of the tooth (↑) of a vertebrate (p. 74). It is composed mainly of crystals of the salts, carbonates, phosphates, and fluorides of calcium held together by small amounts of an organic compound (p. 15).

dentine (*n*) the hard substance that makes up the bulk of the tooth (↑) in mammals (p. 80). It is similar to bone but has a higher mineral content and no cells.

cementum (*n*) the hard substance that covers the root of the tooth (↑) in mammals (p. 80). It is similar to bone but has a higher mineral content and lacks Haversian canals (p. 89).

pulp cavity the substance within the centre of a tooth (↑) which contains the blood vessels (p. 127) and nerves (p. 149) which supply the tooth, together with connective tissue (p. 88). It connects to the tissue (p. 83) into which the tooth is fixed.

gum (*n*) the tissue (p. 83) that surrounds and supports the roots of the teeth (↑) and covers the jaw (↓) bones. It contains nerves (p. 149) and many blood capillaries (p. 127) giving it the characteristic pink colour when it is healthy. Also known as **gingiva**.

jaw (*n*) one of the bones in which the teeth (↑) are set. The jaw movements as well as the dentition (↑) of different animals are specialized for different actions, for example, tearing, snatching and chewing movements, such as crushing and grinding.

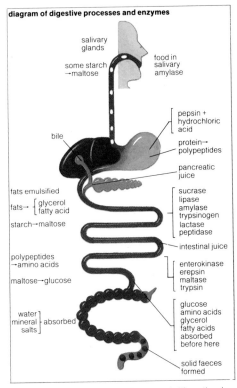

diagram of digestive processes and enzymes

salivary glands

some starch →maltose

food in salivary amylase

bile

pepsin + hydrochloric acid

protein→ polypeptides

pancreatic juice

fats emulsified

fats→ [glycerol / fatty acid]

starch→maltose

sucrase
lipase
amylase
trypsinogen
lactase
peptidase

intestinal juice

polypeptides →amino acids

maltose→glucose

enterokinase
erepsin
maltase
trypsin

water]
mineral ├ absorbed
salts]

glucose
amino acids
glycerol
fatty acids
absorbed
before here

solid faeces formed

secretion (*n*) a material with a special function in an organism which is made within a cell and passed out of the cell to perform its function. **secrete** (*v*).

amylase (*n*) any of a number of enzymes (p. 28) which catalyze (p. 28) the hydrolysis (p. 16) of carbohydrates (p. 17), such as starch (p. 18), into simple sugars. It is secreted (↑) as saliva (p. 99), and in the pancreas (p. 102) and the small intestine (p. 102).

rennin (*n*) an enzyme (p. 28) which coagulates (p. 128) milk. It is secreted (↑) by the gastric glands (p. 87) of the stomach (p. 100).

maltase (*n*) an enzyme (p. 28) which catalyzes (p. 28) the hydrolysis (p. 16) of maltose (p. 18) into two molecules of glucose (p. 17). It is secreted (↑) by the small intestine (p. 102).

lactase (*n*) an enzyme (p. 28) which catalyzes (p. 28) the hydrolysis (p. 16) of the disaccharide (p. 18) lactose (p. 18) into glucose (p. 17) and galactose (p.18). It is secreted (↑) by the small intestine (p.102).

sucrase (*n*) an enzyme (p. 28) which catalyzes (p. 28) the hydrolysis (p. 16) of sucrose (p. 18) into glucose (p. 17) and fructose (p. 17). It is secreted (↑) by the small intestine (p. 102). Also known as **invertase**.

erepsin (*n*) a mixture of enzymes (p. 28) which catalyzes (p. 28) the breakdown of proteins (p. 21) into amino acids (p. 21). It is secreted (↑) by the small intestine (p. 102).

lipase (*n*) an enzyme (p. 28) which catalyzes (p. 28) the hydrolysis (p. 16) of fats into fatty acids (p. 20) and glycerol (p. 20). It is secreted (↑) by the pancreas (p. 102).

enterokinase (*n*) an enzyme (p. 28) which catalyzes (p. 28) the conversion of trypsinogen (p. 108) into trypsin (↓). It is secreted (↑) by the small intestine (p. 102).

chymotrypsin (*n*) an enzyme (p. 28) which catalyzes (p. 28) the conversion of proteins (p. 21) into amino acids (p. 21). It is secreted (↑) by the pancreas (p. 102).

pepsin (*n*) an enzyme (p. 28) which catalyzes (p.28) the hydrolysis (p. 16) of proteins (p. 21) into polypeptides (p. 21) in acid solution (p.118). It is secreted (↑) by the stomach (p. 100) as pepsinogen (↓).

pepsinogen (*n*) the inactive form of pepsin (↑) which is secreted (↑) by the stomach (p. 100) and activated by hydrochloric acid (HCl).

gastrin (*n*) a hormone (p. 130) which stimulates the secretion (↑) of hydrochloric acid (HCl) and pepsin (↑) in the stomach (p. 100). It is activated by the presence of food materials.

trypsin (*n*) an enzyme (p. 28) which catalyzes (p. 28) the hydrolysis (p. 16) of proteins (p. 21) into polypeptides (p.21) and amino acids (p.21). It is secreted (↑) by the pancreas (p. 102) as trypsinogen (p. 108).

trypsinogen (*n*) the inactive form of trypsin (p. 107) which is secreted (p. 106) by the pancreas (p. 102) and converted into trypsin by enterokinase (p. 107).

peptidase (*n*) an enzyme (p. 28) which catalyzes (p. 28) the hydrolysis (p. 16) of polypeptides (p. 21) into amino acids (p. 21) by breaking down the peptide bonds (p. 21). It is secreted (p. 106) by the small intestine (p. 102).

nucleotidase (*n*) an enzyme (p. 28) which catalyzes (p. 28) the hydrolysis (p. 16) of a nucleotide (p. 22) into its component nitrogen bases (p. 22), pentose (p. 17) and phosphoric acid (p. 22). It is secreted (p. 106) by the small intestine (p. 102).

secretin (*n*) a hormone (p. 130) which stimulates the secretion (p. 106) of bile (p. 101) from the liver (p. 103) and digestive (p. 98) juices from the pancreas (p. 102). It is secreted by the duodenum (p. 101).

pancreozymin (*n*) a hormone (p. 130) which stimulates the release of digestive (p. 98) juices from the pancreas (p. 102). It is secreted (p. 106) by the duodenum (p. 101).

microphagous (*adj*) of an animal which feeds on food particles that are tiny compared with the size of the animal so that it must feed continuously to receive enough nutrients (p. 92).

filter feeder a microphagous (↑) feeder which lives in water and filters suspended food particles from the water.

deposit feeder a microphagous (↑) feeder which feeds on particles that have been deposited on and perhaps mixed with the base layer of the environment (p. 218) in which the animal lives.

fluid feeder a microphagous (↑) feeder which feeds by ingesting (p. 98) fluids containing nutrients (p. 92) from living or recently dead animals or plants.

pseudopodial feeder a microphagous (↑) feeder in which cells develop temporary, finger-like projections, pseudopodia (p. 44), to engulf food particles.

mucous feeder a microphagous (↑) feeder in which food particles are trapped in mucus (p. 99) secreted (p. 106) by the organism and moved by ciliary (p. 12) action to the mouth.

types of feeding
microphagous feeder

e.g. Right whale

fluid feeder e.g. mosquito

pseudopodial feeder
e.g. *Amoeba*

mucous feeder   mouth with
e.g. lancelet   tentacles

types of feeding

setous feeder
e.g. *Daphnia*

macrophagous feeders

omnivore
e.g. American oppossum

carnivore
e.g. tiger

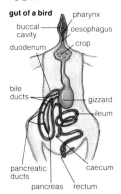

ruminant herbivore
e.g. gazelle

gut of a bird

buccal cavity — pharynx
duodenum — oesophagus
— crop
bile ducts — gizzard
— ileum
pancreatic ducts — caecum
pancreas — rectum

setous feeder a microphagous (↑) feeder in which the food particles are trapped by setae (p. 65) and then moved towards the mouth by beating cilia (p. 12).

macrophagous (*adj*) of an animal which feeds on relatively large food particles and usually, therefore, does not need to feed continuously.

coprophagous (*adj*) of an animal, such as some rodents, which feed on faeces (p. 99) thus improving the digestion (p. 98) of cellulose (p. 19) on second passage.

omnivore (*n*) an animal which feeds by eating a mixed diet of animal and plant food material.

carnivore (*n*) an animal which feeds by eating a diet that consists mainly of animal material. Carnivores may have powerful claws and dentition (p. 104) adapted to tearing flesh.

herbivore (*n*) an animal which feeds by eating a diet that consists mainly of plant material. Herbivores may have dentition (p. 104) and digestion (p. 98) specially adapted to deal with tough plant materials.

ruminant (*n*) one of the group of herbivores (↑) which belong to the order Artiodactyla and in which the stomach (p. 100) is complex and includes a rumen. Food is eaten but not chewed initially, and it is passed to the rumen where it is partly digested (p. 98) and then regurgitated for further chewing before swallowing and passing into the reticulum.

gizzard (*n*) part of the alimentary canal (p. 98) of certain animals. It has a very tough lining surrounded by powerful muscles (p. 143) in which food particles are broken down by grinding action against grit or stones in the gizzard lining or against spines or 'teeth' in the gizzard itself.

crop (*n*) the part of the alimentary canal (p. 98) in animals, such as birds, which either forms part of the gut (p. 98) or the oesophagus (p. 99), and in which food is stored temporarily and partly digested (p. 98).

mandible (*n*) (1) the lower jaw (p. 105) of a vertebrate (p. 74); (2) either of the pair of feeding mouth parts of certain invertebrates (p. 75).

carnivorous plant any plant which supplements its supply of nitrates by capturing, using a variety of means, small animals, such as insects (p. 69) and digesting (p. 98) them with enzymes (p. 28) secreted (p. 106) externally.

parasitism (*n*) an association (p. 227) in which the individuals of one species (p. 40), the parasites, live permanently or temporarily on individuals of another species, the host (↓), deriving benefit and/or nutrients (p. 92) and causing harm or even death to the host.

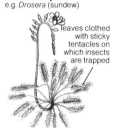

carnivorous plant
e.g. *Drosera* (sundew)

leaves clothed with sticky tentacles on which insects are trapped

parasitism e.g. infestation by hookworms

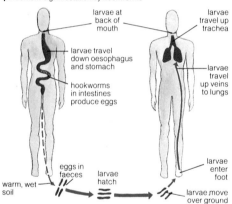

larvae at back of mouth

larvae travel down oesophagus and stomach

hookworms in intestines produce eggs

eggs in faeces

warm, wet soil

larvae hatch

larvae travel up trachea

larvae travel up veins to lungs

larvae enter foot

larvae move over ground

parasite (*n*) *see* parasitism (↑).

endoparasite (*n*) a parasite (↑) which lives within the body of the host (↓) itself, for example, the tapeworm, living within the gut (p. 98) of a vertebrate (p. 74).

ectoparasite (*n*) a parasite (↑) which lives on the surface of the host (↓) and is usually adapted for clinging on to the host on which it often feeds by fluid feeding (p. 108). Ectoparasites usually have special organs for attachment to the host.

intercellular parasite an endoparasite (↑) which lives within the material between the cells of the host (↓).

intercellular (*adj*) between cells.

ectoparasite e.g. dodder (*Cuscuta*) on a bean. The dodder climbs around the host stem and taps into its vascular system via haustoria

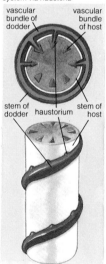

vascular bundle of dodder

vascular bundle of host

stem of dodder

haustorium

stem of host

endoparasite e.g. fungal pathogen. Fungal hyphae weave between cells and tap cells for nutrients with haustoria

host cells

fungal hyphae

haustorium

intracellular parasite an endoparasite (↑)which lives within the cells of the host (↓).

host (*n*) the species (p. 40) of organism in an association (p. 227) within which or on which a parasite (↑) lives and reaches sexual maturity, and which suffers harm or even death as a result.

secondary host a host (↑) on which or within which the young or resting stage of a parasite (↑) may live temporarily. The parasite does not reach sexual maturity on the secondary host. Also known as **intermediate host**.

transmission (*n*) the process by which a substance or an organism is transported from one place to another, e.g. a parasite (↑) is transmitted from one host (↑) to another sometimes via a secondary host (↑) and it may involve considerable risk to the parasite. **transmit** (*v*).

vector (*n*) a secondary host (↑) which is actively involved in the transmission (↑) of a parasite (↑) from one host (↑) to another or an organism which passes infectious disease from one individual to another without necessarily being affected by the disease itself. For example, the blood-sucking mosquito which transmits a malaria-causing blood parasite from one individual on which it feeds to another is a vector.

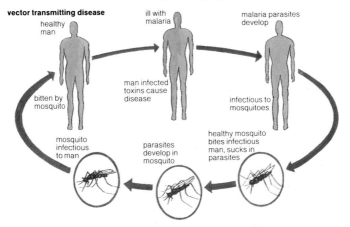

vector transmitting disease

healthy man

ill with malaria

malaria parasites develop

bitten by mosquito

man infected toxins cause disease

infectious to mosquitoes

mosquito infectious to man

parasites develop in mosquito

healthy mosquito bites infectious man, sucks in parasites

respiration (*n*) the process whereby energy is produced in a plant or animal by chemical reactions. In most organisms, this is achieved by taking in oxygen from the environment (p. 218) and, after transportation to the cells, its reaction with food molecules releases carbon dioxide, water and energy that is trapped in ATP (p. 33) – cellular respiration (p. 30).

respiratory quotient RQ. The ratio of the volume of carbon dioxide produced by an organism to the volume of oxygen used up during the same period of respiration.

$$RQ = \frac{\text{carbon dioxide produced}}{\text{oxygen used up}}$$

breathing (*n*) the process of actively drawing air or any other gases into the respiratory (↑) organs for gas exchange (↓). **breathe** (*v*).

gas exchange the process which takes place at a respiratory surface (↓) in which a gas, such as oxygen, from the environment (p. 218) diffuses (p. 49) into the organism because the concentration of that gas in the organism is lower than in the environment, and another gas, such as carbon dioxide, is released from the organism into the environment. In plants, respiratory gas exchange is complicated by the gas exchange that takes place as a result of photosynthesis (p. 93).

respiratory surface the surface of an organ, such as a lung (p. 115), across which gas exchange (↑) takes place. It is usually highly folded to increase the surface area, thin, and, in organisms that live on land, damp.

inspiration (*n*) the process of drawing air into or across the respiratory surface (↑) by muscular (p. 143) action. The pressure within the respiratory organ is reduced to below that of the environment (p. 218) so that air flows in. Also known as **inhalation**. **inspire** (*v*).

expiration (*n*) the process of forcing air and waste gases out of the respiratory (↑) organ by muscular (p. 143) action. The pressure within the respiratory organ is increased so that air flows out. Also known as **exhalation**. **expire** (*v*).

air (*n*) the mixture of gases which forms the atmosphere surrounding the Earth. It is composed of approximately 78 per cent nitrogen, 21 per cent oxygen, 0.03 per cent carbon dioxide, and little under 1 per cent of the so-called noble gases including argon, neon, etc. It also includes water vapour.

gill (*n*) one part of the respiratory surface (↑) found in most aquatic animals, such as fish. Gills are projections of the body wall or of the inside of the gut (p. 98) and may be very large and complex in relation to the animal because they are supported by water. They are very thin and well supplied with blood vessels (p. 127) so that gas exchange (↑) between the water and the blood (p. 90) of the animal is usually very efficient.

two main gill types showing water movements

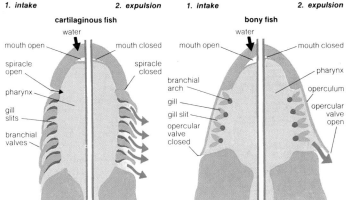

1. intake   *2. expulsion*   *1. intake*   *2. expulsion*

cartilaginous fish   **bony fish**

water — water

mouth open — mouth closed — mouth open — mouth closed

spiracle open — spiracle closed

pharynx — branchial arch — pharynx

gill — operculum

gill slits — gill slit — opercular valve open

branchial valves — opercular valve closed

gill filament one of the numerous flattened lobes which make up a gill and increase its surface area.

gill slit one of a series of openings through the pharynx (p. 99) in fish and some amphibians (p. 77) leading to the gills (↑).

operculum[a] (*n*) a bony plate that covers the visceral clefts (p. 74) and the gill slits (↑) in bony fish (p. 76). It assists in pumping water over the gills (↑) for gas exchange (↑) by inward and outward movements.

counter current exchange system the system found in the gills (p. 113) of bony fish (p. 76) in which the water that is pumped past the gill filaments (p. 113) flows in the opposite direction to the flow of blood (p. 90) within the gill. Gas exchange (p. 112) takes place continuously along the whole length of the gill because the levels of gases never reach equilibrium.

parallel current exchange system the system of gas exchange (p. 112) found in the gills (p. 113) of cartilaginous fish (p. 76) in which the flow of water and the flow of blood (p. 90) are in the same direction. This system is less efficient than a counter current exchange system (↑) because equilibrium is soon reached.

buccal pump one part of the double pump action which causes oxygen-containing water to flow over the gills (p. 113). Muscles (p. 143) in the floor of the buccal cavity (p. 99) cause it to rise and fall as the mouth shuts and opens forcing water over the gills and then drawing in more water through the mouth respectively.

opercular pump one part of the double pump action which causes oxygen-containing water to flow over the gills (p. 113). Muscles cause the operculum (p. 113) to open outwards as water is drawn in through the mouth.

tracheal system in insects (p. 96), a system of gas exchange (p. 112) and transport which is separate from the blood (p. 90) system. Oxygen is carried by air tubes called tracheae (↓) and some of the oxygen in the air diffuses into the body tissues (p. 83). Oxygen is also dissolved in the fluid in the tracheoles (↓).

spiracle (n) the opening to the atmosphere of an insect's (p. 69) trachea(↓).

counter current exchange system

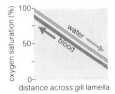

parallel current exchange system

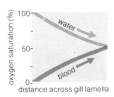

tracheal system

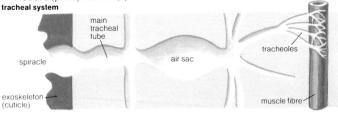

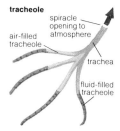

tracheole

spiracle
opening to
atmosphere

air-filled
tracheole

trachea

fluid-filled
tracheole

trachea (*n*) (1) in insects (p. 69), one of a number
of air tubes which lead from the spiracles (↑)
into the body tissues (p. 83); (2) in land-living
vertebrates (p. 74), a tube which leads from the
throat into the bronchi (p. 116). **tracheae** (*pl.*).

tracheole (*n*) one of the very fine tubes into which
the tracheae (↑) of insects (p. 69) branch. They
pass into the muscles (p. 143) and organs of an
insect's body to allow gas exchange (p. 112) to
take place.

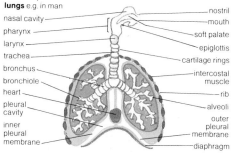

lungs e.g. in man

nasal cavity

pharynx

larynx

trachea

bronchus

bronchiole

heart

pleural
cavity

inner
pleural
membrane

nostril

mouth

soft palate

epiglottis

cartilage rings

intercostal
muscle

rib

alveoli

outer
pleural
membrane

diaphragm

lungs (*n.pl.*) a pair of thin-walled, elastic sacs
present in the thorax (↓) of amphibians (p. 77),
reptiles (p. 78), birds and mammals (p. 80) and
containing the respiratory surfaces (p. 112).

ventilation (*n*) the process whereby the air
contained within the lungs (↑) is exchanged with
air from the atmosphere by regular breathing
(p. 112) in which muscular (p. 143) movements
of the thorax (↓) varies its volume and thus the
volume of the lungs. During inspiration (p. 112)
the volume of the lungs increases and
atmospheric pressure forces air into the lungs.
During expiration (p. 112), the muscles relax
and the volume of the lungs decreases by
virtue of their elasticity so that air is forced out.

thorax (*n*) (1) in arthropods (p. 67), the segments
between the head and the abdomen (p. 116); (2)
in vertebrates (p. 74), the part of the body which
contains the heart (p. 124) and lungs (↑). In mammals
(p. 80) it is separated from the abdomen by the
diaphragm (p. 116) and protected by the rib cage.

thoracic cavity = thorax (↑), in vertebrates (p. 74).

intercostal muscle a muscle (p. 143) which connects adjacent ribs. When the external intercostal muscles contract, the ribs are moved upwards and outwards, increasing the volume of the thoracic cavity (p. 115) and thus the lungs (p. 115) so that air is forced into the lungs for inspiration (p. 112). When the internal intercostal muscles contract the volume of the thoracic cavity decreases and expiration (p. 112) takes place.

diaphragm (*n*) a sheet of muscular (p. 143) tissue (p. 83) which separates the thoracic cavity (p.115) from the abdomen (↓) in mammals (p. 80).

abdomen (*n*) (1) in arthropods (p. 67), the segments at the back of the body; (2) in vertebrates (p. 74), the part of the body containing the intestines, liver, kidney etc.

pleural cavity the narrow space, filled with fluid, between the two layers of the pleural membrane (↓).

pleural membrane the double membrane which surrounds the lungs (p. 115) and lines the thoracic cavity (p. 115). It secretes (p. 106) fluids to lubricate the two layers as the lungs expand and contract during breathing (p. 112).

larynx (*n*) a structure found at the junction of the trachea (p. 115) and the pharynx (p. 99) which contains the vocal cords (↓). During swallowing it is closed off by the epiglottis (p. 99).

vocal cord (*n*) one of the folds of the lining of the larynx (↑) which produce sound as a current of air passes over them.

bronchus (*n*) one of the two large air tubes into which the trachea (p. 115) divides and which enter the lungs (p. 115).

bronchiole (*n*) one of a number of smaller air tubes into which the bronchi (↑) divide after entering the lungs (p. 115). The bronchioles make up the 'bronchial tree' which ends in air tubes called the *respiratory bronchioles*. These divide into *alveolar ducts* (or terminal bronchioles) which give rise to the alveoli (↓).

alveolus (*n*) a pouch-like air sac which occurs in clusters at the ends of the bronchioles (↑) and which contains the respiratory surfaces (p. 112). A network of capillaries (p. 127) covers the thin, elastic epithelium (p.87). **alveoli** (*pl.*)

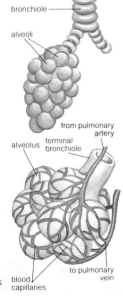

bronchiole and alveoli

bronchiole

alveoli

from pulmonary artery

terminal bronchiole

alveolus

to pulmonary vein

blood capillaries

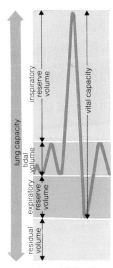

comparison of different lung volumes in man

tidal flow of the system in which inspiration (p. 112) and expiration (p. 112) take place through the same air passages so that the air passes twice over each part of the respiratory surface (p. 112). This is less efficient than a system in which there is a constant throughflow such as that which takes place over the gills (p. 113) of fish.

tidal volume the volume of air which is inspired (p. 112) or expired (p. 112) during normal regular breathing (p. 112). It is considerably less than the lung capacity (↓).

ventilation rate the rate per minute at which the total volume of air is expired (p. 112) or inspired (p. 112).

residual volume the volume of air that always remains within the alveoli (↑) because the thorax (p. 115) is unable to collapse completely. It exchanges oxygen and carbon dioxide with the tidal air.

vital capacity the total amount of air which can be inspired (p. 112) and expired (p. 112) during vigorous activity.

reserve volume the difference in volume between the total lung capacity (↓) and the vital capacity (↑).

lung capacity the total volume of air that can be contained by the lungs when fully inflated.

acclimatization (*n*) the period of time it takes for the respiration (p. 112) of an organism to get used to the reduced partial pressure of oxygen that may occur at high altitudes, for example, where the atmospheric pressure is reduced.

respiratory centre the part of the medulla oblongata (p. 156) which controls the rate of breathing (p. 112) in response to the levels of carbon dioxide dissolved in the bloodstream (p. 90).

oxygen debt the deficit in the amount of oxygen that is available for respiration (p. 112) during vigorous activity so that even when the activity ceases, breathing (p. 112) continues at a high rate until the oxygen debt is made up. Lactic acid builds up in the muscles (p. 143) from lactic acid fermentation (p. 34).

osmosis (*n*) the process by which water passes
through a semipermeable membrane (↓) from
a solution (↓) of low concentration of salts to
one of high concentration thereby diluting it.
Osmosis will continue until the concentrations
of the two solutions are equalized. In living
things osmosis can take place through
membranes (p. 14), e.g. tonoplast (p. 11) or
plasmalemma (p. 14), in either direction. In
plants, the cell walls (p. 8) are elastic so that
they can contain solutions of higher concentration
when osmosis ceases. **osmotic** (*adj*).

osmosis

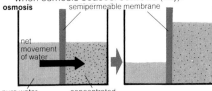

semipermeable membrane

net
movement
of water

pure water   concentrated
or hypotonic   or hypertonic
solution with   solution with
a high osmotic   a low osmotic
pressure   pressure

osmotic potential the tendency of water
molecules to diffuse (↓) through a semipermeable
membrane (↓) from a solution (↓) of low solute
concentration to a solution of high solute
concentration until equilibrium is reached.

semipermeable membrane a membrane (p. 14),
such as a tonoplast (p. 11) or plasmalemma (p. 14),
with microscopic (p. 9) pores (p. 120) through which
small molecules e.g. water will pass but larger
molecules e.g. sucrose (p. 18), or salts will not.

solution (*n*) a liquid (*the solvent*) with substances
(*the solute*) dissolved in it. Substances that will
dissolve are said to be *soluble* and those that
will not, *insoluble*.

isotonic solution (*n*) a solution (↑) in which the
osmotic potential (↑) is the same as that of
another solution so that neither solution either
gains or loses water by osmosis (↑) across a
semipermeable membrane (↑).

hypotonic (*adj*) of one solution (↑) in an osmotic
(↑) system which is more dilute than another.

hypertonic (*adj*) of one solution (↑) in an osmotic
(↑) system which is more concentrated than another.

diffusion (*n*) the process in which molecules move from an area of high concentration to an area of low concentration. Osmosis (↑) is a special type of diffusion restricted to the movement of water molecules.

diffusion pressure deficit the situation which exists between two solutions (↑) on either side of a semipermeable membrane (↑) in which a substance has been added to one of the solutions which cannot pass through the membrane and impedes the passage of water from that solution. The greater the concentration of the solution, the higher the diffusion pressure deficit and the lower the osmotic potential (↑) of that solution.

turgor (*n*) the condition in a plant cell in which water has diffused (↑) into the cell vacuole (p. 11) by osmosis (↑), causing the cell to swell, because the cell fluid was at a lower osmotic potential (↑) than that of its surroundings.

turgid (*adj*) of a plant cell in which the turgor (↑), which is resisted by the elasticity of the cell wall (p. 8), has brought the cell close to bursting and no more water can enter the cell. Turgidity provides the plant with support.

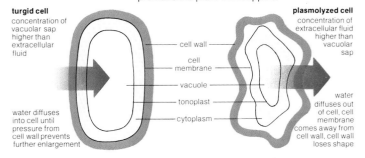

turgid cell
concentration of vacuolar sap higher than extracellular fluid

water diffuses into cell until pressure from cell wall prevents further enlargement

cell wall

cell membrane

vacuole

tonoplast

cytoplasm

plasmolyzed cell
concentration of extracellular fluid higher than vacuolar sap

water diffuses out of cell, cell membrane comes away from cell wall, cell wall loses shape

turgor pressure the pressure exerted by the bulging cell wall (p. 8) during osmosis (↑) into the vacuole (p. 11) of a plant cell.

plasmolysis (*n*) a loss of water, and hence turgidity (↑) from a plant cell when it is surrounded by a more concentrated solution (↑). The cytoplasm (p. 10) loses volume and contracts away from the cell wall (p. 8) causing wilting.

flaccid (*adj*) of cell tissue (p. 83), weak or soft.

wilt (*v*) *of leaves and green stems* to droop.

stoma (*n*) one of the many small holes or pores
(↓) in the leaves (mainly) and stems of plants
through which gas and water vapour exchange
take place. They are able to open and close by
means of their neighbouring guard cells (↓).
stomata (*pl.*).

pore (*n*) a small opening in a surface.

guard cell one of the pair of special, crescent-
shaped cells which surround each stoma (↑)
and which enable the stoma to open or close in
response to light intensity by osmosis (p. 118).
When the guard cells are turgid (p. 119) the
stoma is open.

stomata
surface view of leaf

epidermal cells   stoma

pore   guard cells

guard cells
opening and
closing of
stoma

guard cells

chloroplasts

nucleus of
guard cell

narrow canal
of central region
of guard cell

subsidiary
cell

subsidiary
cell nucleus

stomatal pore

rigid thickened walls
of central region
of guard cell

open   closed

epidermal cells

substomatal chamber the space below the
stoma (↑).

transpiration (*n*) the loss of water from a plant
through the stomata (↑). It is controlled by the
action of the stomata. It provides a flow
(transpiration stream (p. 122)) of water through
the plant and also has a cooling effect as the
water evaporates from the plant's surface. It is
affected by temperature, relative humidity (↓),
wind speed. As air and leaf temperatures
increase so does the rate of transpiration. The
lower the humidity of the atmosphere, the
faster is the rate of transpiration. Increasing
wind speed normally increases the transpiration
rate provided the cooling effect is not greater.

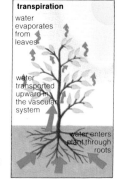

transpiration

water
evaporates
from
leaves

water
transported
upward in
the vascular
system

water enters
plant through
roots

guttation

high humidity

droplets of
water exuded
from hydathodes
(ends of veins at leaf margin)

**movement of water from
soil to centre of root**

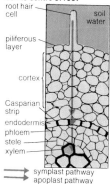

root hair
cell

soil
water

piliferous
layer

cortex

Casparian
strip
endodermis
phloem
stele
xylem

→ symplast pathway
⇒ apoplast pathway

**symplast and apoplast
pathways**

plasmodesmata

apoplast	symplast
substances translocated through cell walls and inter- cellular spaces	substances translocated through living cells and plasmodesmata

relative humidity the percentage of water vapour contained in the air. When the relative humidity is 100 per cent the air is saturated.

guttation (*n*) the process which takes place in some plants in conditions of high relative humidity (↑) in which water is actively secreted (p. 106) in liquid form by special structures called hydathodes (found at the end of the veins on the leaves) rather than being lost as water vapour. This takes place because of osmotic (p. 118) absorption (p. 81) of water by the roots.

atmospheric pressure the pressure which is exerted on the surface of the Earth by the weight of the air in the atmosphere.

vacuolar pathway a pathway for the passage of water by osmosis (p. 118). Vacuoles (p. 11) contain a fluid with a lower osmotic potential (p. 118) than water so that the vacuole will take in water until it becomes turgid (p. 119).

symplast pathway a pathway for the transport of water through a plant by diffusion (p. 119) from one cell to the next through the cytoplasm (p. 10) along the threads called plasmodesmata (p. 15) which link adjacent cells.

apoplast pathway a pathway for the transport of water in a plant, particularly across the root cortex (p. 86), by diffusion (p. 119) along adjacent cell walls (p. 8).

mass flow a hypothesis (p. 235) developed by Munch in 1930 to explain the transport of substances in the phloem (p. 84). Mass flow takes place in the sieve tube (p. 84) lumina as water is taken up by osmosis (p. 118) in actively photosynthesizing (p. 93) regions where concentration is high and flows to areas where water is lost as the products of photosynthesis are being used up or stored and, therefore, concentration is low. Water is carried in the opposite direction in the xylem (p. 84) by the transpiration stream (p. 122).

root pressure the pressure in a plant which causes water to be transported from the root into the xylem (p. 84) by the plant's osmotic (p. 118) gradient.

Casparian strip a thickened waterproof layer which covers the radial and the transverse cell walls (p. 8) of the endodermis (p. 86) so that all water which is transported from the root cortex (p. 86) to the xylem (p. 84) must pass through the cytoplasm (p. 10) of the endodermis cells.

transpiration stream the continuous flow of water which takes place in a plant through the xylem (p. 84) as water is lost to the atmosphere by transpiration (p. 120) and taken up from the soil by the root hairs (p. 81).

cohesion theory the theory (p. 235) which explains that a column of water may be held together by molecular forces of attraction permitting the ascent of sap up a tall stem without falling back or breaking. There is stress or tension in the column of water as water is lost from the xylem (p. 84) vessels by osmosis (p. 118). Similarly, molecular forces of adhesion will cause water to cling to other substances and thereby rise up a stem by capillarity.

translocation[1] (*n*) the transport of organic material through the phloem (p. 84) of a plant. The material includes carbohydrates (p. 17) such as glucose (p. 17), amino acids (p. 21) and plant growth substances (p. 138).

active transport the method by which, with the use of energy, molecules are transported across a cell membrane (p. 14) against a concentration gradient. It probably involves the use of molecular carriers.

transcellular strand hypothesis a hypothesis (p. 235) developed by Thaine to explain why transport rates in plants appear to be greater than would be possible by diffusion (p. 119). It suggests that active transport (↑) takes place along fibrils (p. 11) of protein (p. 21) that pass through the sieve tube (p. 84).

electro-osmotic hypothesis a hypothesis (p. 235) developed by Spanner to explain why transport rates in plants appear to be greater than would be possible by diffusion (p. 119). It suggests that electro-osmotic forces exist across the sieve plates (p. 85)

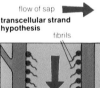

representations of two theories of phloem transport

flow of sap ▶

transcellular strand hypothesis

fibrils

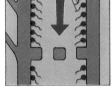

electro-osmotic hypothesis

electro-magnetic force

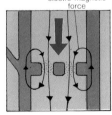

double circulatory system of a mammal

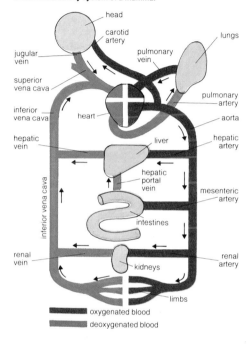

oxygenated blood
deoxygenated blood

single circulation of a fish

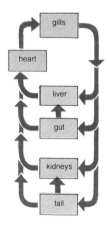

circulatory system a system in which materials can be transported around the body of an animal, needed because the volume of the animal is usually too great for transport to be effective by diffusion (p. 119).

single circulation a circulatory system (↑), such as that which occurs in fish, in which the blood (p. 90) passes through the heart (p. 124) once on each complete circuit.

double circulation a circulatory system (↑), such as that which occurs in birds and mammals (p. 80), in which the blood (p. 90) passes through the heart (p. 124) twice on each complete circuit and so maintains the system's blood pressure. In this system, the heart is divided into left and right sides.

open circulatory system a circulatory system (p. 123), e.g. in arthropods (p. 67), in which the blood (p. 90) is free in the body spaces for most of its circulation. The organs lie in the haemocoel (p. 68) and blood from the arteries (p. 127) bathes the major tissues (p. 83) before diffusing (p. 119) back to the open ends of the veins (p. 127). There are no capillaries (p. 127).

closed circulatory system a circulatory system (p. 123), e.g. in vertebrates (p. 74), in which the blood (p. 90) is contained within vessels (p. 127) for the greater part of its circulation.

heart (n) a muscular (p. 143) organ or specialized blood (p. 90) vessel (p. 127) which pumps around the circulatory system (p. 123).

open circulatory system
e.g. an insect

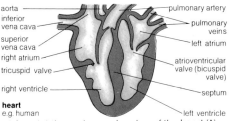

heart
e.g. human

atrium (n) the region or chamber of the heart (↑) which receives the blood (p. 90). The heart of a mammal (p. 80) has a left and a right atrium which are the receivers of oxygenated (p. 126) blood from the lungs (p. 115) and deoxygenated (p. 126) blood from the body respectively. Also known as **auricle**. **atria** (pl.).

ventricle (n) a muscular (p. 143) region or chamber of the heart (↑) which by regular contractions pumps the blood (p. 90). The heart of a mammal (p. 80) has a left and a right ventricle which pump oxygenated (p. 126) blood to the body and deoxygenated (p. 126) blood to the lungs respectively.

cardiac cycle the cycle in which, by rhythmical muscular (p. 143) contractions, blood (p. 90) flows into the atria (↑) of the heart (↑) and is pumped out of the ventricles (↑).

systole (n) contraction phase of cardiac cycle (↑).

diastole (n) relaxation phase of cardiac cycle (↑).

position of heart in various invertebrates

earthworm

crustacean

spider

valve (*n*) a flap or pocket which only allows a liquid, e.g. blood (p. 90), to flow in one direction.

atrioventricular valve the valve which separates the left ventricle (↑) and atrium (↑) preventing blood (p. 90) from flowing back into the atrium by the closure of two membranous (p. 14) flaps. Also known as **mitral valve**.

bicuspid valve = atrioventricular valve (↑).

tricuspid valve the valve which separates the right ventricle (↑) and atrium (↑).

tendinous cords the tough connective tissue (p. 88) in the heart (↑) which prevents the atrioventricular (↑) and tricuspid valves(↑) from turning inside out during contraction.

pocket valves valves between the ventricles (↑) and the pulmonary artery (p. 128) and aorta (↓) which, when closed, prevent the return of blood to the ventricles.

semi-lunar valves = pocket valves.

aorta (*n*) the major artery (p.127) carrying oxygenated (p. 126) blood (p. 90) from the heart (↑).

myogenic muscle (*n*) muscle (p. 143), e.g. cardiac muscle (p. 143), which may contract without nervous (p. 149) stimulation although its rate of contraction is controlled by such stimulation.

heartbeat (*n*) the rhythmic contraction of the myogenic muscle (↑) of the heart (↑). Also known as **cardiac rhythm**.

sino-atrial node a group of cells in the right atrium (↑) which is responsible for maintaining the heartbeat (↑) by nervous (p. 149) stimulation relayed by it.

pacemaker (*n*) = sino-atrial node (↑).

atrio-ventricular node a second group of cells in the right atrium (↑) which receives the nervous (p. 149) stimulation from the sino-atrial node (↑).

Purkinje tissue nervous (p. 149) tissue (p. 83) which conducts the nervous stimulation from the sino-atricular node (↑) to the tip of the ventricle (↑) ensuring that the ventricle contracts from its tip upwards to force blood (p. 90) out through the arteries (p. 127).

sympathetic nerve a motor nerve (p. 149) which arises from the spinal nerve and releases adrenaline (p. 152) into the heart muscle (p. 143) to increase the heartbeat (↑).

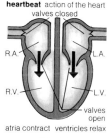

heartbeat action of the heart
valves closed

R.A.

L.A.

R.V.

L.V.

valves open

atria contract ventricles relax

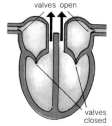

valves open

valves closed

ventricles contract atria relax

changes in volume and pressure during a mammalian cardiac cycle

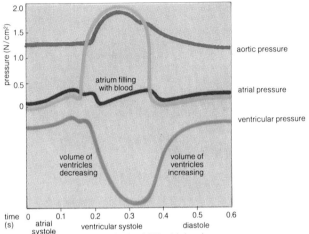

vagus nerve a motor nerve (p. 149) which arises from the medulla oblongata (p. 156) and releases acetylcholine (p. 152) into the heart muscle (p.143) to decrease the heartbeat (p.125).

pulse (*n*) a wave of increased blood (p. 90) pressure which passes through the arteries (↓) as the left ventricle (p. 124) pumps its contents into the aorta (p. 125).

pigment (*n*) a coloured substance. For example, *myoglobin* is a variety of haemoglobin (↓) found in muscle (p. 143) cells, and *chlorocruorin* is a respiratory pigment containing iron which is found in the blood (p. 90) of some polychaetes (p. 65). *See also* chlorophyll (p. 12).

haemoglobin (*n*) a red pigment (↑) and protein (p. 21) containing iron, which is found in the cytoplasm (p. 10) of the red blood cells (p. 91) of vertebrates (p. 74). It combines readily with oxygen to form oxyhaemoglobin and in this form oxygen is transported to the tissues (p. 83) from the lungs (p. 115).

oxyhaemoglobin (*n*) see haemoglobin (↑).

oxygenated (*adj*) containing or carrying oxygen.

deoxygenated (*adj*) not oxygenated (↑).

the pattern of excitation that accompanies contraction of the heart

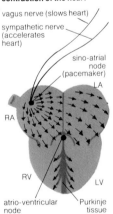

a network of capillaries

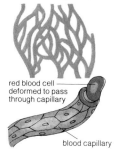

red blood cell
deformed to pass
through capillary

blood capillary

artery

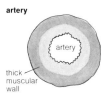

artery

thick
muscular
wall

vein

vein

thin muscular
wall

valve in a vein

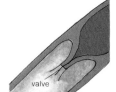

valve

haemocyanin (*n*) a blue pigment (↑) and protein (p. 21) containing copper, which is found in the plasma (p. 90) of certain invertebrates (p. 75). It also combines with oxygen to transport it to the tissues (p. 83).

Bohr effect the effect of increasing the likelihood of the dissociation of oxygen from oxyhaemoglobin (↑) as the level of carbon dioxide is increased so that with increased activity more oxygen is passed to the body tissues (p. 83).

vascular system the system of vessels (↓) which transport fluid throughout the body of an organism.

capillary (*n*) any of the very large numbers of tiny blood vessels (↓) which form a network throughout the body. They present a large surface area and are thin-walled to aid gas exchange (p. 112).

sphincter muscle any of the muscles (p. 143) which, by contraction, close any of the hollow tubes, organs, or vessels (↓) in an organism.

vessel[a] (*n*) a channel or duct with walls e.g. blood (p. 90) flows through a blood vessel.

vein[a] (*n*) any one of the tubular vessels (↑) that conveys blood (p. 90) back to the heart (p. 124). Veins are quite large in diameter but thinner walled than arteries (↓) and the blood is carried under relatively low pressure. Veins have pocket valves (p. 125) which ensure that the blood is carried towards the heart only.

venule (*n*) a small blood vessel (↑) that receives blood from the capillaries (↑) and then, with other venules forms the veins (↑).

artery (*n*) any one of the tubular vessels (↑) that conveys blood (p. 90) from the heart (p. 124). They are smaller in diameter than veins (↑) but the walls are thicker and more elastic and the blood is carried at relatively high pressure. With the exception of the aorta (p. 125) and the pulmonary artery (p. 128), arteries have no pocket valves (p. 125).

arteriole (*n*) a small artery (↑).

sinus (*n*) any space or chamber, such as the sinus venosus which is a chamber found within the heart (p. 124) of some vertebrates (p. 74) especially amphibians (p. 77), and lies between the veins (↑) and the atrium (p. 124).

pulmonary circulation the part of the double circulation (p. 123) in which deoxygenated (p. 126) blood (p. 90) is pumped from the heart (p. 124) to the lungs (p. 115).

pulmonary artery the artery (p. 127) which carries the deoxygenated (p. 126) blood (p. 90) pumped from the heart (p. 124) to the lungs (p.115).

pulmonary vein the vein (p. 127) which carries the oxygenated (p. 126) blood (p. 90) from the lungs (p. 115) back to the heart (p. 124).

systemic circulation the part of the double circulation (p. 123) in which blood (p. 90) is pumped from the heart (p. 124) throughout the body of the animal.

arterio-venous shunt vessel a small blood vessel (p. 127) which bypasses the capillaries (p. 127) and carries blood (p. 90) from the arteries (p.127) to the veins (p. 127) and therefore regulates the amount of blood which enters the capillaries.

lymph (*n*) a milky or colourless fluid which drains from the tissues (p. 83) into the lymphatic vessels (↓) and is not absorbed (p. 81) back into the capillaries (p. 127). It is similar to tissue fluid and contains bacteria (p. 42) but does not contain large protein (p. 21) molecules.

lymphatic vessel any one of the vein-like (p. 127) vessels (p. 127) that carry lymph (↑) from the tissues (p. 83) into the large veins that enter the heart (p. 124).

lymph node a swelling in the lymphatic vessel (↑), especially in areas such as the groin or armpits, which contain special white blood cells (p. 91) known as macrophages (p. 88).

platelet (*n*) any of the fragments of cells present in the blood (p. 90) plasma (p. 90) which are formed in the red bone marrow (p. 90) and which prevent bleeding by combining at the point of an injury and releasing a hormone (p. 130) which stimulates blood clotting (↓). They also release other substances which cause blood vessels (p. 127) to constrict so that they also prevent capillary (p 127) bleeding.

coagulate (*v*) = clot (↓).

anticoagulant (*n*) a substance that stops blood (p. 90) clotting (↓).

lymphatic system

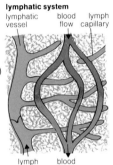

lymphatic vessel   blood flow   lymph capillary

lymph flow   blood flow

cells bathed in tissue fluid

lymph vessels

blood capillaries oxygenated blood

blood capillaries deoxygenated blood

platelet

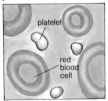

platelet

red blood cell

clot (*v*) *of liquids* to become solid, for example,
blood (p. 90) clots in air. **clot** (*n*).

blood groups in humans, there is a system of
multiple alleles (p. 205) which gives rise to four
main different blood groups with different
antigens (p. 234) or proteins (p. 21) on the
surface of the red blood cells (p. 91). The A allele,
B allele and O allele (producing no antigens)
may combine to give any of the following blood
group combinations: AA, AO, BB, BO, AB or
OO. A and B alleles are both dominant (p. 197) to
O so that there are four groups, A, B, O and AB.

rhesus factor an antigen (p. 234) which is present
in the blood (p. 90) of rhesus monkeys and
most, but not all, humans. During pregnancy
(p. 195) or following transfusion of blood
containing the rhesus factor (Rh+) into blood
lacking it (Rh−), breakdown of the red blood
cells (p. 91) can occur with dangerous results.

the four main blood groups

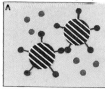 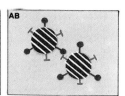

 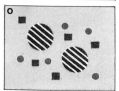

●—● A antigen ◾—◾ B antigen ◼ A antibody ● B antibody

	blood group	antigens on red cells	antibodies in serum	can receive blood type	can donate blood to
	A	**A**	**B**	groups **A** and **O**	groups **A** and **AB**
	B	**B**	**A**	groups **B** and **O**	groups **B** and **AB**
universal recipients	**AB**	**A** and **B**	none	groups **A, B, AB** and **O**	group **AB**
universal donors	**O**	none	**A** and **B**	groups **O** only	groups **A, B, AB** and **O**

homeostasis (*n*) the maintenance of constant internal conditions within an organism, thus allowing the cells to function more efficiently, despite any changes that might occur in the organism's external environment (p. 218).

endocrine system a system of glands (p. 87) in animals which produce hormones (↓). This system and the nervous system (p. 149) combine to control the functions of the body.

endocrine gland a gland (p. 87) which produces hormones (↓).

hormone (*n*) a substance made in very small amounts in one part of an organism and transported to another part where it produces an effect. (1) In plants, the hormones can be referred to as growth substances (p. 138). (2) In animals, hormones are secreted (p. 106) by the endocrine glands (↑) into the blood stream (p. 90), where they circulate to their destination.

adrenal glands in mammals (p. 80), a pair of endocrine (↑) glands (p. 87) near the kidneys (p. 136). They are divided into two parts; the *medulla*, the inner part which secretes (p. 106) adrenaline (p. 152) and noradrenaline (p. 152), and the *cortex*, the outer part which secretes various steroid (p. 21) hormones (↑).

homoiothermic (*adj*) of an organism which maintains its body temperature at a constant level in changing external circumstances. These organisms, including mammals (p. 80), for example, are usually regarded as 'warm blooded' because their body temperature is usually above that of the surroundings.

endothermic (*adj*) = homoiothermic (↑).

poikilothermic (*adj*) of an organism whose body temperature varies with and is roughly the same as that of the environment (p. 218). These organisms, not including birds and mammals (p. 80), are usually regarded as 'cold blooded' although their body temperature may be higher or lower than that of the environment depending on such factors as wind speed or the sun's radiation (↓). They have a lower metabolic rate (p. 32) as their body temperature falls.

exothermic (*adj*) = poikilothermic (↑).

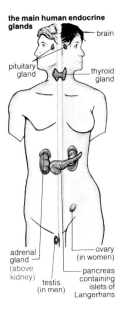

the main human endocrine glands

brain

pituitary gland

thyroid gland

adrenal gland (above kidney)

testis (in men)

ovary (in women)

pancreas containing islets of Langerhans

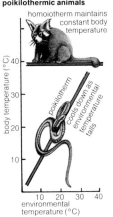

comparison of homoiothermic and poikilothermic animals

homoiotherm maintains constant body temperature

poikilotherm

cools down as environmental temperature falls

body temperature (°C)

10 20 30 40
environmental temperature (°C)

heat gains and losses in a reptile (poikilothermic) animal

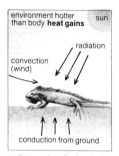

environment hotter than body **heat gains**
sun
radiation
convection (wind)
conduction from ground

environment cooler than body **heat losses**

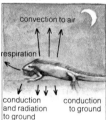

convection to air
respiration
conduction and radiation to ground
conduction to ground

radiation (*n*) the transfer of heat from a hot object, such as the sun, to a cooler object, such as the earth or the body of an organism, through space without increasing the temperature of the space.

evaporation (*n*) the change from a liquid to a vapour or gas that takes place when the liquid is warmed to a temperature at or below its boiling point.

conduction (*n*) the transfer of heat through a solid.

convection (*n*) the transfer of heat in a fluid as the warmed portion of the fluid rises and the cool portion falls.

hair (*n*) a single-celled or many celled outgrowth from the dermis (↓) of a mammal (p. 80) made up of dead material and including the substance keratin. Among other functions, a coat of hair insulates the mammal's body from excessive warming or cooling especially if the hairs are raised to trap a layer of insulating air around the body. Hair is a characteristic of mammals.

skin (*n*) the outer covering of an organism which insulates it from excessive warming or cooling, prevents damage to the internal organs, prevents the entry of infection, reduces the loss of water, protects it from the sun's radiation (↑) and it also contains sense organs which make the organism aware of its surroundings.

dermis (*n*) = skin (↑).

epidermis (*n*) the outer layer of the skin (↑). The epidermis is made up of three main layers of cells: the continuous *Malpighian layer* is able to produce new cells by division and so replace epidermal layers as they are worn away; the *granular layer* grades into the harder, outer *cornified layer* which is composed of dead cells only and forms the main protective part of the skin. This cornified layer may become very hard and thick in areas that are constantly subject to wear, such as the soles of the feet.

sebaceous gland one of the many glands (p. 87) contained within the skin (↑) that open into the follicles (p. 132) and which secrete (p. 106) an oily antiseptic substance which repels water and keeps the skin flexible.

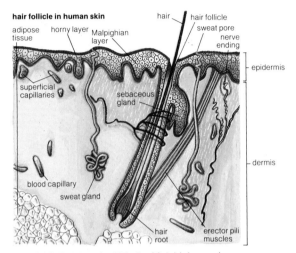

hair follicle in human skin

adipose tissue | horny layer | Malpighian layer | hair | hair follicle | sweat pore | nerve ending | epidermis

superficial capillaries | sebaceous gland | dermis

blood capillary | sweat gland

hair root | erector pili muscles

hair follicle (*n*) a pit within the Malphigian and granular layers of the epidermis (p. 131) which contains the root of a hair and from which the hair grows by cell division. Sebaceous glands (p. 131) open into the hair follicles and also contain muscles (p. 143) to erect the hairs for additional insulation as well as nerve (p. 149) endings for sensitivity.

sweat gland a coiled tubular gland (p. 87) within the epidermis (p. 131) which absorbs moisture containing some salts and minerals from surrounding cells and releases it to the surface through a tube causing the skin (p. 13) to be cooled as the moisture evaporates into the atmosphere.

hibernation (*n*) a process by which certain organisms respond to very low external temperatures in a controlled fashion. The core temperature of the animal falls to near that of the environment (p. 218) and there is a resulting drop in the metabolic rate (p. 32) although the nervous system (p 149) continues to operate so that should the temperature fall near to the lethal temperature, the animal increases its metabolic rate to cope with it.

hyperthermia (*n*) overheating. The condition in which, as a result of vigorous activity, disease, or heating by radiation (p. 131), the body temperature of an organism rises above its normal level. As a result of nerve impulses (p. 150) sent to the hypothalmus (p. 156) counter measures are taken including dilation of the blood vessels (p. 127) to allow greater heat loss by radiation, convection (p.131) and conduction (p. 131), and sweating to cool the body surface as moisture evaporates.

hypothermia (*n*) overcooling. The condition in which the temperature of an organism falls below its normal level. As a result of nerve impulses (p. 150) sent to the hypothalamus (p. 156) counter measures are taken including a reduction in sweating, constriction of the blood vessels (p. 127) to reduce the amount of heat being lost by radiation (p. 131), conduction (p. 131) and convection (p. 131), rapid spasmodic contraction of the muscles (p. 143) to cause shivering, and an increase in metabolic rate (p.32).

aestivation (*n*) (1) the condition of inactivity or torpor into which some animals enter during periods of drought or high temperatures. Lung fish, for example, bury themselves in mud at the start of the dry season and re-emerge when the rain begins to fall again. (2) the arrangement of parts in a flower bud.

leaf fall the condition into which some plants enter during periods of extreme water shortage by losing some of their leaves to reduce water losses by transpiration (p. 120).

osmoregulation (*n*) the process by which an organism maintains the osmotic potential (p.118) in its body fluids at a constant level e.g. freshwater fish take in large volumes of water through the gills (p.113) by osmosis (p.118) which are then excreted (p.134) as urine (p.135) from the kidneys (p.135). Marine fish, either drink sea water (bony fish (p.76)) so that salts are absorbed (p. 81) by the gut (p.98) and water then follows by osmosis, with the salts eliminated by the gills, or retain urea (p.134) so that their fluids are hypertonic (p. 118) to sea water and then, like freshwater fish, they take in more water through their gills.

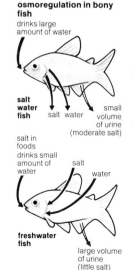

osmoregulation in bony fish

drinks large amount of water

salt water fish

salt water

small volume of urine (moderate salt)

salt in foods

drinks small amount of water

salt

water

freshwater fish

large volume of urine (little salt)

carotid body a small oval structure in the carotid artery (p. 127) containing nerves (p. 149) which respond to the oxygen and carbon dioxide content of the blood (p. 90) and so control the level of respiration (p. 112).

carotid sinus a small swelling in the carotid artery (p. 127) containing nerves (p. 149) which respond to blood (p. 90) pressure and so control circulation (p. 123).

excretion (*n*) the process by which the waste and harmful products of metabolism (p. 26) such as water, carbon dioxide, salts and nitrogenous compounds, are eliminated from the organism.

chemical formulae of nitrogenous wastes

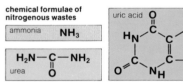

ammonia NH_3

urea $H_2N-\underset{\underset{O}{\|}}{C}-NH_2$

uric acid

excretory organs in invertebrates
Malpighian tubules of insect

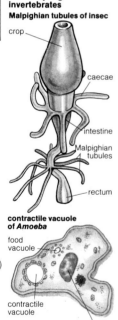

urea (*n*) a nitrogenous organic compound (p. 15) which is soluble in water and which is the main product of excretion (↑) from the breakdown of amino acids (p. 21) in certain animals.

uric acid a nitrogenous organic compound (p. 15), insoluble in water, which is the main product of excretion (↑) from the breakdown of amino acids (p. 21) in certain animals. Because it is insoluble, uric acid is not toxic and can be excreted without losing large amounts of water.

ammonia (*n*) a nitrogenous inorganic compound (p. 15) which is very toxic and may only be found as a product of excretion (↑) in organisms where large amounts of water are available for its removal, such as in aquatic animals.

contractile vacuole a vacuole (p. 11) present in the endoplasm (p. 44) of the Protozoa (p. 44) which is important in the osmoregulation (p. 133) of these organisms. In hypotonic (p.118) solutions (p. 118) the vacuole swells and then seems to contract, releasing to the exterior, water which has entered the cell with food or from the surroundings. The water passes into the solution. In hypertonic (p. 118) and isotonic (p. 118) solutions the vacuole disappears.

contractile vacuole of *Amoeba*

Malpighian tubule one of a number of narrow, blind tubes which are the main organs of osmoregulation (p. 133) and excretion (↑) in insects (p. 69) and other members of the Arthropoda (p. 67). They arise from the gut (p. 98) and in them, uric acid (↑) crystals are produced which can be eliminated with little water loss. Although insects do possess a thin, waterproof cuticle (p. 145) water is still lost through the joints (p. 146) and by respiration (p. 112).

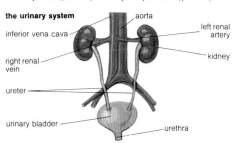

the urinary system

aorta

inferior vena cava

left renal artery

right renal vein

kidney

ureter

urinary bladder

urethra

ureter (n) the tube or duct which carries urine (↓) from the kidney (p. 136) to the bladder (↓).

bladder[a] (n) an extensible sac into which the ureter (↑) passes which fills with urine (↓) secreted (p. 106) continuously by the kidneys (p. 136). When full, it is opened by a sphincter muscle (p. 127), contracts, and releases the urine.

urine (n) the fluid which is finally expelled from the kidneys (p. 136). It contains urea (↑), or uric acid (↑) together with other materials and water.

anti-diuretic hormone a hormone (p. 130) synthesized by the hypothalamus (p. 156) and secreted (p. 106) by the pituitary gland (p. 157). It increases the reabsorption of water in the kidney (p. 136) tubules thus increasing the concentration of the urine (↑).

aldosterone (n) a hormone (p. 130) secreted (p. 106) by the adrenal glands (p. 130) which stimulates the reabsorption of sodium from the kidneys (p. 136) and increases the excretion (↑) of potassium thus tending to increase the concentration of sodium in the blood (p. 90) while decreasing the concentration of potassium.

kidney (*n*) in mammals (p. 80), one of a pair of organs which form the main site of excretion and may also be involved in osmoregulation (p. 113). Blood (p. 90) is pumped by the heart (p. 124) under pressure through the kidneys and, by reabsorptions and secretions (p. 106), useful substances are returned to the blood while wastes are eliminated in the urine. (↓).

nephron (*n*) one of the main structures of excretion (p. 134) in the kidneys (↑). It is a microscopic (p. 9) tubule which is made up of a Malpighian corpuscle and a drainage duct. It is in the nephrons that the processes of filtration and reabsorption take place.

kidney

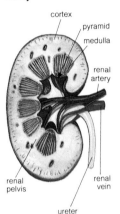

cortex
pyramid
medulla
renal artery
renal vein
renal pelvis
ureter

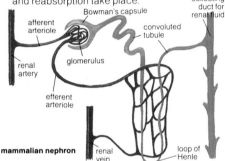

collecting duct for renal fluid
Bowman's capsule
afferent arteriole
convoluted tubule
renal artery
glomerulus
efferent arteriole
mammalian nephron
renal vein
loop of Henle

Bowman's capsule part of the Malpighian corpuscle in the nephron (↑). It is a cup-shaped swelling surrounding the glomerulus (↓) forming part of the structure through which blood (p. 90) is forced under pressure and by which it is 'purified' by a process known as ultra-filtration (↓).

loop of Henle a U-shaped tubule in the nephron (↑) into which isotonic (p. 118) renal fluid (↓) is pumped. As the fluid passes in the counter direction along the other arm of the tubule, sodium ions are actively transported (p. 122) across into the first arm so that there is a high concentration of sodium ions in the bend of the tube. From here there are collecting ducts which open into the renal pelvis from where water is drawn out by osmosis (p. 118). Thus, in the collecting ducts, the renal fluid becomes hypertonic (p. 118).

mesophyte e.g. beech

sheds leaves in autumn

glomerulus (*n*) a knot of capillaries (p. 127) which form part of the Malpighian corpuscle and into which blood (p. 90) is pumped via the renal artery (p. 127) and arterioles (p. 127). Blood is forced through the walls of the capillaries into the Bowman's capsule (↑).

renal fluid a fluid consisting mostly of blood plasma (p. 90) and the soluble materials contained within it, which, the process of ultrafiltration (↓) passes through the kidneys (↑).

ultrafiltration (*n*) the process by which a large proportion of the blood plasma (p. 90) and the soluble materials contained within it are forced under pressure through the walls of the glomerulus (↑), through the walls of the Bowman's capsule (↑) and into the lumen (↓) of the nephron (↑).

lumen (*n*) a space within a tube or sac.

hydrophyte e.g. *Nuphar*

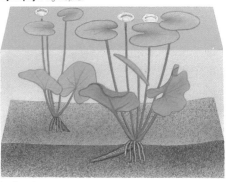

xerophyte
e.g. cactus

thick succulent   hot
stems with   dry
thick cuticle   air

spines

dry desert sand

hydrophyte (*n*) a plant which is adapted to grow in water or in very wet conditions. Leaves and stems often contain air spaces to aid the flotation of the whole plant or part of it.

mesophyte (*n*) a plant which is adapted to grow in habitats (p. 217) with normal supplies of water. Typically, they have large, flattened leaves that are lost during leaf fall (p. 133).

xerophyte (*n*) a plant which is adapted to grow in very dry environments (p. 218).

growth substance (*n*) a hormone (p. 130) which,
in very small quantities, can increase, decrease,
or otherwise change the growth of a plant or
part of a plant.
indole-acetic acid IAA. A growth substance (↑)
which causes plant cells to grow longer and
causes cells to divide. IAA is the most common
of the group of growth substances called
auxins.
auxin (*n*) *see* indole-acetic acid (↑).

indole-acetic acid (IAA)

gibberellin (*n*) a growth substance (↑) which may
effect the elongation of cells in stems and
which may cause the area of leaves to increase.
They also promote a variety of effects in plant
growth such as seed germination (p. 168),
flowering and the setting of fruit.

gibberellin
e.g. gibberellic acid 1

cytokinin (*n*) a growth substance (↑) which, in
association with IAA, affects the rate of cell
division, promotes the formation of buds, and is
essential for the growth of healthy leaves. Also
known as **kinin**.

cytokinins
e.g. kinetin

HC——————CH
HN——CH₂——C   CH
              O
C
HN
N   N
C   C
N   CH
HC   C
N   NH

absicin (*n*) a growth substance (↑) which inhibits plant growth, prevents germination (p. 168), and tends to promote buds to become dormant. Absicin seems to work against normal growth substances by preventing the manufacture of proteins (p. 21) etc.

absicin

H₃C   CH₃   CH₃
H₂C C   C
     C   H   C CH
      OH H
O C   COOH
   C   C CH₃
      H

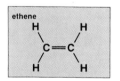

ethene (*n*) a growth substance (↑) which is produced as the result of normal metabolism (p. 26) in plants and which may cause leaves to fall and fruit to ripen. Also known as **ethylene**.
florigen (*n*) a growth substance (↑) which, although it has never been isolated, is believed to promote the production of flowers.
tropism (*n*) the way in which the direction of growth of a plant responds to external stimuli. **trophic** (*adj*).

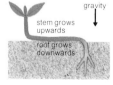

geotropism (*n*) a tropism (↑) in which the various parts of a plant grow in response to the pull of the Earth's gravity. For example, primary roots grow downwards and are referred to as being positively geotrophic while the main stems grow upwards and are referred to as being negatively geotrophic.

statolith (*n*) a large grain of starch (p. 18) which is found in plant cells and which is thought to respond to the effects of gravity causing the effects of geotropism (p. 139).

phototropism (*n*) a tropism (p. 139) in which various parts of the plant grow in response to the direction from which light is falling on the plant. Stems tend to grow towards the light and are referred to as being positively phototrophic. The roots of some plants e.g. those of climbers grow away from the source of light and are referred to as being negatively phototrophic.

phototropism

auxin in phototropism

1 shoot tip in dark, auxin evenly concentrated

2 exposed to light from one side, auxin concentration increases on dark side and decreases on light side

3 increased relative auxin concentration on dark side causes cells on dark side to elongate, and the shoot bends towards the light

hydrotropism (*n*) a tropism (p. 139) in which the roots of plants grow towards a source of water. Hydrotropism will usually override the effects of geotropism (p. 139). **hydrotrophic** (*adj*).

chemotropism (*n*) a tropism (p. 139) in which the roots of a plant or the hyphae (p. 46) of a fungus (p. 46) may grow towards a source of food materials. **chemotrophic** (*adj*).

thigmotropism (*n*) a tropism (p. 139) in which, by the stimulus of touch, certain parts of particular plants, such as the stems of climbing plants, may coil around a support. **thigmotrophic** (*adj*).

nastic movements growth movements, such as the opening and closing of flowers, which although they occur as a result of external stimuli, such as the presence or absence of light, do not take place in a particular direction.

photonasty (*n*) a nastic movement (↑) which is a response to the presence or absence of light or even to light levels e.g. the flowers of daisies close at night and only open during the daylight.

thigmonastic movements of the leaves of a 'sensitive plant' after it has been touched

apical dominance

apex

axillary buds

intact plant: auxin translocated from apex inhibits growth of axillary buds into lateral shoots

apex removed: lateral shoots grow

thermonasty (*n*) a nastic movement (↑) which is a response to the surrounding temperature. For example, the flowers of some plants will open when the weather is warm.

thigmonasty (*n*) a nastic movement (↑) in which the response is to touch. For example, the leaves of the South American plant, commonly known as the 'sensitive plant', fold back when touched.

taxic movements the movement of an organism in which the response takes place in relation to the direction of the stimulus.

phototaxis (*n*) a taxic movement (↑) in which the movement may be away from or towards the direction from which the light is coming. For example, certain insects (p. 69) may hide from the light and are referred to as negatively phototaxic while many algae (p. 44) will move towards the light and are described as positively phototaxic.

thermotaxis (*n*) a taxic movement (↑) in which the movement may be away from or towards regions of higher or lower temperature. For example, a mammal (p. 80) may seek the shade of a tree during the heat of the day to prevent overheating.

chemotaxis (*n*) a taxic movement (↑) in which an organism may move towards a chemical stimulus. For example, a spermatozoid (p. 178) may swim towards a female organ which secretes (p. 106) a substance, such as sucrose (p. 18).

hygroscopic movements movements which take place as the parts of organisms dry out and thicker parts move differently from thinner parts.

autonomic movements movements which take place in an organism without any external stimulus. The stimulus comes from within the organism itself and may include movements such as the coiling of the tendrils of climbing plants such as peas.

apical dominance the state which may occur in plants in which the bud at the tip of a plant stem grows but the lateral ones do not. If the apical bud is removed, lateral branches may grow.

vernalization (*n*) the process whereby certain plants, such as cereal crops, need to be subjected to low temperatures, such as that which occurs through overwintering, during an early part of their growth before they will be induced to flower.

phytochrome (*n*) a light-sensitive pigment (p. 126) present in plant leaves which exists in two forms that can be converted from one to the other. One absorbs red light, the other far red light. In the absence of light, the latter slowly changes back to the former. Phytochromes initiate the formation of hormones (p. 130).

etiolation (*n*) plant growth which takes place in the absence of light. The plants may lack chlorophyll (p. 12) so that they will be yellow or even white in colour. The leaves will be reduced in size and the stems tend to grow much longer.

photoperiodism (*n*) the process in which certain activities, such as flowering or leaf fall (p. 133), respond to seasonal changes in day length.

long-day plants plants, such as cucumber, which only usually flower during the summer months in temperate climates when the hours of daylight exceed about fouteen in twenty-four.

short-day plants plants, such as chrysanthemums, which only usually flower during the spring or autumn months in temperate climates when the hours of daylight are less than about fourteen in twenty-four.

day-neutral plants plants, such as the pea, in which the hours of daylight have no influence on the flowering period.

interconversion of two forms of phytochrome

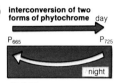

etiolation

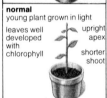

photoperiodism

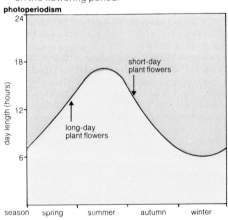

long-day plant

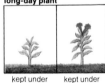

short-day plant

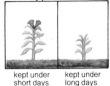

fibre of voluntary muscle

striped band

structure of striated muscle

endomysium

nucleus

A band
I band

H-line
Z-line
myofibril

structure of cardiac muscle

connective tissue   branched fibre

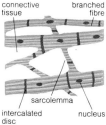

sarcolemma
intercalated disc   nucleus

locomotion (*n*) the ability of an organism to move all or part of its body independent of any outside force. An animal can usually move its whole body whereas a plant may only be able to move certain parts, such as petals (p. 179) or leaves, in response to changes in the environment (p. 218).

muscle (*n*) tissue (p. 83) which is made up of cells or fibres that are readily contracted.

fibre (*n*) a thread-like structure.

skeletal muscle muscle (↑) tissue (p. 83) consisting of elongated cells with many nuclei (p. 13) and cross striations in the cytoplasm (p. 10). It usually occurs in bundles and is under voluntary control of the central nervous system (p. 149) so that it contracts when stimulated to do so. These muscles are attached to parts of the skeleton (p. 145) and their contractions cause these parts to move. Skeletal muscle which has a striped look is known as striated muscle. This consists of long, narrow muscle fibres bounded by a membrane (p. 14) and containing many nuclei. The muscle fibres are bound together into bundles. They contract when stimulated.

voluntary muscle = skeletal muscle (↑).

striated muscle = skeletal muscle (↑). Also known as **striped muscle**.

unstriated muscle = involuntary muscle (↓).

involuntary muscle the muscle (↑) which is found in the internal organs and blood vessels (p. 127) and consists of simple tubes or sheets. It is under the involuntary control of the autonomic nervous system (p. 155). Also known as **smooth muscle**.

visceral muscle a smooth or unstriated (↑) muscle (↑) tissue (p. 83) made up of elongated cells held together by connective tissue (p. 88) and activated involuntarily. It is found in all internal organs as well as blood vessels (p. 127) with the exception of the heart (p. 124).

cardiac muscle muscle (↑) tissue (p. 83) found only in the heart (p. 124) walls. It consists of fibres containing cross striated (↑) myofibrils (p. 144). It contracts rhythmically and automatically (i.e. without nervous (p. 149) stimulation).

muscle fibre the elongated cells which make up striated muscles (p. 143) and which consist of a number of myofibrils (↓).

myofibril (*n*) the very fine threads which make up the muscle fibres (↑) and are found in smooth, striated (p. 143), and cardiac muscles (p. 143). They contain the contractile proteins (p. 21) myosin (↓), actin (↓) and tropomyosin (↓).

sarcomere (*n*) the part of the myofibril (↑) which is responsible for the contraction. It is made up of a dark central A band composed of myosin (↓) on either side of which are I bands composed of actin (↓). Each sarcomere is joined to the next by the Z membrane (p. 14). During contraction the I band shortens while the A bands stay more or less the same length so that the muscle filaments slide between one another.

thick filaments the filaments of a myofibril (↑) which are composed of myosin (↓).

thin filaments the filaments of a myofibril (↑) which are composed of actin (↓).

actin (*n*) the contractile protein (p. 21) which comprises one of the main elements in muscle (p. 143) myofibrils (↑). When stimulated actin and myosin (↓) join together to form actomyosin (↓).

myosin (*n*) the contractile protein (p. 21) which comprises the most abundant element in muscle (p. 143) myofibrils (↑). When stimulated actin (↑) and myosin join together to form actomyosin (↓).

actomyosin (*n*) a complex of the two proteins (p. 21) actin (↑) and myosin (↑) which, when they interact to form the complex, result in the contraction of the muscle (p. 143).

tropomyosin (*n*) the third protein found in myofibrils (↑) which may be responsible for controlling the contractions of muscle (p.143).

sliding filament hypothesis the theory (p. 235) which suggests that when a muscle (p. 143) contracts, the individual filaments do not shorten but that they slide between one another because it can be seen under the electron microscope (p. 9) that the pattern of striations (p. 143) changes during the contraction.

sliding filament hypothesis

thin filament (actin)
relaxed myofibril
thick filament (myosin)
Z-line   Z-line

contracted myofibril

sarcoplasmic reticulum a smooth endoplasmic reticulum (p. 11) which is responsible for absorbing (p. 81) the calcium that is necessary for muscle (p. 143) contraction.

muscle spindle a modified muscle fibre (↑) which is receptive to stimulation and controls the way in which a muscle (p. 143) contracts.

skeleton (*n*) a supporting structure. It can be jointed (p. 146) and the joints (p. 146) are connected by muscles (p. 143) that, when they contract against the limbs as levers, enable the animal to operate the limbs, e.g. the legs, allowing movement on land.

exoskeleton (*n*) the external skeleton (↑) of organisms, such as insects (p. 69), which provides protection for the internal organs and is the structure to which the muscles (p. 143) are attached. For example, the shell of a cockle would be referred to as an exoskeleton.

apodeme (*n*) one of a number of projections on the inside of the exoskeleton (↑) where there are joints (p. 146) and to which the muscles (p. 143) for the movement of those joints are attached.

cuticle[a] (*n*) the outer layer of the endoskeleton (↓) which in animals, such as insects (p. 69), acts as the skeleton (↑) itself. It may be composed of chitin (p. 49) but in shellfish may be hardened with lime-rich salts. It is secreted (p. 106) by the epidermis (p. 131) and is non-cellular. The outer epicuticle is waxy and waterproof in insects and other arthropods (p. 67).

hydrostatic skeleton a form of skeleton (↑) found in soft-bodied animals such as earthworms (p. 66) in which the body fluids themselves provide the structure against which the muscles (p. 143) act.

endoskeleton (*n*) a bony or cartilaginous (p. 90) structure contained within the body of vertebrates (p. 74) which is usually jointed (p. 146) to allow movement and to which the muscles (p. 143) are attached to provide the mechanisms for movement.

musculo-skeletal system the system which enables the animal to move by providing a jointed (p. 146) skeleton (↑) against which the muscles (p. 143) can act to cause operation of the joints using the limbs as levers.

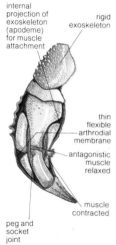

claw of crab showing exoskeleton

internal projection of exoskeleton (apodeme) for muscle attachment

rigid exoskeleton

thin flexible arthrodial membrane

antagonistic muscle relaxed

muscle contracted

peg and socket joint

swimming in fish   waves of lateral undulations pass from the head back along the body

swimming (*n*) the process by which an organism such as a fish propels itself through or on the surface of the water by the action of fins (p. 75) or flexing of the whole body. In fish, when the muscles (p. 143) contract, the body cannot shorten so that it moves from side to side to provide propulsion.

caudal fin the tail fin. The main organ by which a fish propels itself through the water. It is a membrane (p. 14) supported by fin rays attached to the vertebral column (p. 74) of the fish.

myotome muscle one of a number of blocks of striated muscle (p. 143) that completely enclose the vertebral column (p. 74) of fish.

joint (*n*) the region at which any two or more bones of a skeleton (p. 145) come into contact. For example, the elbow joint in a human.

ball and socket joint a movable joint (↑) between limbs in which one bone terminates in a knob-shaped structure which fits into a cup-shape in the meeting bone allowing some movement in all directions, e.g. the joint between the femur and socket of the pelvic girdle (↓).

hinge joint a movable joint (↑) between bones in which the movement can take place in one plane or direction only e.g. the knee joint.

pivot hinge a movable joint (↑) between bones in which movement can take place in all directions by a twisting or rotating movement. For example, the joints in the neck.

ligament (*n*) the strong, elastic connective tissue (p. 88) which, for example, holds together the limb bones at a joint (↑) and which helps to control the movement of the joint.

tendon (*n*) the thick cord of connective tissue (p. 88) which connects a muscle (p. 143) to a bone. It is non-elastic so that when the muscle contracts, it pulls against the bone forcing it to move at the joint (↑).

cross sections of fish body

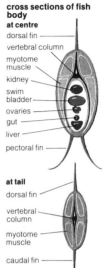

at centre
dorsal fin
vertebral column
myotome muscle
kidney
swim bladder
ovaries
gut
liver
pectoral fin

at tail
dorsal fin
vertebral column
myotome muscle
caudal fin

structure of a joint

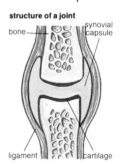

bone
synovial capsule
ligament
cartilage

joints
ball and socket joint   socket

hinge joint

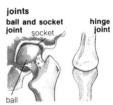

ball

muscles of hind leg of a mammal

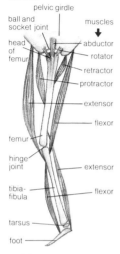

vertebrae
vertebral column (backbone)

joint between vertebrae

a vertebra

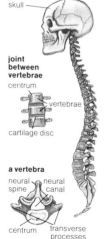

protractor muscle a muscle (p. 143) which on contraction draws the limb bone forwards.

retractor muscle a muscle (p. 143) which on contraction draws the limb bone backwards.

adductor muscle a muscle (p. 143) which on contraction draws the limb bone inwards.

abductor muscle a muscle (p. 143) which on contraction draws the limb bone outwards.

rotator muscle a muscle (p. 143) which on contraction rotates a limb bone outwards or inwards.

flexor muscle a muscle (p. 143) which on contraction draws two limb bones together.

extensor muscle a muscle (p. 143) which on contraction draws two limb bones apart.

vertebra (*n*) one of a number of bones, or, in some cases, segments of cartilage (p. 90), which make up the vertebral column (p. 74). Each vertebra is usually hollow and has muscles (p. 143) attached to it.

pelvic girdle the part of the skeleton (p. 145) of a vertebrate (p. 74) to which the hind limbs are attached. It is rigid and provides the main support for the weight of the body

limb (*n*) any part of the body of an animal, apart from the head and the trunk, including, for example, the arms, the legs or the wings.

flight (*n*) that form of locomotion (p. 143) such as is found in most birds and many insects, whereby the animal is borne through the air either by gliding on the wind using outstretched membranes (p. 14) or by using the lift generated by the special shape and the power provided by wings.

feather (*n*) one of a very large number of structures which provide the body covering of birds. They distinguish birds from all other animals. Feathers insulate the bird s body, repel water, streamline it, and aid in the power of flight.

down (*n*) the soft, fluffy feathers (↑) which form the initial covering of very young birds and which are also found on the undersides of adult birds. The individual barbs (p. 148) are not joined together so that down provides better insulation than flight feathers (p. 148) by trapping a layer of air close to the bird's body.

flight feather one of a number of feathers (p. 147)
giving birds their streamlined effect and which
are elongated to provide the flight surface.

shaft (*n*) the central rod-like but flexible stem of a
feather (p. 147) which is the continuation of the
quill (↓) that is attached to a feather (like hair)
follicle (p. 132) in the skin of a bird. The skin of
a bird has no sweat glands (p. 132).

quill (*n*) the hard tube-like part of the feather
(p. 147) that is attached to the feather follicle
(p. 132) and which is also connected by
muscles (p. 143) that are able to alter the angle
at which the feather lies in relation to the body
of the bird. For example, in cold weather, the
bird would raise its feathers to trap extra layers
of air for more efficient insulation.

vane (*n*) the flat, blade-like part of the feather
(p. 147) which is composed of the shaft (↑) and
its attached web of barbs (↓) and barbules (↓)

barb (*n*) a hook-like process which projects from
the shaft (↑) of a feather (p. 147). Barbs are
interlocked by the barbules (↓).

barbule (*n*) one of the tiny barbs (↑) attached to
the barb of a feather (p. 147) and which link the
barbs by a system of hooks and troughs to
make up the web or vane (↑) of the feather.

pectoralis muscle one of the large, powerful
muscles (p. 143) which pull on the wings of a
bird.to force it upwards and downwards
providing the power for flight. Pectoralis
muscles are attached to the sternum (↓) of the
bird The pectoralis major is the muscle which
depresses the wing while the pectoralis minor
is responsible for raising it.

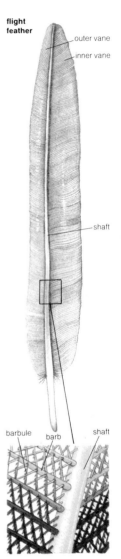

flight
feather

outer vane

inner vane

shaft

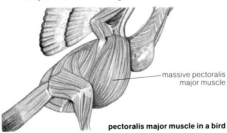

massive pectoralis
major muscle

pectoralis major muscle in a bird

barbule

barb

shaft

sternum (*n*) the breast bone of tetrapods.

gliding (*n*) the process of flight (p. 147) whereby the animal holds the wings outstretched so that they function as aerofoils and the animal soars on a cushion of supporting air.

irritability (*n*) the ability of an organism to respond to changes in its environment (p. 218) e.g. the movement of animals in response to noise or being touched.

nervous system the system within the body of an organism which permits the transmission of information through the body so that its various parts are able rapidly to respond to any stimuli.

central nervous system CNS. That part of the nervous system (↑), which in vertebrates (p. 74) includes the brain (p. 155) and the spinal cord (p. 154) which receives nerve impulses (p. 150) from all parts of the body, internal and external, and responds by delivering the appropriate commands to the various organs and muscles (p. 143) to react accordingly.

peripheral nervous system that part of the nervous system (↑) excluding the CNS (↑). It consists of a network of nerves (↓) running through the body of the organism and connected with the CNS.

neurone (*n*) one of the many specially modified cells which make up the nervous system (↑). Each neurone is connected via synapses (p. 151) to others by a single thread-like axon (↓) or nerve fibre and numerous dendrons (↓) which transmit nerve impulses (p. 150) from neurone to neurone.

nerve cell = neurone (↑).

cell body the part of the neurone (↑) with the nucleus (p. 13). Also known as **centron**.

dendron (*n*) a branching process of cytoplasm (p. 10) from the body of a neurone (↑) which ends in a synapse (p. 151). They may branch into dendrites.

Nissl's granules granules found in the cytoplasm (p. 10) of a neurone (↑). They are rich in RNA (p. 24).

axon (*n*) the long process of a neurone (↑) filled with axoplasm which normally conducts nerve impulses (p. 150) away from the cell body (↑). The axon is enclosed in a myelin (p. 150) sheath which is bounded by the thin membrane (p. 14), the *neurilemma*, of the *Schwann cell*.

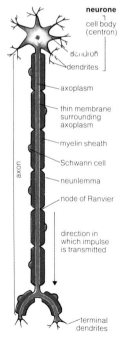

neurone

cell body (centron)

dendron
dendrites

axoplasm

thin membrane surrounding axoplasm

myelin sheath

Schwann cell

neurilemma

node of Ranvier

direction in which impulse is transmitted

terminal dendrites

axon

myelin (*n*) a fatty substance which insulates the axon (p. 149) and speeds up the transmission of nerve impulses (↓). In vertebrates (p. 74) not all axons are myelinated. The myelin sheath is broken at intervals by constrictions called *nodes of Ranvier*.

neuroglia (*n*) specialized cells which protect and support the central nervous system (p. 149).

nerve impulse one of an interspaced succession of impulses or signals that are carried between the neurones (p. 149) via the exchange of sodium ions and changes in the electrical state of the neurone. The impulses travel at a constant speed throughout the nervous system (p. 149) and the energy for the impulse is not provided by the stimulus itself.

transmission of nerve impulse along nerve

direction of impulse

resting potential

membrane polarized: inside negative, outside positive. Sodium ions expelled by sodium pump mechanism

action potential

membrane depolarized: sodium ions enter axon; inside positive, outside negative

resting potential

membrane repolarized

resting potential the state which occurs when a neurone (p. 149) is inactive so that the neurone carries a greater negative charge within the cell and a greater positive charge outside.

action potential the state in which an electrical charge moves along the membrane (p. 14) of the axon (p. 149).

sodium pump mechanism the mechanism by which sodium ions are pumped out of a neurone (p. 149) as soon as the nerve impulse (↑) has passed.

polarization (*n*) the process in which sodium ions are pumped out of the neurone (p. 149) by the sodium pump mechanism (↑) so that the inside of the cell is restored to its resting potential (↑).

all or nothing law

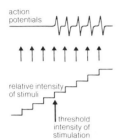

depolarization (*n*) the process in which the membrane (p. 14) of the neurone (p. 149) becomes permeable to the passage of sodium ions which then enter the cell so that the cell becomes positively charged.

stimulus (*n*) any change in the external environment (p. 218) or the internal state of an organism which, (via the nervous system (p. 149) in animals), provokes a response to that change without supplying the energy for it.

threshold intensity the level of stimulus below which there is no nervous (p. 149) response of the stimulated organism.

'all or nothing law' the law which states that an organism will respond to a stimulus in only two ways: that is either no nervous (p. 149) response at all or a response which is of a degree of intensity which does not vary with the intensity of the stimulus.

refractory period the length of time which passes between the passage of a nervous impulse (↑) through a neurone (p. 149) and its return to the resting potential (↑). During this period the neurone cannot further be stimulated.

absolute refractory period a refractory period (↑) in which a further stimulus of any intensity will result in the passage of no further nerve impulse (↑).

relative refractory period a refractory period (↑) in which another, unusually intense stimulus will result in the passage of a further nerve impulse (↑).

transmission speed the speed at which a nervous impulse (↑) travels and which is dependent upon the diameter of the neurone (p. 149).

synapse (*n*) the gap which exists between neurones (p. 149) and which is bridged during the action potential (↑) by a substance secreted (p. 106) by the neurone.

synaptic transmission the process by which nervous impulses (↑) are transmitted between neurones (p. 149) via the synaptic knob (p. 152). The action potential (↑) stops at the synapse (↑) but it causes a substance to be released which travels across the synapse and generates a new action potential in the neighbouring neurone.

synapse

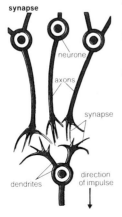

synaptic knob (*n*) the knob-like ending of the axon (p. 149) which projects into the synapse (p. 151).

acetylcholine (*n*) one of the substances that are released as the action potential (p. 150) in a neurone (p. 149) arrives at the synapse (p. 151). It is specifically produced between a neurone and a muscle (p. 143) cell. There is a special enzyme (p. 28) called acetylcholine esterase which breaks it down so its effect does not continue.

atropine (*n*) a substance which is found in the plant, deadly nightshade, and which acts as a poison by preventing nerve impulses (p. 150) being transmitted from the neurone (p. 149) to the body tissues (p. 83).

strychnine (*n*) a substance which is obtained from the seed of an east Indian tree and which has a powerful stimulating effect on the central nervous system (p. 149), so much so, that in greater than minute quantities it acts as a poison.

adrenaline (*n*) a substance, similar to noradrenaline (↓), released by the adrenal glands (p. 130), which increases the metabolic rate (p. 32) and other functions when it is released into the bloodstream (p. 90) during stress or in preparation for action.

noradrenaline (*n*) one of the substances which is released as the action potential (p. 150) in a neurone (p. 149) arrives at the synapse (p. 151). It is produced in the autonomic nervous system (p. 155). It is also secreted (p. 106) by the adrenal glands (p. 130) and affects cardiac muscle (p. 143) and involuntary muscle (p. 143) etc.

summation (*n*) the process in which the additive effect of nerve impulses (p. 150) arriving at different neurones (p. 149) stimulates the impulse in another neurone while the arrival of just one of the impulses produces no effect.

facilitation (*n*) the process in which the stimulation of a neurone (p. 149) is increased by summation (↑).

reflex action the fundamental and innate (p. 164) response of an animal to a stimulus. For example, the automatic escape reaction away from a source of threat or pain, such as withdrawing the hand from a hot object.

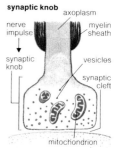

synaptic knob axoplasm
nerve impulse
myelin sheath
synaptic knob
vesicles
synaptic cleft
mitochondrion
membrane of post-synaptic dendrite

simple reflex arc

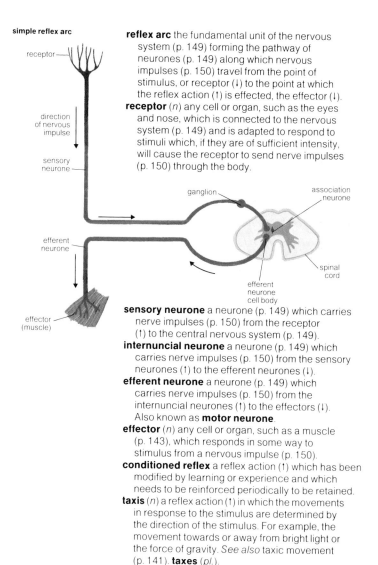

reflex arc the fundamental unit of the nervous system (p. 149) forming the pathway of neurones (p. 149) along which nervous impulses (p. 150) travel from the point of stimulus, or receptor (↓) to the point at which the reflex action (↑) is effected, the effector (↓).

receptor (*n*) any cell or organ, such as the eyes and nose, which is connected to the nervous system (p. 149) and is adapted to respond to stimuli which, if they are of sufficient intensity, will cause the receptor to send nerve impulses (p. 150) through the body.

sensory neurone a neurone (p. 149) which carries nerve impulses (p. 150) from the receptor (↑) to the central nervous system (p. 149).

internuncial neurone a neurone (p. 149) which carries nerve impulses (p. 150) from the sensory neurones (↑) to the efferent neurones (↓).

efferent neurone a neurone (p. 149) which carries nerve impulses (p. 150) from the internuncial neurones (↑) to the effectors (↓). Also known as **motor neurone**.

effector (*n*) any cell or organ, such as a muscle (p. 143), which responds in some way to stimulus from a nervous impulse (p. 150).

conditioned reflex a reflex action (↑) which has been modified by learning or experience and which needs to be reinforced periodically to be retained.

taxis (*n*) a reflex action (↑) in which the movements in response to the stimulus are determined by the direction of the stimulus. For example, the movement towards or away from bright light or the force of gravity. *See also* taxic movement (p. 141). **taxes** (*pl.*).

kinesis (*n*) a reflex action (p. 152) in which the
rate of movement is affected by the intensity of
the stimulus and which is unaffected by its
direction. For example, woodlice move faster in
drier surroundings than in damp ones.

spinal cord the part of the central nervous
system (p. 149) in vertebrates (p. 74) which is
contained within a hollow tube running the
length of the vertebral column (p. 74) and runs
from the medulla oblongata (p. 156). It consists
of neurones (p. 149) and nerve fibres with a
central canal containing fluid. Pairs of spinal
nerves (↓) leave the spinal cord to pass into the
body. The spinal cord carries nerve impulses
(p. 150) to and from the brain (↓) and the body.

spinal cord

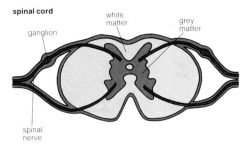

white
matter

grey
matter

ganglion

spinal
nerve

meninges (*n.pl.*) the three membranes (p. 14)
which protect the central nervous system
(p. 149) of vertebrate (p. 74) animals.

pia mater one of the meninges (↑). The soft,
delicate inner membrane (p. 14) next to the
central nervous system (p. 149) which is
densely packed with blood vessels (p. 127).

arachnoid mater the central of the three
meninges (↑) and separated from the pia mater
(↑) by fluid-filled spaces.

dura mater one of the meninges (↑). The stiff,
tough outer membrane which is in direct
contact with the arachnoid mater (↑) and which
contains blood vessels (p. 127).

spinal nerve one of the pairs of nerves (p. 149)
which arise from the spinal cord (↑) in segments.

meninges

skull (bone)   dura mater
arachnoid   cerebro-
mater   spinal fluid
pia mater   brain

grey matter nervous tissue (p. 91) that is grey in colour and is found in the centre of the spinal cord (↑) as well as in parts of the brain (↓). It contains large numbers of synapses (p. 151) and consists mainly of nerve (p. 149) cell bodies (p. 149).

white matter nervous tissue (p. 91) that is whitish in colour and is found on the outer region of the spinal cord (↑) as well as in parts of the brain (↓). It connects different parts of the central nervous system (p.149) and consists mainly of axons (p.149).

autonomic nervous system the part of the central nervous system (p. 149) in vertebrates (p. 74) which carries nerve impulses (p. 150) from receptors (p. 153) to the smooth muscle fibres (p. 144) of the heart (p. 124), gut (p. 98) and other internal organs.

sympathetic nervous system the part of the autonomic nervous system (↑) which increases the heart (p. 124) rate and breathing (p. 112) rate, the secretion (p. 106) of adrenaline (p. 152), the blood (p. 90) pressure, and slows the digestion (p. 98) so that the vertebrate's (p. 74) body is prepared for emergency action in response to stimuli.

parasympathetic nervous system the part of the autonomic nervous system (↑) which effectively works in opposition to the sympathetic nervous system (↑) slowing down the heart (p. 124) beat etc. Both systems act in co-ordination to control the rates of action.

ganglion (*n*) a bundle of nerve (p. 149) cell bodies (p. 149) contained within a sheath which, in invertebrates (p. 75) may form part of the central nervous system (p. 149), and in vertebrates (p. 74) are generally found outside the central nervous system. Also, in the brain (↓) some of the masses of grey matter (↑) are referred to as ganglia (*pl.*).

nerve net an interconnecting network of nerve (p.149) cells found in the bodies of some invertebrates (p. 75) to form a simple nervous system (p. 149).

brain (*n*) the part of the central nervous system (p. 149) which effectively co-ordinates the reactions of the whole body of the organism. It forms as an enlargement of the spinal cord (↑) and is situated at the anterior end of the body.

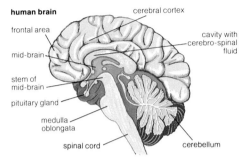

human brain
cerebral cortex
frontal area
cavity with cerebro-spinal fluid
mid-brain
stem of mid-brain
pituitary gland
medulla oblongata
spinal cord
cerebellum

cerebral hemispheres the paired masses of grey
matter (p. 155), beneath which is white matter
(p. 155), that occur at the front end of the
forepart of the brain (p. 155) and by which many
of the animal's activities are controlled. Each
hemisphere controls actions on the opposite
side of the body from which it is situated.

cerebral cortex the highly convoluted grey matter
(p. 155) that forms part of the cerebral
hemispheres (↑).

corpus callosum the band of nerve (p. 149)
fibres which connects the cerebral hemispheres
(↑) allowing their action to be co-ordinated.

medulla oblongata the continuation of the spinal
cord (p. 154) with the hind region of the brain
(p. 155). It contains centres of grey matter
(p. 155) which are responsible for controlling
many of the major functions and reflexes
(p. 153) of the body, for example, the medulla
oblongata contains the respiratory centre (p.117).

cerebellum (*n*) the part of the brain (p. 155) lying
between the medulla oblongata (↑) and the
cerebral hemispheres (↑) which is deeply
convoluted and is responsible for controlling
voluntary muscle (p. 143) action which is
stimulated by the cerebral hemispheres.

hypothalamus (*n*) that region of the forepart of
the brain (p. 155) which is responsible for
monitoring and regulating metabolic (p. 26)
functions such as body temperature, eating,
drinking and excretion (p. 134). It controls the
activity of the pituitary gland (↓).

thalamus (*n*) the part of the brain (p. 155) which carries and co-ordinates nerve impulses (p. 150) from the cerebral hemispheres (↑).

pituitary gland a gland (p. 87) in the brain (p. 155) which secretes (p. 106) a number of hormones (p. 130) that in turn stimulate the secretion of hormones from other glands to affect such metabolic (p. 26) processes as growth, secretion of adrenaline (p. 152), milk production, and so on.

pineal body a gland (p. 87) found as an outgrowth on the top of the brain (p. 155) and which may be responsible for secreting (p. 106) a hormone (p. 130) associated with colour change.

human ear

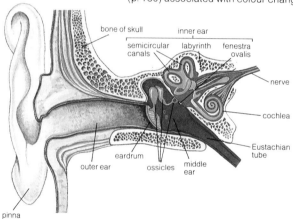

bone of skull · inner ear · semicircular canals · labyrinth · fenestra ovalis · nerve · cochlea · Eustachian tube · eardrum · ossicles · middle ear · outer ear · pinna

ear (*n*) one of the pair of sense organs, situated on either side of the head in vertebrates (p. 74), which are used for hearing (p. 159) and balance (p. 159).

outer ear the tube which leads from the outside of the head to the eardrum (p. 158). In amphibians (p. 77) and some reptiles (p. 78), it is not present because the eardrum is situated at the skin surface.

pinna (*n*) the part of the outer ear (↑), present in mammals (p. 80), situated on the outside of the head and consisting of a flap of skin and cartilage (p. 90), which helps to direct sound into the ear (↑).

inner ear the innermost part of the ear (p. 157,) which is situated within the skull and which detects sound as well as the position of the animal in relation to gravity and acceleration, so enabling the animal to balance. It is fluid filled and is connected to the brain (p. 155) by an auditory nerve (p. 149) so that it is able to convert sound waves into nervous impulses (p. 150). It is made up of a labyrinth of membraneous (p. 14) tubes contained within bony cavities.

middle ear an air-filled cavity situated between the outer ear (p. 157) and the inner ear (↑). It is separated from the outer ear by the eardrum (↓) and, in mammals (p. 80), contains three small bones or ossicles (↓).

eardrum (*n*) a thin, double membrane (p. 14) of epidermis (p. 131) separating the outer ear (p. 157) from the middle ear (↑) and which is caused to vibrate by sound waves. These vibrations are then transmitted through the middle ear, where their force is amplified, into the inner ear (↑). Also known as **tympanic membrane**.

fenestra ovalis a small, oval, membraneous (p. 14) window which connects the middle ear (↑) with the inner ear (↑) allowing vibrations from the eardrum (↑) to be transmitted to the inner ear. It is twenty times smaller than the eardrum so that the force of the vibrations is increased.

fenestra rotunda a small, round, membraneous (p. 14) window which connects the inner ear (↑) with the middle ear (↑) and which bulges into the middle ear as vibrations of the fenestra ovalis (↑) cause pressure increases in the inner ear.

ear ossicle one of usually three small bones that occur in the middle ear (↑) of mammals (p. 80) and which, by acting as levers, transmit and increase the force of vibrations produced by the eardrum (↑) and carry them to the inner ear (↑).

malleus (*n*) a hammer-shaped ear ossicle (↑) which is connected with the eardrum (↑).

incus (*n*) an anvil-shaped ear ossicle (↑) situated between the malleus (↑) and the stapes (↓).

stapes (*n*) a stirrup-shaped ear ossicle (↑) which is connected with the fenestra ovalis (↑).

ear ossicles

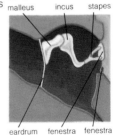

malleus   incus   stapes

eardrum   fenestra   fenestra
            ovalis   rotunda

section through cochlea

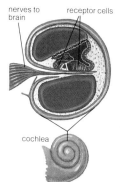

nerves to brain

receptor cells

cochlea

semicircular canals

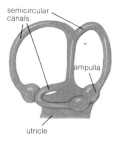

semicircular canals

ampulla

utricle

Eustachian tube a tube which connects the middle ear (↑) with the back of the throat. It is normally closed but opens during yawning and swallowing to balance the pressure on either side of the eardrum (↑) thus preventing the eardrum from bursting.

vestibular apparatus the apparatus contained in a cavity in the inner ear (↑) immediately above and behind the fenestra ovalis (↑) and which contains the organs concerned with the sense of balance and posture.

cochlea (*n*) a spirally coiled tube, which is a projection of the saccule (p. 160), and found within the inner ear (↑). It is concerned with sensing the pitch (↓) of the sound waves entering the ear (p. 157).

hearing (*n*) the sense whereby sound waves or vibrations enter the outer ear (p. 157) and cause the eardrum (↑) to vibrate. In turn, these vibrations are transmitted through the middle ear (↑) and into the inner ear (↑) where they are converted into nervous impulses (p. 150) and transmitted to the brain (p. 155).

intensity (*n*) the degree of loudness or softness of sound. If a sound entering the ear (p. 157) is very loud, muscles (p. 143) attached to the ear ossicles (↑) prevent them from vibrating too much.

pitch (*n*) the degree of height or depth of a sound which depends on the frequency of the sound waves – those of a high frequency are referred to as high and vice versa. Different parts of the cochlea (↑) respond to sounds of different pitch.

balance (*n*) the ability of the animal to orient itself properly in relation to the force of gravity. Animals rely on information received by apparatus within the inner ear (↑) and by the eyes and other senses to achieve balance and posture.

semicircular canal one of three looped tubes positioned at right angles to one another within the inner ear (↑). They contain fluid which flows in response to movement of the head. The movement of the fluid is detected by the sensory hairs in ampullae (p. 160) at the ends of the canals.

ampulla (*n*) a swelling at the end of each semicircular canal (p. 159). It bears a gelatinous cupula (↓), sensory hairs, and receptor (p. 153) cells which are responsible for transmitting information to the brain (p. 155) via the auditory nerve (p. 149). **ampullae** (*pl.*).

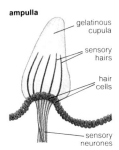

ampulla

gelatinous cupula

sensory hairs

hair cells

sensory neurones

utricle (*n*) a fluid-filled sac within the inner ear (p. 158) from which arise the semicircular canals (p. 159). Within the fluid of the utricle are otoliths (↓) of calcium carbonate. If the head is tilted, the otoliths are pulled downwards by gravity and in turn pull on sensory fibres attached to the wall of the utricle.

saccule (*n*) a lower, fluid-filled cavity in the inner ear (p. 158) from which arises the cochlea (p. 159). Like the utricle (↑), it also contains otoliths (↓) which respond to the orientation of the head with respect to the force of gravity.

cupula (*n*) the gelatinous body which forms part of the ampulla (↑) and which is displaced by fluid that moves in response to the position of the head.

otolith (*n*) one of the granules of calcium carbonate present in the fluid of the utricle (↑) and saccule (↑) which respond to tilting movements of the head by force of gravity.

eye (*n*) the sense organ of sight which is sensitive to the direction and intensity of light and which, in vertebrates (p. 74), is also able to form complex images of the outside world which are transmitted to the brain (p. 155) via the optic nerve (p. 149). The eyes of most animals are roughly spherical in shape, and in vertebrates are contained in depressions within the skull, to which they are attached by muscles (p. 143).

retina (*n*) the light-sensitive inner layer of the eye (↑) which contains rod-shaped receptor (p. 153) cells and cone-shaped receptor cells. Nerve (p. 149) fibres leave the retina and join together to form the optic nerve.

choroid layer a layer of tissue surrounding the eye (↑) between the retina (↑) and the sclerotic layer (↓). It contains pigments (p. 126) to reduce reflections within the eye and blood vessels (p. 127) which supply oxygen to the eye.

human eye

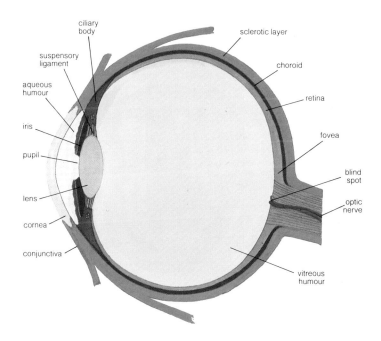

ciliary
body

sclerotic layer

suspensory
ligament

choroid

aqueous
humour

retina

iris

fovea

pupil

blind
spot

lens

optic
nerve

cornea

conjunctiva

vitreous
humour

sclerotic layer the tough, fibrous (p. 143) non-
elastic layer which surrounds and protects the
eye (↑) and is continuous with the cornea (↓).
cornea (*n*) the disc-shaped area at the front of
the eye (↑) which is continuous with the sclerotic
layer (↑) and which is transparent to light. It is
curved so that light passing through it is
refracted and begins to converge before
reaching the lens (p. 162). Indeed, in land-living
mammals (p. 80) this is the main element of the
eye's focusing power.

lens accommodation

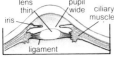

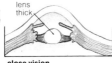

distant vision eye at rest   **close vision**

lens (*n*) a transparent disc which is convex on
both faces and which is attached to the ciliary
body (↓) by suspensory ligaments (↓). It consists
of an elastic, jelly-like material held in a skin.
When the ciliary muscles (↓) contract, the
convexity of the lens is increased so that light
rays entering the eye can be focused on the
retina (p. 160). This allows for the fine focusing
power of the eye (p. 160).

refraction (*n*) the change in direction of light
which takes place as it crosses a boundary
between two substances.

convex (*adj*) of a lens which causes light passing
through it to move closer together.

concave (*adj*) of a lens which causes light passing
through it to move further apart rather
than closer together. *See also* convex (↑).

suspensory ligament the fibrous (p. 143)
ligament (p. 146) which holds the lens (↑) in
position.

ciliary body the circular, thickened outer edge of
the choroid (p. 160) at the front of the eye
which contains the ciliary muscles and to
which the lens (↑) is attached. It also contains
glands (p. 87) which secrete (p. 106) the
aqueous humour (↓).

ciliary muscles *see* ciliary body (↑).

iris (*n*) a ring of opaque tissue (p. 83) that is
continuous with the choroid (p. 160) and which
has a hole or pupil (↓) in the centre through
which light can pass. There are circular
muscles (p.143) and radial muscles surrounding
the pupil which increase or decrease the size of
the pupil in accordance with the intensity of light.

pupil (*n*) the hole in the centre of the iris (↑)
through which light enters the eye (p. 160). It is
usually circular but may be other shapes in
some animals.

refraction

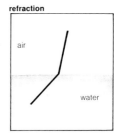

**convex
lens**

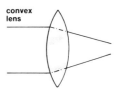

**concave
lens**

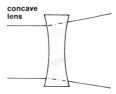

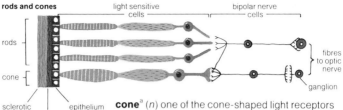

rods and cones

rods

cone

light sensitive cells

bipolar nerve cells

fibres to optic nerve

ganglion

sclerotic layer   choroid   epithelium

normal sight

far   near

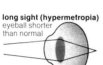

long sight (hypermetropia)
eyeball shorter than normal

convex lens

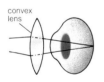

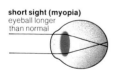

short sight (myopia)
eyeball longer than normal

concave lens

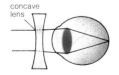

cone[a] (*n*) one of the cone-shaped light receptors (p. 153) present in the retina (p. 160). Cones contain three different pigments (p. 126) which are in turn sensitive to red, green, and blue light so that cones are primarily responsible for colour vision. Cones are concentrated mainly in and around the fovea (↓) and are not present at the edge of the retina. They are also receptors of light of high intensity.

fovea (*n*) a small, central depression in the retina (p. 160) in which most of the receptor (p. 153) cells, especially the cones (↑) are concentrated. It is directly opposite the lens and provides the main area for acute and accurate daylight vision (↓).

daylight vision vision of great sharpness which takes place in bright light since most of the light entering the eye (p. 160) falls on the fovea (↑).

rod (*n*) one of the rod-shaped light receptors (p. 153) present in the retina (p. 160) which are much more sensitive to light of low intensity but are not sensitive to colour. They are not present in the fovea (↑) and increase in numbers towards the edges of the retina. They are also sensitive to movements.

night vision vision which takes place in light of low intensity using the rods (↑).

aqueous humour a watery fluid that fills the space between the cornea (p. 161) and the vitreous humour (↓) and in which lie the lens (↑) and the iris (↑). It is secreted (p. 106) by glands (p. 87) in the ciliary body (↑).

vitreous humour the jelly-like fluid that fills the space behind the lens (↑).

blind-spot the area of the retina (p. 160) from which the optic nerve (p. 149) leaves the eye. It contains no rods (↑) or cones (↑) so that no image is recorded on this part of the retina.

behaviour (*n*) all observable activities carried out by an animal in response to its external and internal environment (p. 218).

ethology (*n*) the study or science of the behaviour (↑) of animals in their natural environment (p. 218).

instinctive behaviour behaviour (↑) which is believed to be controlled by the genes (p. 196) and which is unaffected by experience, e.g. the courtship behaviour of many animals, such as fish and birds, is stimulated by a particular signal provided the animal is sexually mature and has the appropriate level of sex hormones (p. 130) in its body.

innate behaviour behaviour (↑) that does not have to be learned. *See also* instinctive behaviour (↑).

learned behaviour behaviour (↑) in which the response to stimuli is affected by the experience of the individual animal to gain the best advantage from a situation.

habituation (*n*) a learned behaviour (↑) in which the response to a stimulus is reduced by the constant repetition of the stimulus.

associative learning a learned behaviour (↑) in which the animal learns to associate a stimulus with another that normally produces a reflex action (p. 152). For example, dogs respond to the sight of food by salivating. Pavlov's dogs learned to associate the sight of food with the ringing of a bell and would then salivate on hearing the bell without seeing the food.

imprinting (*n*) a learned behaviour (↑) which takes place during the very early stages of an animal's life so that e.g. the animal continues to follow the first object on which its attention is fixed by sight, sound, smell or touch. This is usually the animal's parent.

exploration (*n*) the process through which animals learn about their environment (p. 218) while they are young by play and contact with other animals.

orientation (*n*) the reflex action (p. 152) in which the animal changes the position of part or the whole of its body in response to a stimulus. For example, an animal might turn its head or prick up its ears in response to a sudden or unusual sound.

releaser (*n*) a stimulus which releases instinctive behaviour (↑) in an animal.

incomplete metamorphosis
e.g. locust

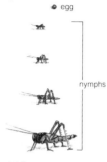

egg

nymphs

adult/imago

complete metamorphosis
e.g. butterfly

egg

larva

pupa

adult/imago

growth (*n*) the permanent increase in size and dry mass of an organism which takes place as the cells absorb (p. 81) materials, expand, and then divide. The temporary take-up of water cannot be considered as growth.

growth rate the amount of growth which takes place in a given unit of time.

incomplete metamorphosis the change which takes place from young to adult form in which the young closely resembles the adult.

instar (*n*) a stage through which an insect (p. 69) passes during incomplete metamorphosis (↑).

ecdysis (*n*) the periodic moulting or shedding of the external cuticle (p. 145) allowing growth, which takes place between instars (↑) during incomplete metamorphosis (↑).

nymph (*n*) the early instar (↑) or young of an insect (p. 69) which is small, sexually immature, and unable to fly.

complete metamorphosis the change which takes place from young to adult form in which the young do not resemble the adult and which can occur through a pupal (↓) stage.

larva (*n*) the immature stage in the life cycle of an animal, for example, an insect (p. 69) which undergoes metamorphosis (p. 70). The larva is usually different in structure and appearance from the adult. It hatches from the egg and is able to fend for itself. **larvae** (*pl.*), **larval** (*adj*).

pupa (*n*) the stage between larva (↑) and adult in an insect (p. 69) in which movement and feeding stop and metamorphosis (p. 70) takes place. **pupae** (*pl.*), **pupate** (*v*).

imago (*n*) the sexually mature, adult stage of an insect's (p. 69) development.

corpora allata a pair of glands (p. 87) in the head of an insect (p. 69) which secrete (p. 106) a hormone (p. 130) that encourages the growth of larval (↑) structures and discourages that of adult structures.

ecdysial glands a pair of glands (p. 87) in the head of an insect (p. 69) which secrete (p. 106) a hormone (p. 130) that stimulates ecdysis (↑) and growth.

morphogenesis (*n*) the process in which the overall form of the organs of an organism is developed, leading to the development of the whole organism.

differentiation (*n*) the process in which cells (unspecialized) change in form and function, during the development of the organism, to give all of the different types of specialized cells that characterize that organism.

neotony (*n*) the retention in some animals of larval (p. 165) or embryonic (↓) features, either temporarily or permanently beyond the stage at which they would normally be lost. It is thought to be important in evolutionary (p. 208) development. For example, humans retain certain resemblances to young apes.

embryology (*n*) the study of embryos (↓).

embryo (*n*) the stage in the development of an organism between the zygote (↓) and hatching, birth, or germination (p. 168). **embryonic** (*adj*).

zygote (*n*) the diploid (p. 36) cell which results from the fusion of a haploid (p. 36) male gamete (p. 175) or spermatozoon (p. 188) and a haploid female gamete or ovum (p. 178 and p. 190).

cleavage (*n*) the process in which the nuclei (p. 13) and cytoplasm (p. 10) of the fertilized (p. 175) zygote (↑) divide mitotically (p. 37) to form separate cells. Also known as **segmentation**.

blastula (*n*) the embryonic (↑) structure or mass of small cells which results from cleavage (↑).

blastocoel (*n*) the cavity which occurs in the centre of a blastula (↑) during the final stages of cleavage (↑).

gastrulation (*n*) the process which follows cleavage (↑) in which cell movements occur to form a gastrula (↓) that will eventually lead to the formation of the main organs of the animal. In simple instances, part of the blastula (↑) wall folds inwards to form a hollow gastrula.

gastrula (*n*) the stage in the development of an animal embryo (↑) which comprises a two-layered wall of cells surrounding a cavity known as the archenteron.

ectoderm (*n*) the external germ layer (↓) of an embryo (↑) which develops into hair, various glands (p. 87), CNS, the lining of the mouth etc.

endoderm (*n*) the internal germ layer (↓) of the embryo (↑) which develops into the lining of the gut (p. 98) and its associated organs.

stages of cleavage in *Amphioxus*

single cell zygote

2-cell stage   nucleus

blastomere

8-cell stage

blastula

blastula in section

blastoderm (one cell thick)   blastocoel

gastrulation

1.

gastrula   infolding of blastoderm

2.   ectoderm

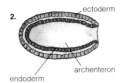

endoderm   archenteron

organogeny development of main organ layers in *Amphioxus*

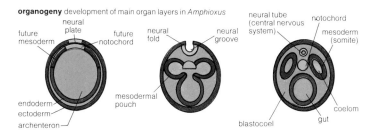

organogeny (*n*) the formation of the organs during growth. **organogenesis** (*n*).

notochord (*n*) the flexible skeletal (p. 145) rod which is present at some stage in the development of all chordates (p. 74). It stretches from the central nervous system (p. 149) to the gut (p. 98) and, in vertebrates (p. 74), it persists as remnants in the backbone throughout the life of the animal although it is present primarily during the development of the embryo (↑).

neural tube the part of the brain (p. 155) and vertebral column (p. 74) which forms first during the growth of the embryo (↑).

mesoderm (*n*) the germ layer (↓) lying between the ectoderm (↑) and endoderm (↑) and which gives rise to connective tissue (p. 88), blood (p. 90), and muscles (p. 143), etc.

germ layer one of the two or three main layers of cells which can be seen in an embryo (↑) after gastrulation (↑). The endoderm (↑), ectoderm (↑) and mesoderm (↑) are all germ layers.

coelom (*n*) a fluid-filled cavity in the mesoderm (↑) of triploblastic (p. 62) animals which, in higher animals, forms the main body cavity in which the gut (p. 98) and other organs are suspended so that their muscular (p. 143) contractions may be independent of those of the body wall.

somite (*n*) any one of the blocks of mesoderm (↑) tissue (p. 83) that flank the notochord (↑) as parallel strips and which develop into blocks of muscle (p. 143), parts of the kidneys (p. 136) and parts of the axial skeleton (p. 145).

myotome (*n*) the part of a somite (↑) that develops into striped muscle (p. 143) tissue (p. 83).

germination (*n*) the first outward sign of the growth of the seeds or spores (p. 178) of a plant which takes place when the conditions of moisture, temperature, light, and oxygen are suitable. **germinate** (*v*).

hydration phase the stage of germination (↑) in which the seed absorbs (p. 81) water and the activity of the cytoplasm (p. 10) begins.

metabolic phase the stage of germination (↑) in which, under enzyme (p. 28) control, the water absorbed (p. 81) during the hydration phase (↑) hydrolyzes (p. 16) the stored food materials into the materials needed for growth.

plumule (*n*) the first apical (↓) leaves and stem which form part of the embryonic (p. 166) shoot of a spermatophyte (p. 57).

radicle (*n*) the first root in an embryonic (p. 166) spermatophyte (p. 57) which later develops into the plant's rooting system.

cotyledon (*n*) the first, simple, leaf-like structure that forms within a seed. Plants, e.g. grasses and cereals, have only one and are called monocotyledons (p. 58) while other flowering plants have two and are called dicotyledons (p.57). Cotyledons contain no chlorophyll (p. 12) at first and may function as food reserves for the germinating (↑) plant but, in most dicotyledons, the cotyledons emerge above ground, turn green and photosynthesize (p. 93).

endosperm (*n*) the layer of tissue (p. 83) which surrounds the embryo (p. 166) in some spermatophytes (p. 57). It supplies nourishment to the developing embryo but, in some plants e.g. peas and beans, it has been absorbed (p.81) by the cotyledons (↑) by the time the seed is fully developed while, in others such as wheat, it is not absorbed until the seed germinates (↑).

testa (*n*) a hard, tough, protective coat which surrounds the seed and shields it from mechanical damage or the invasion of fungi (p. 46) and bacteria (p. 42). Also known as **seed coat**.

epicotyl (*n*) the part of the plumule (↑) which lies above the attachment point of the cotyledons (↑).

hypocotyl (*n*) the part of the plumule (↑) which lies below the attachment point of the cotyledons (↑).

types of seed

cotyledonous
food stored in cotyledons
e.g. bean

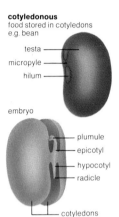

testa
micropyle
hilum

embryo

plumule
epicotyl
hypocotyl
radicle

cotyledons

endospermic
most food stored in endosperm
e.g. maize

endosperm
aleurone layer

testa
embryo
coleoptile
epicotyl
hypocotyl
radicle
cotyledon

hypogeal germination germination (↑) in which
the cotyledons (↑) remain below the surface of
the soil, e.g. in broad beans.
epigeal germination germination (↑) in which the
cotyledons (↑) emerge above the soil surface
and form the first photosynthetic (p. 93) seed
leaves e.g. in lettuces.

epigeal germination

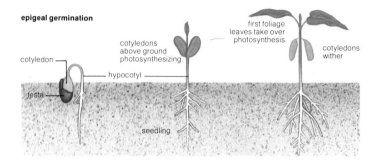

hypogeal germination

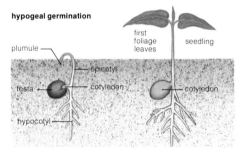

meristem (*n*) the part of an actively growing plant
where cells are dividing and new permanent
plant tissue (p. 83) is formed.
apical meristem a meristem (↑) which occurs at
shoot and root tips. The division of these cells
which are small and contain granular cytoplasm
(p. 10) with small vacuoles (p. 11), results in the
growth of stems and roots.
apex (*n*) the top or pointed end of an object.
apical (*adj*).

lateral meristem a meristem (p. 169), including the vascular cambium (p. 86) and phellogen (p. 172), which occurs along the roots and stems of dicotyledons (p. 57) plants and which is made up of long, thin cells that give rise to xylem (p. 84) and phloem (p. 84).

lateral (*adj*) on, at or about the side of something.

ground meristem the part of the apical meristem (p. 169) from which pith (p. 86), cortex (p. 86), medullary rays (p. 86) and mesophyll (p. 86) are formed.

tunica (*n*) one of the two layers of tissue (p. 83) comprising the apical meristem (p. 169). It is the outer layer of tissue and may itself be made up of one or more layers of cells in which the division takes place at right angles to the surface of the plant (anticlinally).

corpus (*n*) one of the two layers of tissue (p. 83) comprising the apical meristem (p. 169). It is the inner layer of tissue and the division of cells occurs irregularly.

zone of cell division the part of the apex (p. 169) of a root or shoot which includes the apical meristem (p. 169) and the leaf primordium (↓) or the root cap (p. 81).

zone of expansion the part of the apex (p. 169) of a root or shoot which lies behind the zone of cell division (↑) and in which the cells elongate and expand.

zone of differentiation the part of the apex (p. 169) of a root or shoot which lies behind the zone of expansion (↑) and in which the cells differentiate into the form and function of parts of the plant that characterize it.

primordium (*n*) the group of cells in the apex of a shoot or root which differentiates into a leaf etc. **primordia** (*pl.*).

primary growth the growth which takes place only in the meristems (p. 169) which were present in the embryo (p. 166). These include the apical meristems (p. 169) and primary growth results largely in the increase in length.

secondary growth the growth which takes place in the lateral meristems (↑) and which results in increase in girth rather than in length.

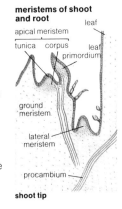

meristems of shoot and root

leaf
apical meristem
tunica corpus leaf
primordium
ground meristem
lateral meristem
procambium

shoot tip

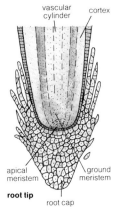

vascular cylinder   cortex
apical meristem   ground meristem

root tip
root cap

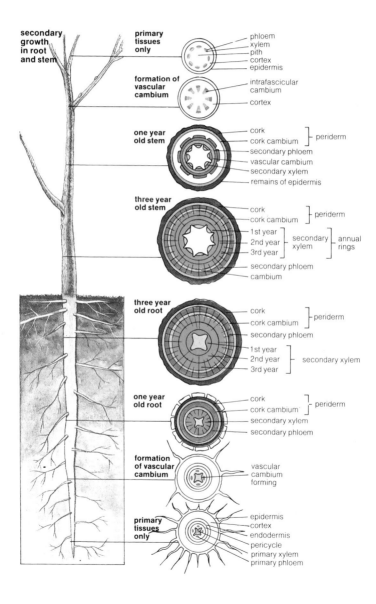

fascicular (*adj*) of meristematic (p. 169) cambium (p. 86) between the xylem (p. 84) and phloem (p. 84) in vascular tissue (p. 83).

interfascicular (*adj*) of meristematic (p. 169) cambium (p. 86) that consists of a single layer of actively dividing cells between the phloem (p. 84) and xylem (p. 84) bundles in stems.

secondary xylem xylem (p. 84) that has been formed by the vascular cambium (p. 86) following the formation of the primary tissues (p. 83).

secondary phloem phloem (p. 84) that has been produced by the cambium (p. 86) following the formation of the primary tissues (p. 83).

annual rings the rings of lighter and darker wood that can be seen in a cross-section of the trunk of a tree living in temperate conditions. Each pair marks the annual increase in the girth of the tree as a result of activity of the cambium (p. 86). The lighter ring is large-celled xylem (p. 84) tissue (p. 83) produced in spring and the darker wood is smaller-celled summer wood.

phellogen (*n*) a layer of cells immediately beneath the epidermis (p. 131) of the stems undergoing secondary growth. It is a lateral meristem (p. 170) whose cells give rise to the phellem (↓) and phelloderm (↓). Also known as **cork cambium.**

bark (*n*) the protective outer layer of the stems of woody plants. It may consist of cork cells only or alternating layers of cork (↓) and dead phloem (p. 84).

cork (*n*) = phellem (↓).

phellem (*n*) an outer layer of dead, waterproof cells formed from the activity of the phellogen (↑) on the stems of woody plants.

phelloderm (*n*) the inner layer of the bark (↑) produced by the activity of the phellogen (↑).

suberin (*n*) a mixture of substances derived from fatty acids (p. 20) and present in the walls of phellem (↑) cells rendering these cells waterproof.

exogenous (*adj*) of branching generated on the outside of the plant.

endogenous (*adj*) of branching generated on the inside of the plant.

annual rings

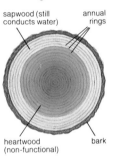

sapwood (still conducts water)

annual rings

heartwood (non-functional)

bark

reproduction (*n*) the means whereby organisms ensure the continued existence of the species (p. 40) beyond the life span of an individual by generating new individuals.

sexual reproduction the generation of new individuals of an organism to continue the life of the species (p. 40) by fusion of haploid (p. 36) nuclei (p. 13) or gametes (p. 175) to form a zygote (p. 166). In most animals a highly motile male spermatozoon (p. 188), generated by the testes (p. 187) and produced in large numbers, unites with a non-motile female ovum (p. 190) produced in small numbers in the ovary (p. 189).

asexual reproduction the generation of new individuals of an organism to continue the life of the species (p. 40) from a single parent by such means as budding (↓) or sporulation (↓). Multiplication is rapid and the offspring are genetically (p. 191) identical to one another and to the parent. For example, binary fission (p. 44) can occur very rapidly and is exponential so that one cell divides into two, two into four, four into eight, and so on, and all the cells are identical to the parent.

motile (*adj*) able to move. **motility** (*n*).

non-motile (*adj*) unable to move.

budding (*n*) asexual reproduction (↑), typical of corals (p. 61) and sponges, in which the parent produces an outgrowth or bud which develops into a new individual.

fragmentation (*n*) asexual reproduction (↑), occurring only in simple organisms such as sponges and green algae (p. 44), in which the parent fragments or breaks up and each fragment develops into a new individual.

sporulation (*n*) asexual reproduction (↑), typical of fungi (p. 46), in which the parent produces often large numbers of small, usually lightweight, single-celled structures, or spores (p. 178), which detach from the parent and may be widely distributed by wind or other mechanisms. Provided they fall into suitable conditions each spore germinates (p. 168) to produce a new individual.

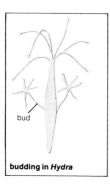

bud

budding in *Hydra*

vegetative propagation asexual reproduction (p. 173), occurring in plants, in which part of the plant, such as a leaf or even a specially produced shoot, detaches from the plant and develops into a new individual.

perennating organ any special structure, found in some biennial (p. 58) and perennial (p. 58) plants, which enables them to survive hostile conditions, such as drought. If more than one structure is produced, asexual reproduction (p. 173) is also achieved. When the conditions deteriorate, the plant dies back, leaving only the perennating organ which, with the onset of suitable conditions, develops into a new individual or individuals.

runner (*n*) an organ in the form of a stem that develops from an axillary (p. 83) bud, runs along the ground, and produces new individuals at its axillary buds or at the terminal bud only.

stolon (*n*) an organ which takes the form of a long, erect branch that eventually bends over, under its own weight, so that the tip touches the ground and roots. At the axillary (p. 83) bud a new shoot grows into a new individual.

rhizome (*n*) a perennating organ (↑) in which the stem of the plant remains below ground and continues to grow horizontally.

bulb (*n*) a perennating organ (↑) which takes the form of an underground condensed shoot with a short stem and fleshy leaves that are close together, overlap and form the food store for the plant. At each leaf base is a bud which can grow into a new bulb. In the growing season, the plant produces leaves and flowers, and exhausts the food store of the bulb. But the new plant makes more food material by· photosynthesis (p. 93) and a new bulb (or bulbs) is formed at the leaf base bud.

corm (*n*) a perennating organ (↑) which takes the form of a special enlarged, fleshy, underground stem that acts as a food store.

tuber (*n*) a perennating organ (↑) similar to a rhizome (↑) in which food is stored in the underground stem or root as swellings which eventually detach from the parent plant.

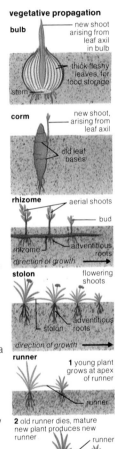

vegetative propagation

bulb
new shoot arising from leaf axil in bulb
thick fleshy leaves, for food storage
stem

corm
new shoot, arising from leaf axil
old leaf bases

rhizome
aerial shoots
bud
rhizome
adventitious roots
direction of growth →

stolon
flowering shoots
adventitious roots
stolon
direction of growth →

runner
1 young plant grows at apex of runner
runner

2 old runner dies, mature new plant produces new runner
runner

sexual reproduction

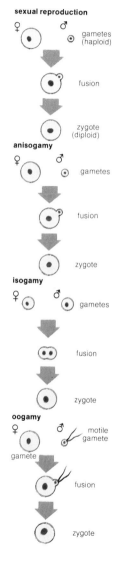

apomixis (*n*) asexual reproduction (p. 173) which superficially resembles sexual reproduction (p. 173) although fertilization (↓) and meiosis (p. 38) do not occur. **apomictic** (*adj*).

fertilization (*n*) the process in sexual reproduction (p. 173) in which the nucleus (p. 13) of a haploid (p. 36) male gamete (↓) fuses with the nucleus of a haploid female gamete to form a diploid (p. 36) zygote (p. 166).

fertile (*adj*) of organisms able to produce young.

mature (*adj*) fully grown, fully developed.

immature (*adj*) not mature (↑).

gamete (*n*) a reproductive (p. 173) or sex cell. Each gamete is haploid (p. 36) and male gametes, or spermatozoons (p. 188) which are small and highly motile, fuse with larger female gametes or ova (p. 190) in the process of fertilization (↑) to form diploid (p. 36) zygotes (p. 166) which are capable of developing into new individuals.

gametangium (*n*) an organ which produces gametes (↑). **gametangia** (*pl.*).

syngamy (*n*) the actual fusion of two gametes (↑) which occurs during fertilization (↑).

isogametes (*n.pl.*) gametes (↑), produced by some organisms, e.g. some fungi (p. 46), which are not differentiated into male or female forms. All the gametes produced by the organism are similar.

anisogametes (*n.pl.*) gametes (↑) which are differentiated in some way either simply by size or by size and form.

heterogametes (*n.pl.*) anisogametes (↑) which are differentiated by size and form into small, highly motile male spermatozoons (p. 188), which are produced in large numbers, and large non-motile female ova (p. 190).

oogamy (*n*) fertilization (↑) which takes place by the union of heterogametes (↑).

dioecious (*adj*) of organisms in which the sex organs are borne on separate individuals which are themselves then described as either males or females.

monoecious (*adj*) of organisms in which the male and female sex organs are borne on the same individual.

hermaphrodite (*n, adj*) = monoecious (↑).

parthenogenesis (*n*) reproduction (p. 173), occurring in some plants and animals such as the dandelion or aphids, in which the female gametes (p. 175) develop into new individuals without having been fertilized (p. 175). The offspring of parthenogenesis are always female and usually genetically (p. 196) identical with the parent and with one another. If the ovum (p. 178 and p. 190) has been produced by meiosis (p. 38), the offspring are haploid (p. 36) while, if it has been produced by mitosis (p. 37) the offspring are diploid (p. 36).

alternation of generations a life cycle of an organism in which reproduction (p. 173) alternates with each generation between sexual reproduction (p. 173) and asexual reproduction (p. 173). It is found, for example, among Coelenterata (p. 60) which have both a polyp (p. 61) and a medusa (p. 61) stage and among bryophytes (p. 52) in which haploid (p. 36) gametes (p. 175) from one stage – gametophyte (↓) – fuse to form a diploid (p. 36) zygote (p. 166) which germinates (p. 168) to form a sporophyte (↓) that, in turn, produces haploid spores (p. 178) by meiosis (p. 38). These develop into a haploid plant body (gametophyte). Each of the generations may be quite different in form.

generation (*n*) a set of individuals of the same stage of development or age, or the time taken for one individual to reproduce (p. 173) and for the progeny (p. 200) to develop to the same stage as the parent.

haplontic (*adj*) of a life cycle, found in some algae (p. 44) and fungi (p. 46), in which a haploid (p. 36) adult form occurs by meiosis (p. 38) of the diploid (p. 36) zygote (p. 166).

diplontic (*adj*) of a life cycle, found in all animals, as well as some algae (p. 44) and fungi (p. 46), in which haploid (p. 36) gametes (p. 175) are produced by meiosis (p. 38) from the diploid (p. 36) adults.

diplohaplontic (*adj*) of a life cycle, found in most plants, in which a diploid (p. 36) sporophyte (↓) generation alternates with a haploid (p. 36) gametophyte (↓) generation.

	gametophyte haploid		sporophyte diploid	
bryophytes				sporophyte dependent on gametophyte
pteridophytes			young sporophyte first leaf first root	sporophyte dependent on gametophyte only in very young stage
gymnosperms	pollen grain ♂ ♀ in ovule			gametophyte dependent on sporophyte
angiosperms	pollen grains ♂ ♀ embryo sac in ovule			gametophyte dependent on sporophyte

alternation of generations and the major plant divisions

sporophyte(*n*) the stage of an alternation of generations (↑), found in most plants, in which the diploid (p. 36) plant produces spores (p. 178) by meiosis (p. 38) which then germinate (p. 168) to produce the gametophyte (↓).

gametophyte (*n*) the stage of an alternation of generations (↑), found in most plants, in which the haploid (p. 36) plant produces gametes (p. 175) by mitosis (p. 37) which fuse to form a zygote (p. 166) that develops into the sporophyte (↑).

archegonium (*n*) the female sex organ of Hepaticae (liverworts) (p. 52), Musci (mosses) (p. 52), Filicales (ferns) (p. 56) and most gymnosperms (p. 57). It is a multicellular (p. 9) structure which is shaped like a flask with a narrow neck and a swollen base which contains the female gamete (p. 175).

oosphere (*n*) the large, unprotected, non-motile female gamete (p. 175) found in an archegonium (↑).

ovum[p] (*n*) the haploid (p.36) female gamete (p.175).

antheridium (*n*) the male sex organ of algae
(p. 44), liverworts (p. 52), mosses (p. 52), ferns
(p. 56) and fungi (p. 46). It may be unicellular
(p. 9) or multicellular (p. 9) and produces small,
motile gametes (p. 175) – the antherozoids (↓).
antheridia (*pl.*).

antherozoid (*n*) the male gamete (p. 175)
produced within the antheridium (↑).

spermatozoid (*n*) = antherozoid (↑).

sporogonium (*n*) the sporophyte (p. 177)
generation of mosses (p. 52) and liverworts
(p. 52) which produces the seed and parasitizes
(p. 110) the gametophyte (p. 177) generation.

sporangium (*n*) (1) the organ in which, in the
sporophyte (p. 177) generation, the haploid
(p. 36) spores (↓) are formed after meiotic
(p. 38) division of the spore mother cells (↓): (2)
in fungi (p. 46) a swelling occurring at the ends
of certain hyphae (p. 46) in which protoplasm
(p. 10) breaks up to form spores during asexual
reproduction (p. 173). **sporangia** (*pl.*).

spore (*n*) a tiny, asexual, unicellular (p. 9) or
multicellular (p. 9) reproductive (p. 173) body
which is produced in vast numbers by fungi
(p. 46) or the sporangia (↑) of plants.

spore mother cell a diploid (p. 36) cell that gives
rise to four haploid (p. 36) cells by meiosis
(p. 38). Also known as **sporocyte**.

microsporangium (*n*) a sporangium (↑) present
in heterosporous (p. 54) plants, which produces
and disperses the microspores (↓).

microspore (*n*) the smaller of the two different
kinds of spore (↑) produced by ferns (p. 56) and
spermatophytes (p. 57) and which gives rise to
the male gametophyte (p. 177) generation.

microsporophyll (*n*) a modified leaf that bears
the microsporangium (↑).

megasporangium (*n*) a sporangium (↑) present
in heterosporous (p. 54) plants, which produces
and disperses the megaspores (↓).

megaspore (*n*) the larger of the two different
kinds of spore (↑) produced by ferns (p. 56) and
spermatophytes (p. 57) and which gives rise to
the female gametophyte (p. 177) generation.

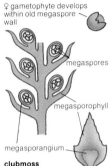

heterosporous plants
the production of megaspores

♀ gametophyte develops
within old megaspore
wall

megaspores

megasporophyll

megasporangium

clubmoss

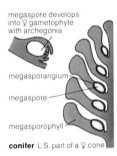

megaspore develops
into ♀ gametophyte
with archegonia

megasporangium

megaspore

megasporophyll

conifer L.S. part of a ♀ cone

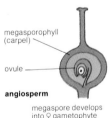

megasporophyll
(carpel)

ovule

angiosperm

megaspore develops
into ♀ gametophyte
(embryo sac)

megaspore

young
ovule

flower

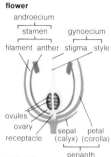

perianth

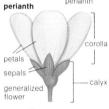

gynoecium

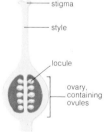

megasporophyll (*n*) a modified leaf that bears the megasporangium (↑). They can be grouped in a strobilus (p. 55).

flower (*n*) the structure concerned with sexual reproduction (p. 173) in angiosperms (p. 57). It is a modified vegetative shoot.

corolla (*n*) the part of the flower which is made up of all the petals (↓). It varies considerably in size, shape, form and colour and often attracts insects (p. 69) to visit the flower, pollinating (p. 183) the plant in the process.

petal (*n*) one of the often brightly coloured and scented individual elements which make up the corolla (↑). They are thought to be modified leaves. Those flowers which are pollinated (p. 183) by the wind have petals which are greatly reduced in size or absent.

calyx (*n*) the outermost part of a flower which comprises a number of sepals (↓) that protect the flower while it is developing in the bud stage.

sepal (*n*) the usually green, often hairy, leaf-like structures which make up the calyx (↑).

perianth (*n*) the part of the flower, comprising the corolla (↑) and calyx (↑), which surrounds the stamens (p. 181) and carpels (↓).

gynoecium (*n*) the female reproductive (p. 173) structure of a flower which is made up of the carpels (↓).

carpel (*n*) one of the single or more individual female reproductive (p. 173) structures of a plant which make up the gynoecium (↑). Each carpel is made up of an ovary (p. 180), a style (p. 181) and a stigma (p. 181). If there is more than one carpel, they may be fused together or separate.

types of ovary

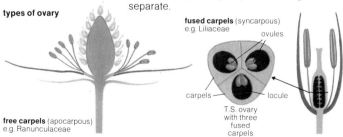

free carpels (apocarpous) e.g. Ranunculaceae

fused carpels (syncarpous) e.g. Liliaceae

T.S. ovary with three fused carpels

ovary[p] (*n*) the part of the carpel (p. 179) which contains the ovules (↓).

ovule[p] (*n*) the structure containing the female gametes (p. 175) and which, after fertilization (p. 175), becomes the seed.

funicle (*n*) the stalk which attaches the base of the ovule (↑) to the wall of the carpel (p. 179).

placenta[p] (*n*) the part of the wall of the ovary (↑) to which the ovules (↑) are attached.

apocarpous (*adj*) of a gynoecium (p. 179) in which the carpels (p. 179) are not fused.

syncarpous (*adj*) of a gynoecium (p. 179) in which the carpels (p. 179) are fused.

placentation (*n*) the position and arrangement of the placentas (↑) in a syncarpous (↑) gynoecium (p. 179).

parietal (*adj*) of placentation (↑) in which the carpels (p. 179) are fused only by their margins with the placentas (↑) becoming ridges on the inner side of the wall of the ovary (↑).

axile (*adj*) of placentation (↑) in which the carpels (p. 179) fold inwards at their margins, fuse, and become a central placenta (↑).

free central of placentation (↑) in which the placenta (↑) grows upwards from the base of the ovary (↑).

nucellus (*n*) the central tissue (p. 83) of the ovule (↑) enclosing the megaspore (p. 178) or ovum (p. 178).

micropyle (*n*) a canal through the integument (↓) near the apex of the ovule (↑) which, in the seed, becomes a pore (p. 120) through which water may enter to enable germination (p. 168).

integument (*n*) the outermost layer of the ovule (↑) which forms the seed coat.

chalaza (*n*) the base of the ovule (↑) to which the funicle (↑) is attached. It is situated·at the point where the nucellus (↑) and integuments (↑) merge.

embryo sac a large, oval-shaped cell, surrounded by a thin cell wall (p. 8), in the nucellus (↑) in which fertilization (p. 175) of the ovum (p. 178) takes place and the embryo (p. 166) develops.

polar nuclei a pair of haploid (p. 36) nuclei (p. 13) found towards the centre of the embryo sac (↑).

egg cell the female gamete (p. 175).

ovule structure

embryo sac
(♀ gametophyte)

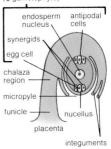

endosperm nucleus · antipodal cells
synergids
egg cell
chalaza region
micropyle
funicle · nucellus
placenta
integuments

placentation types
ovaries cut through to show internal structure

axile — locules

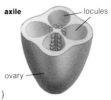

ovary

parietal — locule

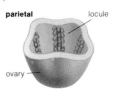

ovary

free-central — locule

ovary

male floral parts
pollen sac   pollen grains

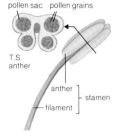

T.S.
anther

anther
                    } stamen
filament

**pollen grain of
angiosperm**

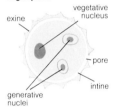

exine

vegetative
nucleus

pore

intine

generative
nuclei

actinomorphic flower
(radial symmetry)

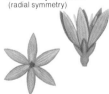

zygomorphic flower
(bilateral symmetry)

synergid (*n*) one of the two haploid (p. 36) cells
which occur at the micropyle (↑) end of the
embryo sac (↑) near the egg cell (↑).

antipodal cell one of the three haploid (p. 36)
cells that move to the end of the embryo sac (↑)
nearest the chalaza (↑). They do not take part in
fertilization (p. 175).

style (*n*) the part of the carpel (p. 179) which joins
the ovary (↑) and the stigma (↓).

stigma (*n*) the receptive tip of the carpel (p. 179)
to which pollen (↓) becomes attached during
pollination (p. 183).

androecium (*n*) the male reproductive (p. 173)
structure of a flower which is made up of the
stamens (↓).

stamen (*n*) one of the male reproductive (p. 173)
structures which make up the androecium (↑). A
stamen consists of an anther (↓) and a filament (↓).

anther (*n*) the tip of the stamen (↑) which produces
the pollen (↓) contained in pollen sacs (↓).

filament (*n*) (1) the stalk of the stamen (↑) to
which the anther (↑) is attached; (2) in plants, a
chain of cells, e.g. some green algae (p. 44) are
filamentous; (3) in animals, any fine threadlike
structure.

pollen sac the chamber in which pollen is formed.

pollen (*n*) grain-like microspores (p.178) produced
in the pollen sac (↑) in huge numbers by meiotic
(p. 38) division of their spore mother cells
(p. 178). They contain the male gametes (p. 175).

tapetal cell one of the layer of cells whicn
surrounds the spore mother cells (p. 178) and
which provide nutrients for the spore mother
cells and the developing spores (p. 178).

generative nucleus one of the two nuclei (p. 13)
found in each grain of pollen (↑), both of which
are transferred to the ovule (↑) via growth of the
pollen tube (p. 184).

receptacle (*n*) the part of the stem of a flower
which is often expanded and which bears the
organs of the flower.

zygomorphic (*adj*) of a flower, such as a
snapdragon, which is bilaterally symmetrical (p.62).

actinomorphic (*adj*) of a flower, such as a
buttercup, which is radially symmetrical (p. 60).

inflorescence
e.g. capitate inflorescence of a composite

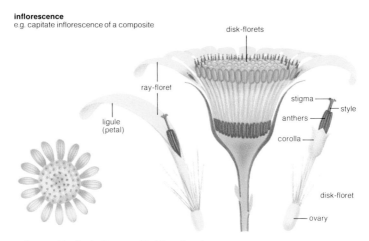

disk-florets

ray-floret

stigma

style

ligule
(petal)

anthers

corolla

disk-floret

ovary

unisexual (*adj*) of a flower, which has the stamens
(p. 181) and carpels (p. 179) on separate
flowers. Unisexual flowers may be either
monoecious (p. 175) or dioecious (p. 175).

nectary (*n*) a glandular (p. 87) swelling found on
the receptacle (p. 181) or other parts of some
flowers, which produces nectar (↓).

nectar (*n*) a sweet, sugary solution (p. 118)
produced by the nectaries (↑). Many insects
(p. 69) visit flowers which produce nectar to feed
on it and, in so doing, pollinate (↓) the flower.

inflorescence (*n*) a group of flowers sharing the
same stem.

inflorescence types

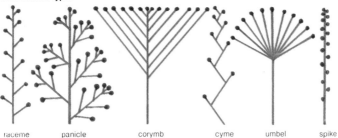

raceme   panicle   corymb   cyme   umbel   spike

spikelet

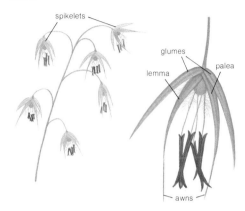

spikelets

glumes

palea

lemma

awns

spathe

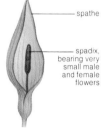

spathe

spadix, bearing very small male and female flowers

cross-pollination

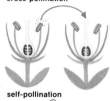

self-pollination

spikelet (*n*) the inflorescence (↑) of a grass.

spathe (*n*) a leaf-like structure or bract which encloses the spadix (↓) of certain monocotyledonous (p. 58) flowers.

spadix (*n*) the inflorescence (↑) of certain monocotyledonous (p. 58) flowers, bearing unisexual (↑) or hermaphrodite (p. 175) flowers.

floral formula a 'shorthand' way of describing the structure of a flower. It is given by a combination of capital letters and numbers as follows: K = calyx (p. 179); C = corolla (p. 179); A = androecium (p. 181); and G = gynoecium (p. 179). Thus, a flower with the floral formula K6 C6 G1 A5 would have six sepals, six petals, one carpel and five stamens.

pollination (*n*) the process in which pollen (p. 181) is transferred from the anther (p. 181) to the stigma (p. 181). **pollinate** (*v*).

self-pollination (*n*) pollination (↑) within the same flower or flowers from the same plant.

cross-pollination (*n*) pollination (↑) between flowers of different plants.

wind pollination pollination (↑) in which the pollen (p. 181) is carried from one flower to the next by the wind.

anemophily = wind pollination (↑).

insect pollination pollination (p. 183) in which the pollen (p. 181) is transferred from one flower to the next on the bodies of insects (p. 69) which are attracted to the flowers by the brightly coloured petals (p. 179), the scent, and the promise of nectar (p. 182).

entomophily = insect pollination (↑).

pollen tube a tubular outgrowth which forms when the pollen (p. 181) grain germinates (p. 168) and which is the means through which male gametes (p. 175) are carried to the egg.

double fertilization in flowering plants, the union of one generative nucleus (p. 181) with an ovum (p. 178) to form a zygote (p. 166) and the other with the two polar nuclei (p. 180) to form the primary endosperm (p. 168) nucleus (p. 13) which is triploid (p. 207). Subsequent division of the primary endosperm nucleus produces the endosperm.

seed (*n*) the structure which develops after the fertilization (p. 175) of the ovule (p. 180) and which is made up of the testa (p. 168) surrounding the embryo (p. 166). In suitable conditions, each seed may germinate (p. 168) and form a fully independent plant. Seeds in flowering plants may be contained within a fruit.

fruit (*n*) the ripened ovary (p. 180) wall of a flower which contains the seeds. Depending upon the method by which the seeds of the plant are distributed, the fruit may be fleshy (distributed by animals) or dry (distributed by wind or water).

pericarp (*n*) the outer wall of the ovary (p. 180) which develops into the fruit.

endocarp (*n*) the inner layer of the pericarp (↑) which develops into the stony covering of the seed of a drupe (↓), such as a cherry.

mesocarp (*n*) the middle layer of the pericarp (↑) which can form the fleshy part of a drupe (↓), such as a cherry, or the hard shell of a nut, like an almond.

exocarp (*n*) the tough outer 'skin' of a fruit.

epicarp (*n*) = exocarp (↑).

aleurone layer the outermost, protein-rich layer of the endosperm (p. 168) of the seeds of grasses.

fertilization in angiosperms

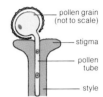

1 pollen grain lands on stigma, pollen tube grows through tissues of style carrying the male gametes

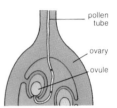

2 pollen tube grows through ovary wall and into micropyle of ovule

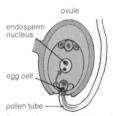

3 one male gamete fertilizes egg cell, the other fertilizes the endosperm nucleus forming endosperm mother cell

fruits and fruit structure

berry e.g. tomato

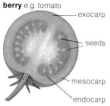

— exocarp

— seeds

— mesocarp

— endocarp

drupe e.g. apricot

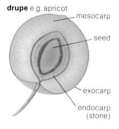

— mesocarp

— seed

— exocarp

— endocarp (stone)

legume e.g. pea

— seeds

pod —

achene e.g. strawberry

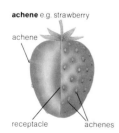

achene —

receptacle   achenes

scutellum (*n*) the part of the embryo (p. 166) of a grass seed, which is situated next to the endosperm (p. 168).

coleoptile (*n*) the protective sheath with a hard pointed tip which protects the plumule (p. 168) in a germinating (p. 168) grass seedling.

dehiscence (*n*) the process in which the mature fruit wall opens, sometimes violently, to release the seeds. **dehisce** (*v*).

berry (*n*) a fruit, such as a blackberry, which, unlike a drupe (↓) does not have a stony endocarp (↑) so that the seeds are surrounded by a fleshy mesocarp (↑) and endocarp.

drupe (*n*) a fruit, such as a plum, formed from a single carpel (p. 179) which has a stony endocarp (↑) that surrounds the seed.

follicle[P] (*n*) a dry fruit, such as a delphinium, which has formed from a single carpel (p. 179) and in which, during dehiscence (↑), the fruit or pod splits along one line to release the seed.

legume (*n*) a dry fruit, such as a pea, which has formed from a single carpel (p. 179) and in which, during dehiscence (↑), the fruit or pod splits along two sides to release the seed.

siliqua (*n*) a dry, elongated fruit or special type of capsule (p. 53), such as that found in the cabbage family, formed from two carpels (p. 179) which are fused together but separated by a false septum or wall. During dehiscence (↑), the siliqua splits as the carpel walls separate leaving the seeds attached to the septum.

silicula (*n*) a type of siliqua (↑), found in plants such as the shepherd's purse, which is short and broad in shape.

achene (*n*) a dry fruit, such as that of the buttercup, which is formed from a single carpel (p. 179), contains only one seed, has a leathery pericarp (↑), and has no particular method of dehiscence (↑).

cypsela (*n*) a dry fruit, such as that of the dandelion, which is formed from two carpels (p. 179) of an inferior ovary (p. 180) which retains a plumed calyx (p. 179) to aid in wind dispersal (p. 186).

caryopsis (*n*) a dry fruit, such as that of grasses, which is similar to an achene (p. 185) except that the pericarp (p. 184) is united with the testa (p. 168).

nut (*n*) a dry fruit, such as that of the hazel, which is similar to an achene (p. 185), except that the pericarp (p. 184) is stony.

samara (*n*) a dry fruit, such as that of the elm, which is similar to an achene (p. 185), except that part of the pericarp (p. 184) forms wings that aid in wind dispersal (↓).

false fruit a fruit that includes other parts of the flower, such as the inflorescence (p. 182), as well as the ovary (p. 180).

pome (*n*) a fleshy false fruit (↑), such as that of the apple. The main flesh is made of the swollen receptacle (p. 181).

fruit dispersal the various methods by which a flower distributes its seeds and which may include the fruit or just the seeds alone.

mechanical dispersal fruit dispersal (↑) in which the fruit itself is responsible for distributing the seed by opening explosively when the seed is mature and scattering it widely.

wind dispersal fruit dispersal (↑) in which the seed is carried on the wind either because it is small and lightweight or by bearing wing-like structures which give them extra lift. In some cases, such as that of the poppy, the capsule (p. 53) itself sways in the wind to distribute the seed, like a censer.

animal dispersal fruit dispersal (↑) in which the seed is distributed by being transported by animals, including humans. The seed or fruit may have hooks or spines which stick to the animals' coats or their fruits may be palatable while the seeds may be indigestible so that the fruits are eaten and the seeds pass through undamaged. Indeed, some seeds can only germinate (p. 168) when they have passed through the digestive (p. 98) system of certain animals.

water dispersal fruit dispersal (↑) in which the fruit or seed is specially adapted to be carried in running water.

caryopsis e.g. wheat

nut e.g. hazelnut

samara e.g. sycamore

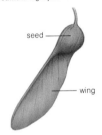

seed

wing

pome e.g. apple

gonad (*n*) the male or female organ of reproduction (p. 173) in sexual animals that produce gametes (p. 175). In some cases, the gonads also produce hormones (p. 130).

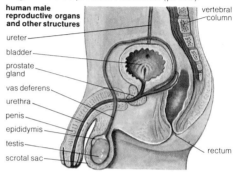

human male reproductive organs and other structures

vertebral column — ureter — bladder — prostate gland — vas deferens — urethra — penis — epididymis — testis — scrotal sac — rectum

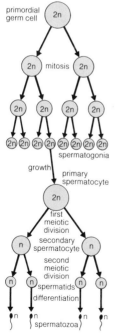

spermatogenesis

primordial germ cell — 2n

2n — mitosis — 2n

2n 2n 2n 2n

2n 2n 2n 2n 2n 2n 2n 2n

spermatogonia

growth — primary spermatocyte

2n

first meiotic division

n — secondary spermatocyte — n

second meiotic division

n n spermatids n n

differentiation

n n spermatozoa n n

testis (*n*) the male reproductive (p. 173) organ which produces spermatozoa (p. 188) by spermatogenesis (↓). In vertebrates (p. 74) there are two testes which usually lie in a sac of skin, the scrotum or scrotal sac (p. 188), outside the main body cavity and behind the penis (p. 189). In vertebrates, the testes also produce androgens (p. 195).

testicle (*n*) = testis (↑).

seminiferous tubule one of several hundred tiny coiled tubes which compose the testis (↑) and in which all stages of spermatogenesis (↓) take place.

Sertoli cell one of a number of large, specialized cells made of germinal epithelium (p. 87) and found in the testis (↑) which are thought to nourish the spermatids (p. 188) to which they are attached.

spermatogenesis (*n*) the process by which spermatozoa (p. 188) are produced in the testes (↑). A germ cell (p. 36) divides a number of times during a multiplication phase to produce spermatogonia (p. 188) each of which then grows into a primary spermatocyte (p. 188). In turn, the spermatocyte undergoes two phases of meiotic (p. 38) division to produce spermatids (p. 188) which then differentiate into the spermatozoa.

spermatogonium (*n*) one of the large numbers of cells found in the testes (p. 187) and which grows into a primary spermatocyte (↓) during spermatogenesis (p. 187). **spermatogonia** (*pl.*).

spermatocyte (*n*) one of the large numbers of reproductive cells found in the seminiferous tubules (p. 187) and which is produced by the growth of a spermatogonium (↑).

spermatid (*n*) one of the large numbers of reproductive (p. 173) cells found in the seminiferous tubules (p. 187) during spermatogenesis (p. 187). It is produced as a result of two phases of meiosis (p. 38) of a spermatocyte (↑). Each spermatid, nourished by the Sertoli cells (p. 187), differentiates and matures into a spermatozoon (↓).

spermatozoon (*n*) the small, differentiated, highly motile mature male gamete (p. 175) or reproductive (p. 173) cell. Spermatozoa (*pl.*) are produced continuously in large numbers in the seminiferous tubules (p. 187). Locomotion (p. 143) takes place by movements of a flagellum (p. 12).

vas efferens one of the small channels through which spermatozoa (↑) are transported from the seminiferous tubules (p. 187) to the epididymis (↓).

epididymis (*n*) a muscular (p. 143), coiled tubule between the vas efferens (↑) and the vas deferens (↓) which functions as a temporary storage vessel for spermatozoa (↑) until they are released during mating.

vas deferens one of the pair of muscular (p. 143) tubules, with mucous (p. 99) glands (p. 87), which leads from the epididymis (↑) and through which spermatozoa (↑) are released into the urethra (↓) during mating.

urethra (*n*) a duct which leads from the bladder (p. 135) to the exterior and through which urine (p. 135) is excreted (p. 134). In males it also connects with the vas deferens (↑).

scrotal sac the external sac of skin which is divided into two, each carrying one testis (p. 187). Thus, the testes are maintained at a lower temperature than the rest of the body to ensure the best conditions for the development of spermatozoa (↑).

spermatozoon
e.g. human sperm

acrosome (contains agent which dissolves egg membrane during fertilization)

head

nucleus (rich in DNA)

neck

centriole

axial filament

middle piece

mitochondria

centriole

axial filament

tail

tail sheath

end piece

prostate gland a gland (p. 87) surrounding the urethra (↑) which, under the control of androgens (p. 195), secretes (p. 106) alkaline substances that reduce the urine's (p. 135) acidity and aid in the motility of spermatozoa (↑).

seminal vesicle one of the two organs connected to the vas deferens (↑) in most male mammals (p. 80). It is under hormonal (p. 130) control and secretes (p. 106) fluid which makes up the bulk of the semen (p. 191) improving the motility of the spermatozoa (↑).

Cowper's gland one of two glands (p. 87) connected to the vas deferens (↑) which secretes (p. 106) fluid for the semen (p. 191).

penis (n) the organ through which the urethra (↑) connects with the exterior and which functions, during mating, to transport spermatozoa (↑) to the female reproductive (p. 173) organs. It contains spongy tissue (p. 83) which fills with blood (p. 90) during mating to become more rigid or erect.

ovary[a] (n) one of the pair of female reproductive (p. 173) organs in which ova (p. 190) are produced during oogenesis (↓). Female hormones (p. 130) are also produced in the ovaries.

oogenesis (n) the process by which ova (p. 190) are produced in the ovaries (↑). A germ cell (p. 36) divides by mitosis (p. 37) to form a number of oogonia (↓) each of which grows to give rise to a primary oocyte (↓). By two phases of meiotic (p. 38) division – with the second phase usually following fertilization (p. 175) – an ovum is produced together with additional polar bodies (↓).

oogonium (n) a specialized cell found within the ovary (↑) which is produced by mitotic (p. 37) division of the germ cell (p. 36) and which grows to give rise to a primary oocyte (↓) during oogenesis (↑). **oogonia** (pl.).

oocyte (n) a reproductive (p. 173) cell found within the ovary (↑) during oogenesis (↑). It results from the growth of an oogonium (↑).

polar body a tiny cell produced during oogenesis after the second meiotic (p. 38) division when the ovum (p. 190) is formed. The polar body contains a nucleus (p. 13) but virtually no cytoplasm (p. 10).

oogenesis

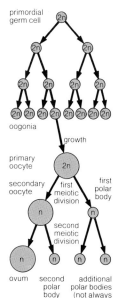

primordial germ cell

oogonia

growth

primary oocyte

secondary oocyte

first meiotic division

first polar body

second meiotic division

ovum

second polar body

additional polar bodies (not always formed)

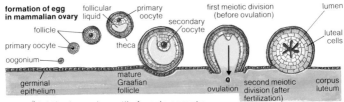

formation of egg in mammalian ovary follicular liquid, follicle, primary oocyte, primary oocyte, oogonium, theca, secondary oocyte, first meiotic division (before ovulation), lumen, luteal cells, germinal epithelium, mature Graafian follicle, ovulation, second meiotic division (after fertilization), corpus luteum

ovum[a] (n) the large, immotile female gamete
(p. 175) produced in the ovary (p. 189) during
oogenesis (p. 189). If it is fertilized (p. 175) by a
spermatozoon (p. 188) it develops into a new
individual. Fertilization may take place at the
oocyte (p. 189) stage following the first meiotic
(p. 38) division. **ova** (pl.).

Graafian follicle a fluid-filled, spherical mass of
cells with a cavity that is found in the ovary
(p. 189) and contains an oocyte (p. 189)
attached to its wall. It is the site of development
of the ovum (↑) and grows from one of the large
number of follicles within the ovary.

corpus luteum a gland (p. 87) which forms
temporarily in the Graafian follicle (↑) after
rupture during ovulation (p. 194). It secretes
(p. 106) the hormone (p. 130) progesterone
(p. 195) which, if the ovum (↑) is fertilized
(p. 175), continues to be released to prepare
the female reproductive (p. 173) tract for
pregnancy (p. 195). If fertilization does not take
place, the corpus luteum degenerates. **corpora
lutea** (pl.).

oviduct (n) a muscular (p. 143) tube lined with
cilia (p. 12) by which ova (↑) are transported
from the ovaries (p. 189) to the exterior.

uterus (n) a thick-walled organ in which the
embryo (p. 166) develops. It is muscular (p. 143)
and the smooth muscle increases in amount
during pregnancy (p. 195) so that it is able to
expel the young at birth. The size of the uterus
as well as the thickness of its wall, which
provides a point of attachment and nourishment
for the developing embryo, varies cyclically
and with sexual activity or inactivity under the
influence of reproductive (p.173) hormones
(p. 130). Also known as **womb**.

human female reproductive organs

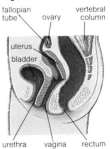

fallopian tube, ovary, vertebral column, uterus, bladder, urethra, vagina, rectum

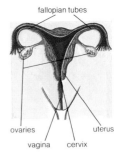

fallopian tubes, ovaries, uterus, vagina, cervix

cervix (*n*) a ring of muscle (p. 143) between the uterus (↑) and the vagina (↓) which also contains mucous glands (p. 87).

vagina (*n*) the muscular (p. 143) duct which connects the uterus (↑) to the exterior and which receives the penis (p. 189) during mating.

copulation (*n*) the sexual union of male and female animals during mating in which, in mammals (p. 80), the penis (p. 189) is received by the vagina (↑) and ejaculation (↓) takes place. Also known as **coitus**.

semen (*n*) a fluid containing spermatozoa (p. 188) produced by the testes (p. 187) and other liquids produced by the prostate gland (p. 189). During copulation (↑) semen is passed from male to female.

ejaculation (*n*) the rhythmic and forcible discharge of semen (↑) from the penis (p. 189).

orgasm (*n*) the climax of sexual excitement which takes place during mating and involves a complex series of reactions of the reproductive (p. 173) organs and other parts of the body including the skin.

implantation (*n*) following fertilization (p. 175), the process in which the developing zygote (p. 166) embeds itself in the wall of the uterus (↑).

foetus in uterus

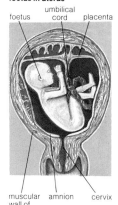

foetus · umbilical cord · placenta

muscular wall of uterus · amnion · cervix

foetus (*n*) an embryo (p. 166) with an umbilical cord (p. 192) which is sufficiently developed to show the main features that the mammal (p. 80) will possess after birth.

foetal membrane any one of those membranes (p. 14) or structures which are developed by the embryo (p. 166) for nourishment and protection but which do not form part of the embryo itself.

amnion (*n*) the fluid-filled sac in which the embryo (p. 166) develops in mammals (p. 80). The amnion offers the embryo protection from any pressure exerted on it by the organs of the mother and a liquid environment (p. 218) in which to develop (important for land animals). The sac wall consists of two layers of epithelium (p. 87) and sometimes only the inner layer is referred to as the amnion. **amniotic** (*adj*), **amniote** (*adj*).

amniotic cavity the amnion (↑) or the fluid-filled cavity within the amnion which contains the developing embryo (p. 166).

allantois (*n*) a sac-like extension of the gut (p. 98) which is present in the embryos (p. 166) of reptiles (p. 78), birds and mammals (p. 80) and which grows out beyond the embryo itself. The connective tissue (p. 88) which covers it is liberally supplied with blood vessels (p. 127) and functions for gas exchange (p. 112) of the embryo as well as for storing the products of excretion (p. 134).

chorion (*n*) the outermost membrane (p. 14), the outer epithelium (p. 87) of the amnion (p. 191) wall which surrounds the embryo (p. 166) of mammals (p. 80) and which unites with the allantois (↑) to develop into the placenta (↓).

placenta[a] (*n*) a disc-shaped organ which develops within the uterus (p. 190) during pregnancy (p. 195) and which is in close association with the embryo (p. 166) and with tissues (p. 83) of the mother. The placenta serves for attachment and nourishment over its large surface area.

umbilical cord a cord which connects the placenta (↑) to the navel of the foetus (p. 191) allowing interchange of materials via two arteries (p. 127) and a vein (p. 127).

viviparity (*n*) the condition in which embryos (p. 166) develop within a uterus (p. 190), are attached to a placenta (↑), and are born alive. **viviparous** (*adj*).

gestation period the time which elapses between fertilization (p. 175) of the ovum (p. 190) and the birth of the young in viviparous (↑) animals. It varies from species (p. 40) to species.

parturition (*n*) the process of giving birth to live young in viviparous (↑) animals by rhythmic contractions stimulated by the secretion (p. 106) of certain hormones (p. 130).

lactation (*n*) the production of milk in the mammary glands (p. 87) to nourish the young in mammals (p. 80).

puberty (*n*) the sexual maturity of a mammal (p. 80).

menopause (*n*) the period in females during which the menstrual cycle (p. 194) becomes irregular with increasing age of the individual before ceasing totally.

embryonic membranes of a mammal

embryo — chorion — embryonic gut — amnion — yolk sac — yolk sac placenta villi — allanto-chorionic placenta villi — allantois

embryonic membranes in a reptilian egg

amniotic fluid — chorion — amnion — shell — yolk — yolk sac — allantois — embryo

sexual cycle the sequence of events which occurs in the females of animals that reproduce (p. 173) sexually and which, in humans, takes place on a monthly pattern with menstruation (p. 194) alternating with ovulation (p. 194).

oestrus cycle the rhythmic sexual cycle (↑) which occurs in mature females of most mammals (p. 80) assuming that the female does not become pregnant (p. 195). There are four main events in the oestrus cycle of which the most important is oestrus (p. 194) itself. In the *follicular phase*, there is growth of the Graafian follicles (p. 190), a thickening of the lining of the uterus (p. 190) and an increase in the production of oestrogen (p. 194). This is followed by *oestrus*. Then comes the *luteal phase* during which a corpus luteum (p. 190) grows from the Graafian follicle which secretes (p. 106) progesterone (p. 195) with a reduction in the secretion of oestrogen. If fertilization (p. 175) and pregnancy occur then the cycle is interrupted and the fourth phase does not follow. If fertilization does not occur, then the corpus luteum diminishes, hormone (p. 130) levels fall, and a new Graafian follicle begins to grow.

relationships between hormones secreted by pituitary, the oestrus cycle and pregnancy in a human

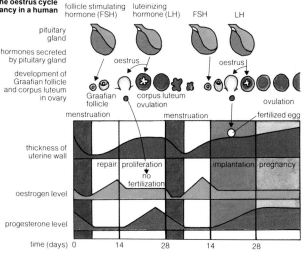

menstrual cycle in humans and some other primates a modified version of the oestrus cycle (p. 193) in which the oestrus (↓) is not obvious so that the female is continuously attractive and receptive to males. There is a regular discharge of blood (p. 90) and the lining of the uterus (p. 190) (menstruation) which occurs, following ovulation (↓) when fertilization (p. 175) does not occur.

ovulation (*n*) the release from the Graafian follicle (p. 190) of an immature ovum (p. 190) or oocyte (p. 189). It takes place, under the influence of a hormone (p. 130) released by the pituitary gland (p. 157) at regular intervals (approximately every 28 days in humans) and in the presence of oestrogen (↓). **ovulate** (*v*).

oestrus (*n*) a short period during the sexual cycle (p. 193) of animals in which the female ovulates (↑) and is also sexually attractive to males so that copulation (p. 191) takes place.

follicle-stimulating hormone FSH. A hormone (p. 130) produced by the pituitary gland (p. 157), following the completion of the oestrus cycle (p. 193) or pregnancy (↓) which stimulates the growth of the Graafian follicles (p. 190) and the ova (p. 190) in females and spermatogenesis (p. 187) in males.

oestrogen (*n*) a female sex hormone (p. 130), produced in the Graafian follicle (p. 190), which stimulates the production of a suitable environment (p. 218) for fertilization (p. 175) and then growth of the embryo (p. 166) by repairing the walls of the uterus (p. 190) following menstruation (↑). During the first part of the oestrus cycle (p. 191), it builds up until it stimulates the production of luteinizing hormone (↓) by the pituitary gland (p. 157). It is also involved in the development of other female organs associated with the sexual cycle.

luteinizing hormone LH. A hormone (p. 130) secreted (p. 106) by the pituitary gland (p. 157) under the influence of oestrogen (↑). It stimulates ovulation (↑) and development of a Graafian follicle (p. 190) into a corpus luteum (p. 190) which produces progesterone (p. 195).

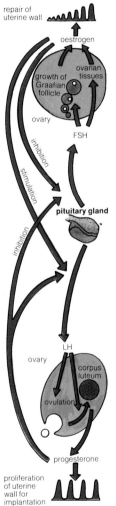

interaction of hormones in the female sexual cycle

repair of uterine wall

oestrogen

ovarian tissues

growth of Graafian follicle

ovary

FSH

inhibition

stimulation

inhibition

pituitary gland

LH

ovary

corpus luteum

ovulation

progesterone

proliferation of uterine wall for implantation

progesterone (*n*) a hormone (p. 130) secreted (p. 106) by the corpus luteum (p. 190) which stops further Graafian follicles (p. 190) from developing by preventing the secretion of follicle-stimulating hormone (↑). It also prepares the uterus (p. 190) for implantation (p. 191) of the ova (p. 190), and assists in the development of the placenta (p. 192) and mammary glands (p. 87).

pregnancy (*n*) the condition which occurs in a female following successful fertilization (p. 175) and implantation (p. 191). The oestrus cycle (p. 193) is suspended in the luteal phase. The production of hormones (p. 130) is altered so that they are produced by the placenta (p. 192) as well as the pituitary gland (p. 157) to ensure that parturition (p. 192) and lactation (p. 192) take place properly. **pregnant** (*n, adj*).

oxytocin (*n*) a hormone (p. 130) produced by the pituitary gland (p. 157) at the end of pregnancy (↑) which stimulates the contraction of uterine (p. 190) muscles (p. 140) during labour and prepares the mammary glands (p. 87) for the production of milk during lactation (p. 192).

prolactin (*n*) a hormone (p. 130) produced by the pituitary gland (p. 157) which stimulates and controls the production of milk during lactation (p. 192).

breeding season in animals in which the oestrus cycle (p. 193) does not occur continuously throughout the year, the time during which it does take place and which is usually under the influence of climate or other environmental (p. 218) factors.

androgens (*n.pl.*) the male sexual hormones (p. 130), such as testosterone (↓), produced essentially by the testes (p. 187), and which stimulate and control spermatogenesis (p. 187) as well as other male characteristics, such as the growth of facial hair.

testosterone (*n*) an androgen (↑) produced by male vertebrates (p. 74).

interstitial cell-stimulating hormone a luteinizing hormone (↑) which stimulates the secretion (p. 106) of androgens (↑) by the testes (p. 187) in males.

genetics (*n*) the study or science of inheritance concerning the variations between organisms and how these are affected by the interaction of environment (p. 218) and genes (↓).

inherit (*v*) to receive genetic (↓) material from one's parents or ancestors. **inheritance** (*n*).

genotype (*n*) the actual genetic (↓) make-up of an organism which may, for example, define the limits of its growth that are then affected by the environment (p. 218).

phenotype (*n*) the total characteristics and appearance of an organism. Organisms may have the same genotype (↑) while the phenotypes may be different because of the effects of the environment (p. 218).

genome (*n*) the genetic (↓) material.

gene (*n*) the smallest known unit of inheritance that controls a particular characteristic of an organism, such as eye colour. A gene may be considered to be a complex set of chemical compounds sited on a chromosome (p. 13). A gene may replicate to produce accurate copies of itself or mutate (p. 206) to give rise to new forms. **genetic** (*adj*).

Mendelian genetics the system of genetics (↑) developed by the Austrian monk, Gregor Mendel (1822-84), in which he studied inheritance by a series of controlled breeding experiments with the garden pea. He studied simple characteristics, controlled by a single gene (↑), and, using statistics, analyzed the results of cross breeding. In this way he showed that phenotypes (↑) did not result from a blending of genotypes (↑) but that the phenotypes were passed on in different ratios.

first filial (F₁) generation the first generation of offspring resulting from cross breeding pure lines (↓) or parentals (↓) of a single species (p. 40).

second filial (F₂) generation the generation of offspring resulting from the cross breeding between individuals of the first filial generation (↑).

pure line the succession of generations which results from the breeding of a homozygous (↓) organism so that they breed true and produce genetically (↑) identical offspring.

phenotype the actual appearance

genotype the actual genetic makeup as determined by the chromosomes

chromosome

parental (*n*) the succession of generations which leads to filial generations (↑).

monohybrid inheritance the result of cross breeding from pure lines (↑) with one pair of contrasting characteristics to give offspring with one of the characteristics, such as Mendel's cross of tall and dwarf garden peas to give a tall monohybrid.

dominant (*adj*) of (1) a gene (↑) which gives rise to a characteristic that always appears in either a homozygous (↓) or a heterozygous (p. 198) condition e.g. in Mendel's cross of tall and dwarf garden peas, all the F_1 generation (↑) were tall while, in the F_2 generation (↑), tall individuals were in a ratio of 3:1 to dwarf individuals. Thus, the dominant gene was for tallness. (2) a plant species (p. 40) which, in any particular community (p. 217) of plants, is the most common and characteristic species of that community in its numbers and growth. The dominant species has a direct effect on the other plants in the community.

recessive (*adj*) of a gene (↑) which gives rise to a characteristic that can only appear in a homozygous (↓) condition and is suppressed by the dominant (↑) gene in the heterozygous (p. 198) condition. For example, in Mendel's cross of tall and dwarf garden peas, the recessive gene was for dwarfness.

allele (*n*) one of the alternative forms of a gene (↑) e.g. from the pair of genes designated BB giving rise to brown eyes and the pair of genes designated bb giving rise to blue eyes, the genes B and b are said to be alleles of the same gene and B is dominant (↑) while b is recessive (↑).

homozygous (*adj*) of an organism which has the same two alleles (↑) for a particular characteristic, such as eye colour. If a homozygote is crossed with a similar homozygote, it breeds true for that characteristic. If an organism is homozygous for every characteristic and breeds with a genetically (↑) identical organism, the offspring will be identical to the parents. This gradually occurs with constant inbreeding so that, while the organisms may well be adapted to their particular environment (p. 218), if it should change, they would be slow to respond.

alleles

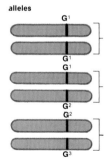

some possible combinations of 3 alleles on a chromosome pair

heterozygous (*adj*) of an organism which has two different alleles (p. 197) for a particular characteristic, such as eye colour, so that the dominant (p. 197) allele is expressed in the phenotype (p. 196). If a heterozygote breeds with a genetically (p. 196) identical heterozygote, some recessive (p. 197) characteristics will appear in some of the offspring. Heterozygous organisms are more adaptable to changing conditions than homozygous (p. 197) ones.

law of segregation Mendel's first law. One of the two laws formulated by the Gregor Mendel, to explain the way in which inheritance occurred. It states that in two alleles (p. 197) on a gene (p. 196) for a pair of characters, only one can be carried in a single gamete (p. 175).

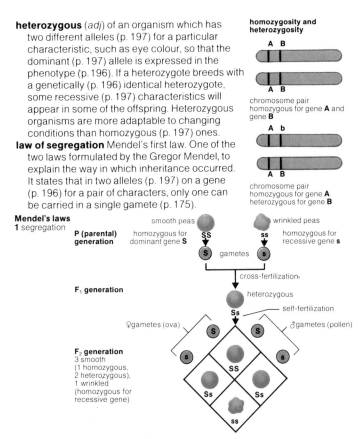

homozygosity and heterozygosity

A B

A B

chromosome pair homozygous for gene **A** and gene **B**

A b

A B

chromosome pair homozygous for gene **A** heterozygous for gene **B**

Mendel's laws
1 segregation

P (parental) generation   homozygous for dominant gene **S**   smooth peas   **Ss**   **ss**   wrinkled peas   homozygous for recessive gene **s**

S   gametes   **s**

cross-fertilization

F₁ generation   heterozygous

Ss   self-fertilization

♀gametes (ova)   **S**   **S**   ♂gametes (pollen)

F₂ generation
3 smooth
(1 homozygous,
2 heterozygous),
1 wrinkled
(homozygous for
recessive gene)

s   **s**

SS

Ss   **Ss**

ss

test cross a test to show whether or not an organism, which shows a characteristic associated with a dominant (p. 197) gene (p. 196), is heterozygous (↑) or homozygous (p. 197) for that characteristic, by crossing it with a double recessive (↓) for the characteristic. If the organism under test is homozygous, all the offspring will show the characteristic of the dominant gene while, if it is heterozygous, half show the dominant character and half the recessive (p. 197).

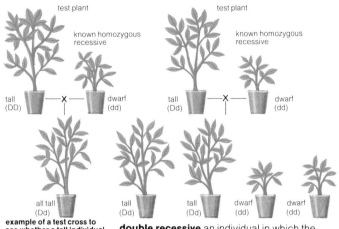

test plant

known homozygous
recessive

tall
(DD) —X— dwarf
(dd)

test plant

known homozygous
recessive

tall
(Dd) —X— dwarf
(dd)

all tall
(Dd)

tall
(Dd)

tall
(Dd)

dwarf
(dd)

dwarf
(dd)

**example of a test cross to
see whether a tall individual
is heterozygous or
homozygous. If
homozygous the progeny
are all tall; if heterozygous
half are tall and half dwarf**

double recessive an individual in which the
alleles (p. 197) of a particular gene (p. 196) are
identical for a recessive (p. 197) characteristic
so that the recessive characteristic is
expressed in the phenotype (p. 196).

carrier (*n*) an organism which may carry a
recessive (p. 197) gene (p. 196) for a
characteristic which may be harmful and which
is not expressed in the carrier because it is
masked by the dominant (p. 197) gene for that
characteristic.

dihybrid inheritance the result of cross breeding
from pure lines (p. 196) of homozygous (p. 197)
organisms with two different alleles (p. 197) for
different characteristics, such as Mendel's cross
of yellow round and wrinkled green garden
peas to give a yellow round dihybrid in which
the genes (p. 196) for yellow and round are
dominant (p. 197) and suppress the genes for
wrinkled and green which are recessive (p. 197).

dihybrid cross the result of dihybrid inheritance
(↑). If the offspring of dihybrid inheritance are
self crossed, the characteristics are expressed
in the ratio 9:3:3:1, in other words, in the Mendel
cross, nine plants are yellow and round, three
are yellow and wrinkled, three are green and
round, and one is green and wrinkled.

law of independent assortment Mendel's
second law. One of the two laws of inheritance
formulated by the Gregor Mendel, which states
that each member of one pair of alleles (p. 197)
is as likely to be combined with one member of
another pair of alleles as with any other member
because they associate randomly (and
independently).

Mendel's laws
2 independent assortment

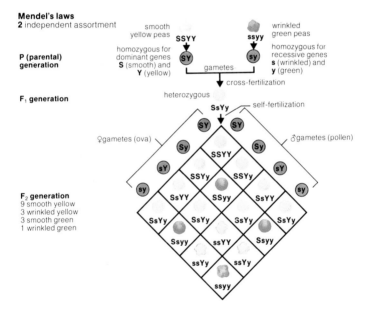

P (parental) generation

F₁ generation

F₂ generation
9 smooth yellow
3 wrinkled yellow
3 smooth green
1 wrinkled green

progeny (*n.pl.*) the offspring which result from
reproduction (p. 173).
linkage (*n*) the situation in which genes (p. 196)
on the same chromosome (p. 13) are said to be
linked so that they are unable to assort
according to the Law of Independent
Assortment (↑) and are inherited together.
linkage group a group of linked (↑) genes (p. 196)
on the same chromosome (p. 13) which are
inherited together.

sex chromosomes

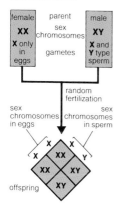

inheritance of colour blindness

X = normal sex chromosome

X^c = sex chromosome with gene for colour blindness

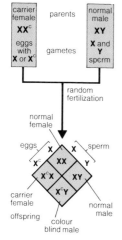

sex chromosomes the chromosomes (p. 13) which control whether or not a given individual of most animals should be male or female. There is a homologous (p. 39) pair of chromosomes in the nucleus (p. 13) of one sex, usually the female, and an unlike or single chromosome in the nucleus of the other, usually the male.

X chromosomes the sex chromosomes (↑) which occur as a like pair XX in the nuclei (p. 13) of the homogametic (↓) sex and usually are responsible for the female sex in most animals. All the gametes (p. 175) of the homogametic sex will contain one X chromosome.

Y chromosomes the sex chromosomes (↑) which occur either as an unlike pair with an X chromosome (↑) or unpaired in the nuclei (p. 13) of the heterogametic (↓) sex and are usually responsible for the male sex in most animals. The gametes (p. 175) of the heterogametic sex are of two kinds, with or without an X chromosome, which are equal in number.

hetersomes (*n.pl.*) homologous chromosomes (p. 39), such as the sex chromosomes (↑) which are not normally identical in appearance.

autosomes (*n.pl.*) homologous chromosomes (p. 39) which are not sex chromosomes (↑) and which are normally identical in appearance.

homogametic sex the sex, usually the female, which contains sex chromosomes (↑) that occur as a like pair of XX chromosomes (p. 13) in the nuclei (p. 13) of an organism.

heterogametic sex the sex, usually the male, which contains sex chromosomes (↑) that occur as an unlike pair of XY chromosomes (p. 13) or unpaired in the nuclei (p. 13) of an organism.

sex-linked (*adj*) of certain characteristics associated with recessive (p. 197) genes (p. 196) which are linked to the sex of the individual because they are attached to the X chromosome (↑).

colour blindness a sex-linked (↑) characteristic in which there is an inability to distinguish between pairs of colours, usually red/green, although the ability to distinguish shade and form is unaffected.

how crossing over and recombination during meiosis shuffles genes and causes variation

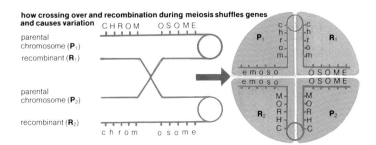

parental chromosome (**P₁**)

recombinant (**R₁**)

parental chromosome (**P₂**)

recombinant (**R₂**)

haemophilia (*n*) a sex-linked (p. 201) characteristic or disease, known only in males, in which the blood (p. 90) is unable to clot (p. 129) properly after wounding.

crossing over the exchange of genetic material (p. 203) during meiosis (p. 39), between male and female parentals (p. 197) in which the chromatids (p. 35) of homologous chromosomes (p. 39) break at the chiasmata (p. 39) and rejoin to allow the assortment of linked (p. 201) genes. Crossing over leads to increased variation (p. 213).

recombinants (*n.pl.*) the gametes (p. 175) which result from crossing over (↑) so that an exchange of genetic material (p. 203) between the parentals (p. 197) gives rise to some characteristics which are not present in either parental. This leads to increased variability in the offspring and greater change of adaptation to changing conditions. **recombine** (*v*).

crossover frequency the number of recombinants (↑) that are likely to occur as a result of the crossing over (↑) between two genes (p. 196) on different parts of the same chromosome (p. 13). It is usually expressed as a percentage of the number of recombinants compared with the total number of offspring produced. The crossover frequency is lower the closer the genes occur on chromosomes.

chromosome map a diagram of the order and distance between the genes (p. 196) on a chromosome (p. 13) worked out by experiments and an analysis of the crossover frequency (↑).

crossing over
during first meiotic division

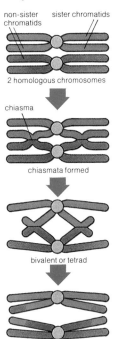

non-sister chromatids   sister chromatids

2 homologous chromosomes

chiasma

chiasmata formed

bivalent or tetrad

genetic material exchanged, chromosomes separate

locus

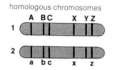

alleles **A**, **B**, **C**, **X**, **Y**, **Z**, occupy
same loci (positions) on
chromosome 1 as alleles **a**, **b**,
c, **x**, **y**, **z**, on chromosome 2

gene locus the precise position of a gene (p. 196)
on the chromosome (p. 13). Alleles (p. 197) of
the same gene occupy the same loci (*pl.*) on
homologous chromosomes (p. 39).

multiple alleles a series of three or more alleles
(p. 197) on the same gene (p. 196) which give
rise to a particular characteristic. Only two of these
alleles in various combinations can occupy
the same gene locus (1) on a pair of homologous
chromosomes (p. 39) at the same time.

lethal alleles alleles (p. 197) which will kill the
individual if they are dominant (p. 197) in a
heterozygous (p. 198) individual or if they are
recessive (p.197) in a homozygous (p.197) one.

partial dominance the situation which occurs
between dominant (p. 197) alleles (p. 197) in
which one may be slightly more dominant than
the other. For example, if the allele for red is
dominant in individuals of the same flower
while the allele for white is dominant in other
individuals, breeding them may produce pink
flowers which will be reddish pink if the allele
for red is more dominant than the allele for
white. Also known as **co-dominance**.

epistasis (*n*) the interaction of non-allelic (p. 197)
genes (p. 196) in which one gene suppresses the
characteristics which would normally be expressed
by another gene. It is similar to recessiveness
(p. 197) and dominance (p. 197) between alleles.

genetic material the organic compounds (p. 15)
which carry the genetic (p. 196) information
from one generation to the next and from cell to
cell. Chromosomes (p. 13) are composed of
proteins (p. 21) and DNA (p. 24) which carries
the genetic information.

genetic code the sequence of the four bases (p.22)
adenine (p. 22), guanine (p. 22), cytosine (p.22),
and thymine (p. 22) on a strand of DNA (p. 24)
represents a code that controls the construction of
proteins (p. 21) and enzymes (p. 28) which make
up the cytoplasm (p. 10) of an organism and directs
its functioning. Triplets of these bases code for
the twenty different amino acids (p. 21), and
groups of these triplets, code for whole proteins.
More than one triplet can code for an amino acid.

amino acids and the genetic code		amino acid general formula
(see previous page)		$$NH_3^+ - \overset{\displaystyle COO^-}{\underset{\displaystyle R}{C}} - H$$
	R = side group	

codon	amino acid	side group (R)	side group (R)	amino acid	codon
AAA AAG }	lysine	$-CH_2CH_2CH_2CH_2NH_3^+$	$-H$	glycine	{ GGU GGC GGA GGG
AAU AAC }	asparagine	$-CH_2CONH_2$	$-CH_2COO^-$	aspartic acid	{ GAU GAC
ACU ACC ACA ACG }	threonine	$-CHOHCH_3$	$-CH_2CH_2COO^-$	glutamic acid	{ GAA GAG
AGU AGC }	serine	$-CH_2OH$	$-CH_3$	alanine	{ GCU GCC GCA GCG
AGA AGG }	arginine	$-CH_2CH_2CH_2NHC\underset{N^+H_2}{\overset{NH_2}{}}$	$CH_3\underset{\|}{CHCH_3}$	valine	{ GUU GUC GUA GUG
AUU AUC AUA }	isoleucine	$CH_3CH_2\underset{\|}{C}HCH_3$			
AUG }	methionine	$-CH_2CH_2SCH_3$	phenylalanine	{ UUU UUC
CCU CCC CCA CCG }	proline				
CAU CAC }	histidine			leucine	{ UUA UUG
				tyrosine	{ UAU UAC
CAA CAG }	glutamine	$-CH_2CH_2CONH_2$		NONSENSE	{ UAA UAG
CGU CGC CGA CGG }	arginine	$-CH_2CH_2CH_2NHC\underset{NH_2}{\overset{N^+H_2}{}}$	$-CH_2SH$	cysteine	{ UGU UGC
CUU CUC CUA CUG }	leucine	$-CH_2CH\underset{CH_3}{\overset{CH_3}{}}$		tryptophan	{ UGG
				NONSENSE	{ UGA
			$-CH_2OH$	serine	{ UCU UCC UCA UCG

phenylalanine side group: $-CH_2C$ (ring) with $HC=CH$ / $HC-CH$

leucine side group: $-CH_2-\underset{CH_3}{\overset{CH_3}{CH}}$

tyrosine side group: $-CH_2C$ (ring) COH

transcription (*n*) the process in which the genetic code (p. 203) is, in the first place, copied from the DNA (p. 24) on to a single strand of RNA (p. 24) in the nuclei (p. 13) of cells.

translation (*n*) the process in which the messenger RNA (p. 24) from the transcription (↑) then leaves the nucleus (p. 13) and passes into the ribosomes (p. 10) in the cytoplasm (p. 10) to function as a pattern from which amino acids (p. 21) are built into proteins (p. 21).

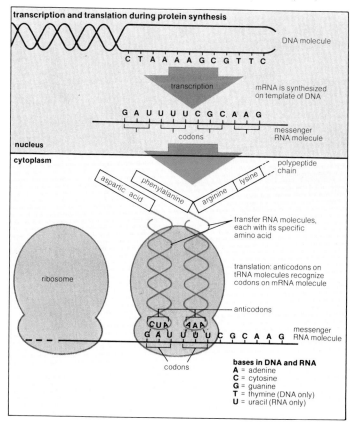

transcription and translation during protein synthesis

DNA molecule

C T A A A A G C G T T C

transcription — mRNA is synthesized on template of DNA

G A U U U U C G C A A G

codons — messenger RNA molecule

nucleus

cytoplasm

aspartic acid

phenylalanine

arginine

lysine

polypeptide chain

transfer RNA molecules, each with its specific amino acid

translation: anticodons on tRNA molecules recognize codons on mRNA molecule

ribosome

anticodons

CUA AAA

G A U U U U C G C A A G — messenger RNA molecule

codons

bases in DNA and RNA
A = adenine
C = cytosine
G = guanine
T = thymine (DNA only)
U = uracil (RNA only)

mutation (*n*) a change in the structure of the genetic material (p. 203) of an organism which will be inherited if it occurs in the cells which produce the gametes (p. 175). It can occur as a result of changes in genes (p. 196) or of changes in the structure or number of chromosomes (p. 13). Most mutations are harmful but some allow the organism to adapt to changing circumstances and as a source of increased variation (p. 213) are the very material of evolution (p. 208). Mutations can be stimulated by increases in certain chemicals or ionizing radiation.

mutant (*n*) the result of a mutation (↑) which is usually recessive (p. 197) in the most common types of mutation.

mutagenic agent some stimulus, such as certain chemicals or ionizing radiation, which is likely to cause a mutation (↑).

chromosome mutation a change or mutation (↑) of the number or arrangement of the chromosomes (p. 13).

deletion (*n*) a chromosome (p. 13) mutation (↑) which occurs if a segment of a chromosome breaks away and is lost during nuclear division (p. 35) with a resulting loss of genetic material (p. 203).

inversion (*n*) (1) a chromosome (p. 13) mutation (↑) which occurs if a segment of a chromosome breaks away during nuclear division (p. 35) and rejoins the chromosome the wrong way round to reverse the sequence of genes (p. 196); (2) a gene mutation (↓) in which the order of the bases (p. 22) in a strand of DNA (p. 24) is changed.

translocation[2] (*n*) a chromosome (p. 13) mutation (↑) which occurs if a segment of the chromosome breaks away during nuclear division (p. 35) and rejoins the original chromosome in a different place or joins another chromosome.

duplication (*n*) a chromosome (p. 13) mutation (↑) in which a segment of the chromosome is duplicated either on the same or on another chromosome.

deletion

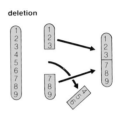

inversion

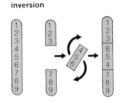

translocation

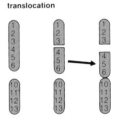

duplication

how changes of single
nucleotides in a triplet
(substitution) can cause
mutants to occur

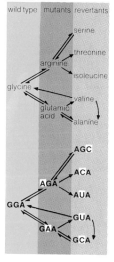

gene mutation a mutation (↑) in which the
   sequence of bases (p. 22) is not copied precisely
   in replicating a strand of DNA (p. 24) resulting
   in a change in the formation of the proteins
   (p. 21). Once it has occurred, it is replicated
   in the formation of further strands of DNA.
substitution (n) a gene mutation (↑) in which one
   DNA (p. 24) base (p. 22) is replaced by another.
insertion (n) a gene mutation (↑) in which another
   base (p: 22) is inserted in the existing sequence
   of bases in the strand DNA (p. 24).
sickle-cell anaemia a disease which is inherited and
   exhibits partial dominance (p. 203). A sickle cell
   contains a mutant (↑) gene (p. 196) which
   crystallizes haemoglobin (p. 126) in the
   erythrocytes (p. 91) of human blood (p. 90) and
   distorts them causing the blood vessels (p. 127)
   to clog. It is found usually among negroid people
   and is thought to offer some resistance to malaria.
polyploidy (n) the condition in which the cells of
   an organism contains at least three times the
   normal haploid (p. 36) number of chromosomes
   (p. 13). **polyploid** (adj).
triploid (adj) of a polyploid (↑) cell in which there is
   three times the normal haploid (p. 36) number of
   chromosomes (p. 13). It results from the fusion
   of a haploid and a diploid (p. 36) gamete (p. 175).
tetraploid (adj) of a polyploid (↑) cell in which there
   is four times the normal haploid (p. 36)
   number of chromosomes (p. 13). It occurs as a
   result of the fusion of two diploid (p. 36) cells.
aneuploidy (n) the condition in which
   chromosome (p. 13) mutation (↑) results in the
   gain or loss of chromosomes from a set.
euploidy (n) the condition in which chromosome
   (p. 13) mutation (↑) results in the gain of a
   whole set of chromosomes.
autopolyploid (adj) of the condition of polyploidy
   (↑) which results from euploidy (↑) in which the
   cell has multiple sets of its chromosomes (p. 13).
allopolyploid (adj) of the condition of polyploidy
   (↑) which results from euploidy (↑) in which the
   cell contains two different sets of chromosomes
   (p. 13) from the hybridization (p. 216) of two
   closely related organisms especially plants.

evolution (*n*) the process whereby all organisms descend from the common ancestors which emerged on the Earth. Over successive generations throughout geological time, populations (p. 214) are modified in response to changes in environment (p. 218) by such processes as natural selection (↓) so that new species (p. 40) are formed which are all related, however distantly, by common descent.

Darwinism (*n*) the mechanism first put forward by the British naturalist, Charles Darwin (1809–82), following careful observation of animals and plants all over the world, e.g. Darwin's finches, to explain how organisms changed slowly, over millions of years, to evolve new forms. He suggested that in any given population (p. 214) of an organism there was considerable variation (p. 213) between individuals. Some would exhibit different characteristics which would be better fitted to their circumstances and environment (p. 218) than others. Thus, these individuals would be more likely to survive to maturity and breed so that their offspring would also exhibit these characteristics. Those individuals that were less well suited to their conditions would have less chance of breeding success so that eventually the population would contain more and more of the individuals that were better suited to their environment and the character of the species (p. 40) would change as a whole and result in a new species. Darwin was unable to explain, however, how the variations were produced in the first place. He called this process the theory of natural selection (↓).

natural selection one of the central deductions of Darwinism (↑). If the variations (p. 213), which occur among individuals within a population (p. 214) of animals, gives to those individuals a better chance of surviving, they are more likely to reach sexual maturity and breed so that their offspring will also inherit those advantageous characteristics. Eventually, the inheritance of variations in a particular direction over generations will lead to a new species (p. 40).

Darwin's finches on the Galapagos, Darwin observed that there were many species of finches which he thought had evolved from one species after it had arrived on the islands

C. crassirostris
vegetarian

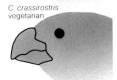

C. psittacula
insect feeding

Camarhynchus pallidus
woodpecker finch

G. fortis
ground feeding

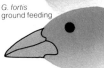

Geospiza scandens
cactus feeding

Pinaroloxias inornata
warbler-like

survival of the fittest the idea behind natural selection (↑), which suggests that only the animals which are best fitted to their circumstances will survive in the struggle for existence while those that are less well fitted will tend to perish.

Neodarwinism (*n*) the modern, modified version of Darwinism (↑) which, with the aid of the theories (p. 235) of genetics (p. 196) based on the work of Gregor Mendel (p. 196), seeks to explain the mechanisms for the existence of advantageous variations (p. 213) which may occur naturally in a population (p. 214) of organisms which, because of lack of knowledge at the time, Darwin was unable to account for.

origin of species the theory (p. 235) leading from Darwinism (↑), which was developed by Charles Darwin in a paper published by him in 1859 and entitled *On the Origin of Species by Means of Natural Selection and the Preservation of Favoured Races in the Struggle for Life*. The theory suggests that within a population (p. 214) of one species (p. 40), various factors exist, such as geographical barriers (rivers, oceans, mountains, etc) or specific differences in behaviour (p. 164) which can isolate (p. 214) breeding groups within the population. This tends to maintain the integrity of the genes (p. 196) which carry the variations (p. 213) within the breeding group that are advantageous for the local environment (p. 218). In this way, the genetic differences between one group and another can build up, and over many generations lead to the development of new species, each fitted to its own conditions. This is referred to as speciation (p. 213). *See also* natural selection (↑).

Lamarckism (*n*) a theory (p. 235) which stemmed from the observations of the French biologist, Jean de Lamarck (1744–1829), who noticed that particular organs of an animal could fall into use or disuse if they were needed or not. From this he suggested that these acquired characteristics could be inherited. Modern genetic (p. 196) studies, however, have been unable to discover any mechanism whereby characters developed during an individual's lifetime could be passed on to its offspring so that the theory has fallen into disuse.

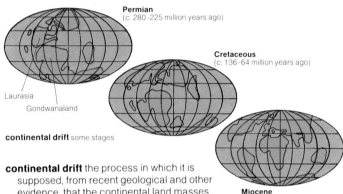

Permian
(c. 280-225 million years ago)

Cretaceous
(c. 136-64 million years ago)

Laurasia

Gondwanaland

continental drift some stages

Miocene
(c. 26-7 million years ago)

continental drift the process in which it is supposed, from recent geological and other evidence, that the continental land masses have not always occupied their present position on the globe and that, powered by processes within the Earth itself, they are slowly and continuously on the move. In this way, new land masses are created and destroyed, split apart and joined, over geological time. This is one of the processes that can lead to the geographical isolation (p. 214) of a breeding population (p. 214) and so to speciation (p. 213).

Pangea (n) the single landmass or 'supercontinent' which itself was formed during Devonian times, some 395 to 345 million years ago, by collision of the two original continents, known as Gondwanaland and Laurasia. The present continents evolved from Pangea by the process of continental drift (↑).

plate tectonics the theory (p. 235) which has been developed recently to provide a mechanism for continental drift (↑). It supposes that the surface layers of the Earth fit together, rather like a spherical jigsaw puzzle, and that the individual pieces are on the move in relation to one another. In this way, pieces may slide past one another, collide with one piece being forced beneath the other, or separate with new crust being formed as they move apart. It is at the boundaries between the individual plates that the majority of volcanic eruptions and earthquakes take place.

analogous structures

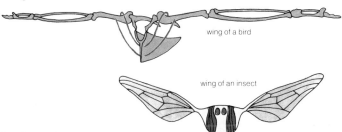

wing of a bird

wing of an insect

homologous structures

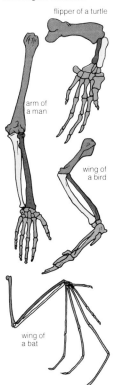

flipper of a turtle

arm of a man

wing of a bird

wing of a bat

analogous (*adj*) of structures or organs which occur in different species (p. 40) of organisms and that have similar functions but a different evolutionary (p. 208) and embryological (p. 166) origin so that their structure is also different. For example, the wings of birds and those of the insects both enable the animals to fly but their origins and form are quite different.

homologous (*adj*) of structures or organs which occur in different species (p. 40) of organisms but which have similar evolutionary (p. 208) and embryological (p. 166) origins even though their functions may have been modified. For example, the limbs of all tetrapod (p. 77) vertebrates (p. 74) are based on the pattern of five digits. This suggests evolutionary relationships between different species.

divergent (*adj*) of evolution (p. 208) in which homologous (↑) structures have become adapted to perform different functions. For example, the flippers of sea mammals (p. 80), such as seals, are homologous with the limbs of land-based vertebrates (p. 74) but, are used in a different way as the seals have become better adapted to their marine environment (p. 218).

convergent (*adj*) of evolution (p. 208) in which analogous (↑) structures have become adapted to perform the same function. For example, the eye of a cephalopod (p. 72) performs the same function as that of a vertebrate (p. 74) but has a quite different origin and structure.

vestigial (*adj*) of a structure or organ which
originally performed a useful function but, through
evolution (p. 208), has become reduced to a
remnant of its former self and no longer functions.
For example, the appendix (p. 102) in humans.

primitive (*adj*) of a structure or organism that is at
an early stage in evolution (p. 208), or is like an
organism at an early stage.

phylogenetic (*adj*) of a classification (p. 40) which
is based on the apparent evolutionary (p. 208)
relationships between organisms.

palaeontology (*n*) the science or study of ancient
life forms through their remains as fossils (↓).

fossil (*n*) any remains or trace of a once-living
organism that has been preserved in some way
such as in the rocks or in ice.

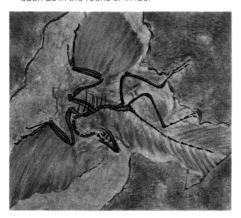

fossil e.g.
Archaeopteryx

fossil record the continuing record of the origins,
development and existence of life on Earth as
expressed through the finds of fossils (↑)
preserved in the rocks from the origins of the
planet to the present day.

geological column a tabular time scale which
has been worked out by geologists on the basis
of the fossil record (↑) and other evidence, such
as radiometric dating, in which the history of
the Earth is broken down into eras, periods, and
epochs.

variation (*n*) the differences in form and structure
which occur naturally among individuals within
the same species (p. 40) and which may result
from genetic (p. 196) changes, such as
mutations (p. 206), or from differences in such
factors as nutrition (p. 92) or the density of the
population (p. 214).

diversity (*n*) the state of things being different
from each other.

isolating mechanisms factors, such as the
existence of geographical barriers, behaviour
(p.164), or the timing of the breeding season (p.195),
which tend to separate groups of individuals
into reproductive (p. 173) communities (p. 217).

gene pool the total number and type of genes
(p. 196) that exist at any given time within a
breeding population (p. 214) that has been
separated by various isolating (p. 214)
mechanisms. The genes within a given gene
pool may then be intermixed randomly by
interbreeding within the group.

speciation (*n*) the process by which two or more
new species (p. 40) evolve (p. 208) from one
original species as breeding groups become
separated by isolating (p. 214) mechanisms
and develop a range of distinctive characters,
as a result of natural selection (p. 208), to the
extent that the isolated populations (p. 214) are
no longer able to breed with one another.

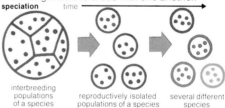

speciation   time

interbreeding
populations
of a species

reproductively isolated
populations of a species

several different
species

differential mortality the basis of natural
selection (p. 208) during periods of increasing
population (p. 214) when those individuals of
the overpopulated community which are best
fitted to their environment (p. 218) survive to breed
while those that are less well fitted die so that
evolution (p. 208) takes place by natural selection.

melanism (*n*) the condition in which such structures as hair, skin and eyes are coloured by the dark-brown pigment (p. 126) melanin. Melanistic skin protects the individual from the harmful effects of prolonged exposure to sunlight. Consequently, humans who have evolved (p. 208) in areas of high sunlight intensity have, by natural selection (p. 208), evolved darker skin colour.

gene frequency the occurrence of one particular gene (p. 196) in a given population (↓) in relation to all its other alleles (p. 197).

Hardy-Weinberg principle a law formulated in 1908 from which the effects of natural selection (p. 208) can be better understood. It suggests that in any population (↓), in which mating takes place at random, the proportion of dominant (p. 197) to recessive (p. 197) genes (p. 196) in the population remains unchanged from one generation to the next. Until the principle was worked out, it was thought, quite reasonably, that the numbers of recessive genes would decline while the dominant genes would increase.

gene flow the process by which genes (p. 196) move within a population (↓) by mating and the exchange of genes.

genetic drift the process by which the genetic (p.196) structure of a small population (↓) of organisms changes by chance rather than by natural selection (p. 208). In a small population the Hardy-Weinberg principle (↑) may not be maintained because the number of pairings will not be random.

isolation (*n*) the process by which two populations (↓) become separated by geographical, ecological (p. 217), behavioural (p. 164), reproductive (p. 173), or genetic (p. 196) factors. After two populations have become genetically or reproductively separated, they will not revert to the same species (p. 40) even if they come together geographically again.

population (*n*) a group of organisms of the same species (p. 40) which occupies a particular space over a given period of time. The actual numbers of individuals within a population may rise and fall as a result in changes of the birth and death rate and such factors as climate, food supply, and disease.

allopatric (*adj*) of two or more populations (↑) of the same or related species (p. 40) which could interbreed if they were not geographically isolated (↑) from one another.

sympatric (*adj*) of two or more related species (p. 40) which are not geographically isolated (↑) from one another and which could interbreed apart from differences in behaviour (p. 164) or the timing of the breeding season (p. 195) etc.

ecological isolation isolation (↑) which occurs within populations (↑) as a result of the different ways in which they relate to their environment (p. 218).

reproductive isolation isolation (↑) which occurs within populations (↑) as a result of differences in their breeding behaviour (p. 164) or timing of their breeding season (p. 195).

sympatric
e.g. two species occurring in the same place

allopatric
e.g. two species occurring in different places

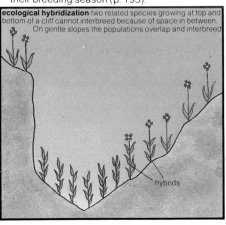

ecological hybridization two related species growing at top and bottom of a cliff cannot interbreed because of space in between. On gentle slopes the populations overlap and interbreed

hybrids

genetic isolation isolation (↑) which occurs within populations (↑) as a result of their genetic (p. 196) incompatibility so that they are unable to produce fertile (p. 175) offspring.

artificial selection the process by which humans make use of the principles of genetics (p. 196) and evolution (p. 208) to create breeds or hybrids (p. 216) which would not be expected to occur as a result of natural selection (p. 208).

inbreeding (*n*) breeding by the mating of closely related individuals, including self-fertilization (p. 175) in plants. It tends to reduce the genetic (p. 196) variability of the population (p. 214) and leads to a greater frequency of expression of recessive (p. 197) characteristics. Humans make use of inbreeding during artificial selection (p. 215) to develop characteristics which are seen as useful.

outbreeding (*n*) breeding by the mating of individuals which are not closely related. The most extreme form of outbreeding is between organisms of different species (p. 40) which leads to the production of non-fertile (p. 175) offspring. Outbreeding normally gives rise to greater genetic (p. 196) variability and vigour and there may be various mechanisms within organisms to encourage it.

hybrid vigour an increase in the vigour of such factors as growth or fertility (p. 175) in the offspring as compared with the parents which results from the cross-breeding of individuals from lines which are genetically (p. 196) different leading to greater heterozygosity (p. 198) and an increased expression of dominant (p. 197) genes.

hybrid (*n*) the offspring of parents from genetically (p. 196) different lines. **hybridization** (*n*).

spontaneous generation the idea, disproved by the French bacteriologist, Louis Pasteur (1822–95) and others, that, in suitable conditions, organisms, especially microorganisms, could be generated from inorganic compounds (p. 15).

special creation a hypothesis (p. 235) which suggests that every form of life that exists or has ever existed was created separately by a deity or other supernatural force. Palaeontological (p. 212) and genetic (p. 196) evidence suggests that this is unlikely and few scientists take the hypothesis seriously today.

steady state a hypothesis (p. 235) which suggests that all organisms were created at some time in the past and have remained unchanged ever since with each generation being identical to its predecessor. Palaeontological (p. 212) evidence suggests that this cannot be the case.

ecology (*n*) the science or study of organisms in relation to one another and the environment (p. 218).

biosphere (*n*) the part of the Earth which includes all of the living organisms on the planet and their environment (p. 218).

biome (*n*) a part of the biosphere (↑) which might be a large, regional community (↓) of interrelated organisms and their environment (p. 218) and would include such habitats (↓) and communities as a tropical rainforest or grassland.

ecosystem (*n*) a self-contained and perhaps small unit or area, such as a woodland, which would include all the living and non-living parts of that unit.

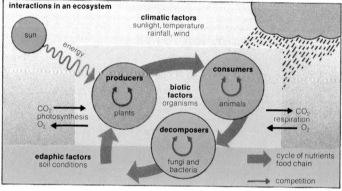

interactions in an ecosystem

community (*n*) a localized group of a number of populations (p. 214) of different species (p. 40) living and interacting with one another within an ecosystem (↑). Communities can be described as open (niches (p. 219) unstable or 'empty' allowing new species into the community) or closed (niches stable and full).

habitat (*n*) a part of an ecosystem (↑), such as a desert, in which particular organisms live because the environmental (p. 218) conditions within the habitat are essentially uniform even though they may vary with the season or between, say, ground level and the tops of the trees.

microhabitat (*n*) a small area within a habitat (↑) such as the underside of a stone.

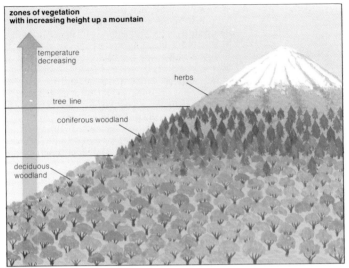

zones of vegetation
with increasing height up a mountain

temperature
decreasing

herbs

tree line

coniferous woodland

deciduous
woodland

zone (*n*) a part of a biome (p. 217) which is characterized by one particular group of organisms depending upon the environmental (↓) conditions present in that area.

climate (*n*) the sum total of all the interrelating weather conditions, such as temperature, pressure, rainfall, sunshine, etc., that exists in a particular region throughout the year and averaged over a number of years.

microclimate (*n*) the climate (↑) which occurs in a small region, such as a town or woodland, which differs in some way from the overall climate of the region due to the effects of other factors within the area. For example, the temperature in a large city may be significantly higher than that of its rural surroundings because of the heat that is trapped by the buildings and re-released.

environment (*n*) the sum total of all the external conditions within which an organism lives.

territory (*n*) any area which is occupied and defended by an animal for purposes of breeding, feeding etc.

niche (*n*) the local physical and biological conditions which an organism fills in an ecosystem (p. 217). If, at some stage, more than one species (p. 40) of organism attempt to occupy the same niche, then they compete with one another until one is eliminated. On the other hand, it is possible for different species to occupy the same niche in geographically separated regions or for one species to evolve (p. 208), by natural selection (p. 208), to occupy different niches.

abiotic (*adj*) of the physical environment (↑) to which organisms are subjected, such as temperature, light intensity, availability of water, etc.

aquatic (*adj*) of a watery environment (↑) or a species (p. 40) which lives primarily in water.

freshwater (*adj*) of an aquatic (↑) environment (↑), such as a river, which does not contain salt and is, therefore, not marine (↓). Also, of a species (p. 40) which lives primarily in fresh water.

marine (*adj*) of an aquatic (↑) environment (↑), such as the ocean, which contains salt. Also describes a species (p. 40) which lives primarily in a marine environment.

littoral (*n*) the zone (↑) of a freshwater (↑) environment (↑) between the water's edge and a depth of about six metres or the zone of a marine (↑) environment between the high and low water marks. A littoral species (p. 40) is one which lives primarily in the littoral zone.

amphibious (*adj*) of an organism which is capable of or spends part of its time living in water and part on land.

terrestrial (*adj*) of those organisms that spend most or all of their lives on land.

subterranean (*adj*) of those organisms that spend most or all of their lives underground, in caves, for example.

arboreal (*adj*) of those organisms that spend most or all of their lives living among the branches of trees.

aerial (*adj*) of those organisms or parts of organisms that spend part or all of their lives in the air. The roots of certain trees grow in the air and are referred to as aerial.

climatic factors those aspects of the environment (p. 218), grouped together as the climate (p. 218), including temperature, rainfall, etc., which affect the distribution of organisms.

edaphic factors those aspects of the environment (p. 218) concerned with the soil and including moisture content, pH (p. 15), etc., which affect the distribution of organisms.

biotic (*adj*) of those biological parts of the environment (p. 218) other than the abiotic (p. 219) factors to which organisms are subjected, and include their relationships with other organisms such as competition (↓) for habitat (p. 217) etc.

predation (*n*) the process by which certain animals gain nutrition (p. 92) by killing and feeding upon other animals. A predator is a secondary consumer (p. 223) and predators do not include parasites (p. 110).

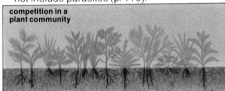

competition in a plant community

leaves compete for light, CO_2, space

roots compete for nutrients and water

competition (*n*) the process in which more than one species (p. 40) or individuals of the same species attempt to make use of the same resources in the environment (p. 218) because there are not enough resources to satisfy the needs of all the organisms. Competition often leads to differential mortality (p. 213).

intraspecific (*adj*) of an action, for example, competition (↑), which takes place between individuals of the same species (p. 40).

interspecific (*adj*) of an action, for example, competition (↑), which takes place between different species (p. 40).

mimicry (*n*) the process in which one organism resembles another and thereby gains some advantage, e.g. a defenceless hoverfly closely resembles the form and colour of a wasp and may, therefore, be avoided by predators (↑).

mimicry

hoverfly

harmless hoverfly mimics unpleasant wasp

wasp

synecology (*n*) the study or science of all communities (p. 217) and ecosystems (p. 217) within an environment (p. 218) and their relationships to one another.

autecology (*n*) the study or science of individuals of one species (p. 40) in relation to one another and to their environment (p. 218).

succession (*n*) a progressive sequence of changes which takes place, after the first colonization (↓) of a particular environment (p. 218), in the organisms which occupy that environment until a stable position is reached where no further changes can take place unless the abiotic (p. 219), edaphic (↑), or climatic factors (↑) themselves are altered. The process takes place rapidly at first and then slows down as stability is approached.

succession

a pioneer species colonizes a habitat

pioneer plants grow and reproduce

growth of plants alters edaphic and biotic factors and more species colonize

climax community with many plant species. Conditions no longer suitable for pioneer

colonization (*n*) the arrival and growth to reproductive (p. 173) age of an organism in an area, i.e. the spread of species (p. 40) to places where they have not lived before. **colonize** (*v*), **colony** (*n*).

pioneer (*n*) a plant species (p. 40) that is found in the early stages of succession (↑).

climax (*adj*) of a community (p. 217) which, following succession (↑), has reached stability.

sere (*n*) a succession (↑) of plant communities (p. 217) which themselves affect the environment (p. 218) leading to the next community and resulting ultimately in the climax (↑) community.

soil (*n*) the material which forms a surface covering over large areas of the Earth and in which organisms gain support, protection, and nutrients (p. 92). It results from the weathering and breakdown of rocks into inorganic (p. 15) mineral particles which are then further acted upon by climatic (p. 220) and biotic (p. 220) factors. The composition depends upon the composition of the original rock.

inorganic component the part of the soil which results from the action of weather on the parent rocks, breaking it down into mineral particles of varying size and composition depending upon the composition of the original rock.

organic component the part of the soil which is derived from the existence and activity of the large numbers of living organisms in the soil.

sand (*n*) the inorganic component (↑) in which the particles range in size from 0.02–2.0 millimetres and are angular. A soil with a high sand content tends to be dry, because of the ease with which water drains away, acidic (p. 15), and low in nutrient (p. 92) content.

clay (*n*) the inorganic component (↑) in which the particles are less than 0.02 millimetres in size and are relatively smooth and rounded. A soil with a high clay content tends to be easily waterlogged, can become compacted, and will harden on drying. It is usually rich in nutrient (p. 92) content, however.

humus (*n*) the organic component (↑) of soil which results from the activity and decomposition (↓) of the living organisms within a soil and which is a mixture of fibrous (p. 143) and colloidal materials made up essentially of carbon, nitrogen, phosphorus and sulphur. Humus improves the structure and texture of a soil, helps it to retain water and nutrients (p. 92), and raises the soil's temperature by absorbing more of the sun's energy because of its dark colour.

erosion (*n*) the process by which the products of weathering of a rock or a soil are worn away by the action of wind, running water, or moving ice, etc.

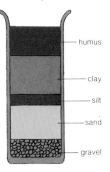

the constituents of loam
soil shaken up with water

— humus
— clay
— silt
— sand
— gravel

soil profile the series of distinct layers that can be observed in a vertical section through soil from the parent rock, through weathered rock and *subsoil* to *topsoil*.

a generalized soil profile

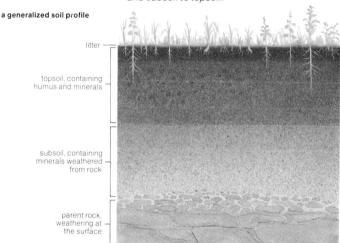

litter

topsoil, containing humus and minerals

subsoil, containing minerals weathered from rock

parent rock, weathering at the surface

pyramid of available energy at the trophic levels of a food web

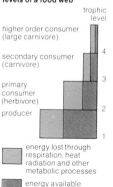

trophic level

higher order consumer (large carnivore)

secondary consumer (carnivore)

4

primary consumer (herbivore)

3

producer

2

1

☐ energy lost through respiration, heat radiation and other metabolic processes

■ energy available as food

producers (*n.pl.*) the organisms, especially green plants and some bacteria (p. 42), which are able to manufacture nutrients (p. 92) from inorganic (p. 15) sources by such processes as photosynthesis (p. 93).

consumers (*n.pl.*) the heterotrophic (p. 92) organisms which obtain their nourishment by consuming the producers (↑) or other consumers.

decomposers (*n.pl.*) the organisms which obtain their nutrients (p. 92) by feeding upon dead organisms, breaking them down into simpler substances and, in so doing, making other nutrients available for the producers (↑).

decomposition (*n*).

trophic level the particular position which an organism occupies in an ecosystem (p. 217) in respect of the number of steps away from plants at which the organism obtains its food. The producers (↑) are at the lowest trophic level while the predators (p. 220) at the highest trophic levels.

carbon cycle

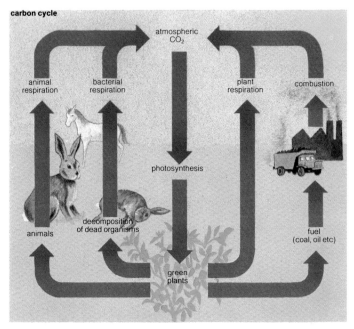

carbon cycle the chain or cycle of events by which carbon is circulated through the environment (p. 218) and living organisms. Plants take in carbon dioxide from the atmosphere and turn it into carbohydrates (p. 17), proteins (p. 21) and fats. Some of the carbon dioxide is returned to the atmosphere during the plants' respiration (p. 112). The plants are eaten by herbivores (p. 105) which, in turn, are eaten by carnivores (p. 105). When the herbivores and carnivores die, they are fed upon by saprophytes (p. 92) and decomposers (p. 223) so that carbon is returned to the soil or to the atmosphere as a product of respiration of bacteria (p. 42) and fungi (p. 46).

oxygen cycle the chain or cycle of events by which oxygen is circulated through the environment (p. 218) and living organisms.

nitrogen cycle the chain or cycle of events by which nitrogen is circulated through the environment (p. 218) and living organisms. Some bacteria (p. 42) and algae (p. 44) can make use of nitrogen directly, and lightning, acting upon atmospheric nitrogen and oxygen, causes it to combine into nitrous and nitric oxide which dissolve in falling rain to enter the soil and form nitrates and nitrites. Most plants make use of nitrogen as nitrates and use them in the manufacture of proteins (p. 21). The plants are fed upon by herbivores (p. 105) which, in turn, are eaten by carnivores (p. 105) which also make use of the nitrogen in the manufacture of animal proteins. When animals and plants die, the nitrogen is returned to the soil by nitrifying bacteria as nitrites, ammonia, and ammonium compounds.

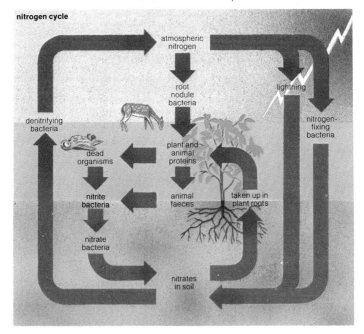

nitrogen cycle

atmospheric nitrogen

root nodule bacteria

lightning

denitrifying bacteria

nitrogen-fixing bacteria

dead organisms

plant and animal proteins

nitrite bacteria

animal faeces

taken up in plant roots

nitrate bacteria

nitrates in soil

water cycle the chain or cycle of events by which water, essential for life, is circulated through the environment (p. 218) and living organisms.

food chain the sequence of organisms from producers (p. 223) to consumers (p. 223) which feed at different trophic levels (p. 223). A simple food chain: grass grows; a cow eats the grass; a human eats the cow or drinks its milk.

food web an interconnected group of food chains (↑). There are few systems as simple as a food chain and many chains may interlink to form a complex web.

a food chain

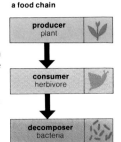

a food web

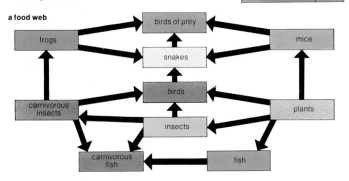

biomass (*n*) the total mass or volume of all the living organisms within a particular area, community (p. 217), or the Earth itself.

pyramid of biomass a diagrammatic representation, which forms the shape of a gently sloping pyramid, to show the biomass (↑) at every trophic level (p. 223).

standing crop the total amount of nutritional (p. 92) living material in the biomass (↑) of a given area at a particular time.

diurnal rhythm the rhythmic sequence of metabolic (p. 26) events, such as the motion of leaves in plants, that take place over a roughly twenty-four hour pattern, and which can be shown to occur in all living organisms even if they are isolated from their normal external environment (p. 218).

circadian rhythm = diurnal rhythm (↑).

annual rhythm the rhythmic sequence of metabolic (p. 26) events, such as germination (p. 168), flowering and fruiting in plants, that take place over a roughly yearly pattern even if they are isolated from their normal external environment (p. 218).

zooplankton

phytoplankton

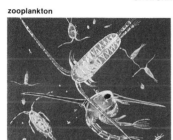

plankton (*n*) any of the various, usually tiny or microscopic (p. 9), organisms that float freely in an aquatic environment (p. 218) that have no visible means of locomotion (p. 143) and depend on the currents in the water for distribution. They are not attached to any other organism or substrate.

phytoplankton (*n*) the plant plankton (↑), especially the diatoms, which are an important source of food for other organisms such as many species (p. 40) of whales.

zooplankton (*n*) the animal plankton (↑) including the larvae (p. 165) of many species (p. 40) of fish.

pelagic (*adj*) of the upper waters of an aquatic, especially marine environment (p. 218), as opposed to the bed of the ocean or lake, and the organisms which inhabit them.

benthic (*adj*) of the bed of an aquatic, especially marine environment (p. 218), and the organisms which live on or in it.

association (*n*) any relationship which exists between organisms to the benefit of one or all of them. In plants, a climax (p. 221) community (p. 217) dominated by one or a small number of species (p. 40) and named after them. *See also* parasitism (p. 110).

symbiosis (*n*) an association (p. 227) between two or more species (p. 40) of organisms to their mutual benefit, such as the association of the mycorrhiza (p. 49) of certain fungi (p. 46) with the roots of trees whereby the tree provides nutrients (p. 92) for the fungus which helps the tree to take up water and supplies nitrates to the roots. **symbiotic** (*adj*).

ectotrophic mycorrhiza

endotrophic mycorrhiza
L.S. root

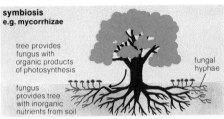

symbiosis
e.g. mycorrhizae

tree provides
fungus with
organic products
of photosynthesis

fungal
hyphae

fungus
provides tree
with inorganic
nutrients from soil

commensalism (*n*) an association (p. 227) in which one species (p. 40) of organism, the commensal, benefits while the other species is neither harmed nor gains benefit. The bacteria (p. 42) in the gut (p. 98) of mammals (p. 80) are commensals.

mutualism (*n*) an association (p. 227) between two or more species (p. 40) in which both benefit. In some cases of mutualism, neither species may be able to survive without the other while, in others, both species may be able to survive independently. It is a form of symbiosis (↑). For example, a species of sea anemone lives on the back of the hermit crab and benefits from being transported to new feeding sites where it feeds on debris from the crab's meals while the crab is protected from predation (p. 220) by the stinging tentacles of the anemone.

epiphyte (*n*) any plant, such as some ferns or lichens (p. 49), which grows on another plant, in a commensal (↑) association (p. 227), using it only for support and not involved in any parasitism (p. 110).

epizoite (*n*) any animal, such as the remora fish which is attached to a shark by a powerful sucker, which lives permanently on another animal, using it for transport etc. and not involved in any parasitism (p. 110).

farming (*n*) the process in which humans exploit naturally occurring plants and animals to provide food for their own needs either by deliberately cultivating wild species (p. 40) or by developing new types of organisms and then sowing, planting, tending and protecting them.

fishery (*n*) the process in which humans catch fish or other aquatic animals for food, and exploit the natural processes of population (p. 214) control to increase the size of the catch.

maximum sustainable yield the maximum size of catch of, for example, fish that can be obtained and sustained over years from a given area of water which is fished in such a way that the stocks are larger than they would be if they were unfished. The adult fish are removed from the water for food so that the young do not have to compete with the adults to the same degree for food and the biomass (p. 226) of the water is increased by their survival.

agriculture (*n*) all of the processes associated with the growing of food in a systematic way, including cultivation of land, tending of stock, development of new types, and destruction of competing (p. 220) species (p. 40), so that the yield from a given area can be increased to cope with the increasing demands of a growing human population (p. 214).

pest (*n*) any species (p. 40) of animal or plant which, in the light of modern methods of agriculture (↑) where vast tracts of land are given over to one species, is not subject to the controls of a natural ecosystem (p. 217) and may increase rapidly in numbers to destroy the crop.

weed (*n*) any species (p. 40) of plant which may be able to grow in an area which has been given over to the cultivation of food plants and which will compete with those food plants for space, light, water and nutrients (p. 92).

biological control a method of reducing the numbers of weeds (↑) or pests (↑) by introducing a natural predator (p. 220) of the pest species (p. 40) If the predator is also able to feed on species which are not regarded as pests, then its numbers will not be reduced when the numbers of pests have fallen. This attempts to maintain a natural equilibrium between the pest and the predator.

biological control
e.g. ladybirds are introduced
to control aphids (pests)

aphid (pest)

ladybird

pesticide (*n*) any agent, usually chemical, which is used to control and destroy pests (p. 229).

herbicide (*n*) any agent, usually chemical, which is used to destroy or control weeds (p. 229).

water purification all of the processes, including storage, straining, filtering and sterilizing which are used by the water authorities to maintain drinking water fit for human consumption. Since drinking water is drawn from rivers, lakes and underground wells, it is also important to ensure that pollutants (↓) from industry or agriculture (p. 229) do not enter the supplies to unacceptable levels.

sewage treatment all of the processes, including the removal of sludge by sedimentation, screening to remove large particles of waste, biological oxidation (p. 32), removal of grit, filtering etc, to ensure that the effluent, which would otherwise contain human waste etc, can be returned to the water cycle (p. 226) without the risk of spreading diseases etc.

conservation (*n*) the use of the natural resources in such a way that they are not despoiled. It is usually taken to include the act of study, management and protection of ecosystems (p. 217), habitats (p. 217) or species (p. 40) of organisms in order to maintain the natural balance of wildlife and its environment (p. 218).

endangered species any species (p. 40) of animal cr plant which, by changes in the natural environment (p. 218) or by human intervention, are threatened with death and extinction.

over-exploitation the use of natural resources in such a way that natural ecosystems (p. 217) may be irreversibly disturbed, habitats (p. 217) destroyed, or organisms threatened with extinction.

pollution (*n*) the act of introducing into the natural environment (p. 218) any substance or agent which may harm that environment and which is added more quickly than the environment is able to render it safe. **pollutant** (*n*), **pollute** (*v*).

water pollution the pollution (↑) of marine and freshwater habitats (p. 217) by the unthinking introduction of human, agricultural (p. 229), and industrial waste into rivers, lakes, and oceans.

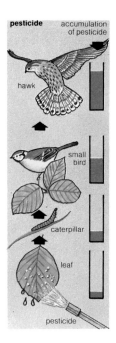

pesticide · accumulation of pesticide

hawk

small bird

caterpillar

leaf

pesticide

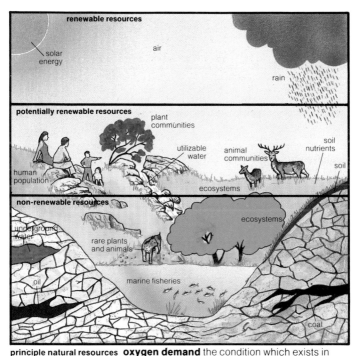

renewable resources

solar energy

air

rain

potentially renewable resources

plant communities

utilizable water

animal communities

soil nutrients

soil

human population

ecosystems

non-renewable resources

underground water

rare plants and animals

ecosystems

oil

marine fisheries

coal

principle natural resources exploited by man

oxygen demand the condition which exists in aquatic environments (p. 218) into which pollutants (↑) have been introduced which promote the growth of aerobic (p. 32) bacteria (p. 42) causing a depletion of the levels of oxygen in the water. Thus, the natural plant life of the environment is reduced and, with it, the animal life that depends upon the plants.

eutrophication (*n*) the situation which occurs when an excess of nutrients (p. 92) is introduced into a freshwater habitat (p. 217) causing a dramatic growth in certain kinds of algae (p. 44). When the nutrients have been used up, the algae die, and the bacterial (p. 42) decomposers (p. 223) which feed on the dead algae use up the oxygen in the water giving rise to an oxygen demand (↑).

algal bloom the dramatically increased
population (p. 214) of algae which occurs in an
aquatic environment (p. 218) which occurs as
a result of eutrophication (p. 231).

air pollution pollution (p. 230) of the atmosphere
which results from burning fossil fuels, such as
coal and oil, with the introduction into the air of
organic (p. 15) and inorganic (p. 15)
compounds, such as carbon dioxide, carbon
monoxide, sulphur dioxide, etc.

smog (*n*) fog which has been polluted (p. 230). A
mixture of smoke and fog.

marine pollution the pollution (p. 230) of the
marine environment (p. 218) primarily by crude
oil as a result of the illegal washing of tanks at
sea or by accidental loss. The damage to
seabird populations (p. 214) is great and well
known but there is also poisoning of marine
plankton (p. 227) which thereby affects the
whole marine food web (p. 226).

radioactive pollution the pollution (p. 230) of the
environment (p. 218) by accidental leakage
from sites of nuclear energy production or from
the dumping of nuclear waste products. The
radioactive materials which find their way into
the environment can lead to chromosome
(p. 13) damage and mutations (p. 206).

terrestrial pollution the pollution (p. 230) of the
land environments (p. 218) by the dumping of
waste materials from mining industries, for
example, or by pesticides (p. 230).

birth control the attempt by humans artificially to
limit the rapid growth which has taken place in
the world human population (p. 214) which may
otherwise place possibly disastrous strains on
food supplies and other non-replaceable
resources. It involves methods of preventing
conception with the use of contraceptives such
as the Pill, vasectomy, etc.

hygiene (*n*) the science which deals with the
preservation of human health by such means
as improvements of sanitation to prevent the
spread of disease. It is thought that hygiene
improvements are among the most important
factors in the increase in human life expectancy.

air pollution smog

pollution of water

disease (*n*) any disorder or illness of the body or organ.

infectious (*adj*) of a disease caused by viruses (p. 43) or other parasitic (p. 92) organisms, such as certain bacteria (p. 42), which can be passed from one individual to another.

contagious (*adj*) of a disease which can be passed from one individual to another by contact, which may be direct touching of the individuals or through objects which have been contaminated by the diseased individual and then handled by another individual.

antiseptic (*adj*) of any agent which destroys the microorganisms which invade the body leading to disease.

aseptic (*adj*) of conditions in which disease-causing microorganisms are not present.

antibiotic (*n*) any substance, produced by a living organism, for example, the fungus (p. 46) *Penicillium*, which is poisonous to other living organisms. Antibiotic substances are used in medicine to destroy disease-causing microorganisms.

antibody (*n*) a protein (p. 21) which is produced by an organism following the invasion of the body fluids by a substance which is not normally present and which may be harmful. The antibody combines with the invading substance thereby removing it from the body.

immunity (*n*) the state in which organisms are protected from the invasion of disease which mainly involves the production of antibodies (↑).

active immunity immunity (↑) in which the body's defensive mechanisms are stimulated by the invasion of foreign microorganisms to produce antibodies (↑).

passive immunity immunity (↑) in which the body's own defensive mechanisms are not stimulated by the invasion of foreign micro-organisms but in which antibodies (↑) have been transferred to it from another animal in which active immunity (↑) has been stimulated.

inherited immunity passive immunity (↑) in which the resistance to certain diseases is inherited genetically (p. 196) from the parents.

the production of antibodies

harmful substances invade animal

defence mechanisms of animal produce antibodies

antibodies combine with harmful substances

combinations of antibody and now harmless substances removed from the body

acquired immunity active immunity (p. 233) by exposure to an infectious (p. 233) disease which is too restricted to cause the symptoms of the disease or passive immunity (p. 233) by the transfer of antibodies (p. 233) from the mother to the offspring across the placenta (p. 192).

vaccination (*n*) the injection into the body of an animal modified forms of the microorganisms which will cause a particular disease so that the body produces antibodies (p. 233) that will resist any possible invasion of the disease itself. The animal gains acquired immunity (↑).

vaccine (*n*) any substance containing antigens (↓) which is injected into an animal's body to produce antibodies (p. 233) and give the animal acquired immunity (↑) to specific diseases.

antigen (*n*) any substance, produced by a microorganism, which will stimulate the production of antibodies (p. 233).

epidemic (*adj*) of a disease which is not normally present in a population (p. 214) and which, therefore, will spread rapidly from individual to individual and infect (p. 233) a large number of the population because there is no natural immunity (p. 233) to the infection.

endemic (*adj*) of a disease which occurs naturally in particular, geographically restricted populations (p. 214).

pandemic (*adj*) of a disease which occurs throughout the population (p. 214) of a whole continent or even the world.

allergy (*n*) the condition in which certain individuals may be particularly sensitive to substances which are quite harmless to other individuals. For example, asthmatic attacks may be stimulated by breathing dust or pollen (p. 181). Allergic reactions may include inflammation or swelling.

symptom (*n*) a sign or condition of the presence of, e.g. a disease.

contract (*v*) (1) *of diseases* to get or to catch. (2) to become smaller or shorter e.g. muscles (p. 143) contract. **contraction** (*n*), **contractile** (*adj*).

acquired immunity

animal with disease-carrying microorganisms

a few microorganisms injected into another animal

antigens produced by the microorganisms cause animal to produce antibodies

animal has gained an acquired immunity to the disease through vaccination

scientific method a means of gaining knowledge of the environment (p. 218) by observation (↓) which leads to the development of a hypothesis (↓). From the hypothesis, predictions (↓) are made which are tested by experiments (↓) that include controls (↓).

observation (*n*) (1) a natural event or phenomenon which is viewed or learned; (2) that which is viewed or learned.

hypothesis (*n*) an idea which has been put forward to explain the occurrence of a natural event or events noted by observation (↑). **hypotheses** (*pl.*).

prediction (*n*) the process of foretelling likely events or phenomena in a given system from those already noted by observation (↑). Predictions follow from the hypothesis.(↑).

experiment (*n*) a means of examining a hypothesis (↑) by testing a prediction (↑) made on the basis of the hypothesis.

control (*n*) an experiment (↑) performed at the same time as the main experiment which differs from it in one factor only. Controls are a means of testing those factors which affect a phenomenon.

theory (*n*) an idea or set of ideas resulting from the scientific method (↑) used as principles (↓) to explain natural phenomena which have been noted by observation (↑).

phenomenon (*n*) any observable fact that can be described scientifically. **phenomena** (*pl.*).

principle (*n*) a general truth or law at the centre of other laws.

adaptation (*n*) a change in structure, function etc, which fits a new use. A particular adaptation may make an organism better fitted to survive (p. 209) in its environment (p. 218). **adapt** (*v*).

structure (*n*) the way in which all the parts of an object, or organism or part of an organism are arranged. The structure of anything is closely related to the function it performs.

function (*n*) the normal action of an object or part of an organism, for example, the function of the ear (p. 157) is to hear (p. 159).

adjacent (*adj*) near by, next to or close to.

amorphous (*adj*) without shape or form, e.g. cells which have not been differentiated.

anterior (*adj*) at, near or towards the front (or
  head) end of an animal, usually the end directed
  forward when the animal is moving (in humans
  the anterior is the ventral (p. 75) part).

articulation (*n*) the movable or non-movable
  connection or joint between two objects.

axis (*n*) a real or imaginary straight line about
  which an object rotates e.g. the axis of
  symmetry (p. 60).

cavity (*n*) a hole or space e.g. the buccal cavity (p. 99).

comatose (*adj*) inactive and in a deep sleep as,
  for example, in a hibernating (p. 132) animal.

comparable (*adj*) of two or more objects, of
  similar quality. **compare** (*v*).

concentration (*n*) the strength or quantity of a
  substance in, for example, a solution (p. 118).

constituent (*n*) a part of the whole. **constituent** (*adj*).

constrict (*v*) to make thinner, for example, a
  narrowing of the blood vessels (p. 127).
  constriction (*n*).

dilate (*v*) to make wider, for example, in blood
  vessels (p. 127). **dilation** (*n*).

convoluted (*adj*) rolled or twisted into a spiral or
  coil, for example, the convoluted tubules in the
  kidney (p. 136). **convolute** (*v*).

co-ordinate (*v*) to cause two or more things, e.g.
  limbs, to work together for the same purpose.
  co-ordination (*n*).

crystallize (*v*) to form crystals (regular shapes).

deficiency (*n*) a shortage or lack of something.
  For example, vitamin deficiency (*see* p. 238).

development (*n*) a stage in growth which includes
  changes in structures and the appearance of
  new organs and tissues (p. 83).

duct (*n*) a tube formed of cells.

equilibrium (*n*) the state in which an object is
  steady or stable because the forces acting
  upon it are equal.

essential (*adj*) very necessary.

external (*adj*) of the outside.

internal (*adj*) of the inside.

extract (*v*) to remove or draw out one substance
  from a particular material.

filter (*n*) an instrument used to take solids and
  other substances out of liquids. **filtration** (*n*).

flex (v) of a joint to bend, of a muscle (p. 143) to contract.

gradient (n) the increase or decrease in a substance over a distance.

increase (v) to become or to make greater in some way, for example, in size, value, concentration etc. **increase** (n).

decrease (v) to become or to make less or fewer in some way, for example, in size, value, concentration etc. **decrease** (n).

insulation (n) any material used to prevent the passage of heat (or electricity), for example, hair insulates the bodies of mammals (p. 80) and feathers (p. 147) the bodies of birds.

intermediate (adj) of an object in the middle, e.g. an intermediate stage in metabolism (p. 26).

lubricate (v) to make smooth or slippery in order to make the movement of parts of a machine or organism easier. **lubrication** (n).

offspring (n) = progeny (p. 200).

parallel (adj) of lines or planes which run in the same direction and never meet.

permeable (adj) of, for example, a membrane (p. 14) which allows a substance to pass through. See also semipermeable membrane (p. 118).

posterior (adj) at, near or towards the back or hind end of an animal, usually the end directed backwards when the animal is moving.

product (n) a substance that is produced.

byproduct (n) a substance that is produced in the course of producing another substance.

protuberance (n) a part or thing which swells or sticks out, for example, a pseudopodium (p. 44) of Amoeba (p. 44).

sedentary (adj) of an animal that remains attached to a surface and does not carry out locomotion (p. 143), for example, a coral polyp (p. 61).

synthesize (v) to make a substance from its parts.

tensile (adj) of a material which is able to be stretched.

transparent (adj) of a material that lets light pass through and through which objects can be clearly seen.

viscous (adj) of a fluid that will not flow i.e. is rather solid.

Vitamins

NAME	LETTER	MAIN SOURCES	FUNCTION	EFFECTS OF DEFICIENCY	FAT (F) OR WATER (W) SOLUBLE
retinol	A	liver, milk, vegetables containing yellow and orange pigments e.g. carrots	light perception, healthy growth, resistance to disease	night blindness, poor growth, infection, drying and degeneration of the cornea	F
calciferol	D	fish liver, eggs, cheese, action of sunlight on the skin	absorption of calcium and phosphorus and their incorporation into bone	bone disorders e.g. rickets	F
tocopherol	E	many plants, such as wheatgerm and green vegetables	cell respiration, conservation of other vitamins	in humans, no proved effect, may cause sterility, muscular dystrophy in rats	F
phylloquinone	K	green vegetables, egg yolk, liver	synthesis of blood clotting agents	haemorrhage, prolonged blood clotting times	F
thiamin	B_1	most meats and vegetables, especially cereals and yeast	coenzyme in energy metabolism	beri-beri, loss of apetite and weakness	W

riboflavin	B_2	milk, eggs, fish, green vegetables	coenzyme in energy metabolism	ulceration of the mouth, eyes and skin	W
niacin	B complex (B_2)	fish, meat, green vegetables, wheatgerm	coenzyme in energy metabolism	pellagra: skin infections, weakness, mental illness	W
pantothenic acid	B_5	most foods, especially yeast, eggs, cereals	coenzyme in energy metabolism	headache, tiredness, poor muscle co-ordination	W
pyridoxine	B_6	most foods, especially meat, cabbage, potatoes	release of energy, formation of amino acids	nausea, diarrhoea, weight loss	W
biotin	B complex (H)	most foods, especially milk, yeast, liver, egg yolk	coenzyme in energy metabolism	dermatitis	W
folic acid	B_c	green vegetables, liver, kidneys	similar to vitamin B_{12}	a form of anaemia	W
cobalamin	B_{12}	meats e.g. liver, heart, herrings, yeast, some green plants	maturing red blood cells, growth, metabolism	a form of anaemia	W
ascorbic acid	C	citrus fruits, green vegetables	collagen formation	scurvy: tooth loss, weakness susceptibility to disease, weight loss	W

Nutrients

carbon dioxide a colourless, odourless gas at normal temperature and pressure with the chemical formula CO_2. It is denser than oxygen and occurs in the atmosphere at lower levels. It is absorbed by plants and is used to make complex organic compounds especially by photosynthesis. It is a waste product of respiration.

oxygen a colourless, odourless gas at normal temperature and pressure with the chemical formula O_2. It is a vital element of the inorganic and organic compounds, such as carbohydrates, proteins and fats, which make up all living organisms. It is taken in by plants as gaseous oxygen in the dark and as carbon dioxide and water and released as a gas from photosynthesis. It is essential for respiration in aerobic organisms.

water a colourless, tasteless liquid, at normal temperatures and pressures, with the chemical formula H_2O. Most nutrients are soluble in water. Water takes part in many of the chemical reactions involved in nutrition and is also an essential fluid in the transport of materials throughout the body of an organism. It is a waste product of respiration and is essential in photosynthesis.

PLANT NUTRIENTS

macronutrients

potassium a macronutrient which is absorbed by plants in the form of potassium salts and which is required as a component of enzymes and amino acids. Potassium deficiency will eventually lead to the plant's death and is indicated by yellow edges to the leaves.

calcium a macronutrient which is absorbed by plants in the form of calcium salts and which is required in cell walls. Calcium deficiency will cause a plant to have stunted roots and shoots because the growing points die.

nitrogen a macronutrient present in the atmosphere as a colourless, odourless gas at normal temperatures and pressures but absorbed by plants in the form of nitrates. It is an essential part of proteins and amino acids etc. Nitrogen deficiency causes the plant to show stunted growth with yellowing of the leaves.

phosphorus a macronutrient which is absorbed by plants as H_2PO_4 and is found in proteins, ATP and nucleic acids. Phosphorus deficiency causes the plant to show stunted growth with dull dark green leaves.

magnesium a macronutrient which is absorbed by plants in the form of magnesium salts and is found in chlorophyll. Magnesium deficiency causes yellowing of the leaves.

sulphur a macronutrient which is absorbed by plants as sulphates and is found in certain proteins. Sulphur deficiency causes roots to develop poorly as well as yellowing of the leaves.

iron a macronutrient which is absorbed by plants as iron salts and is found in cytochromes. Iron deficiency causes yellowing of the leaves.

micronutrients

boron a micronutrient absorbed by plants in the form of borates. It is important after pollination in the stimulation of germination of the pollen grains as well as in the absorption of calcium through the roots. Boron deficiency results in certain diseases of plants, such as internal cork in apples.

zinc a micronutrient which is absorbed by plants in the form of zinc salts. It is important in the activation of certain enzymes and in the production of leaves. Zinc deficiency results in the abnormal growth of leaves.

copper a micronutrient which is absorbed by plants in the form of copper salts. It is required by some enzymes. Copper deficiency results in the growtn of the plant showing certain kinds of abnormality.

molybdenum a micronutrient absorbed by plants in the form of molybdenum salts. It is important in the function of certain enzymes for the reduction of nitrogen. Molybdenum deficiency results in the overall growth of the plants being reduced.

chlorine a micronutrient which is absorbed by plants in the form of chlorides. It is important in osmosis etc. although it cannot easily be shown to have effects if there is a deficiency.

manganese a micronutrient which is absorbed by plants in the form of manganese salts. It is an important activator of certain enzymes. Manganese deficiency results in the yellowing of the leaves as well as grey mottling.

ANIMAL NUTRIENTS

minerals

calcium a mineral, present in milk products, fish, hard water and in bread, which is required for healthy bones and teeth to aid in the clotting of blood and in muscles. The average adult human requires 1.1 grams per day and the total body content is about 1000 grams.

phosphorus a mineral, present in most foods but especially cheese and yeast extract, which is required for healthy bones and teeth, and takes part in the DNA, RNA and ATP metabolism. The average human adult requires 1.4 grams per day and the total body content is about 780 grams.

sulphur a mineral, present in foods containing proteins, such as peas, beans and milk products. It is required as a constituent of certain proteins, such as keratin and vitamins, such as thiamine. The average human adult requires 0.85 grams per day and the total body content is about 140 grams.

potassium a mineral present in a variety of foods, such as potatoes, mushrooms, meats and cauliflower, which is required for nerve transmission acid-base balance. The human requires 3.3 grams per day and the total body content is about 140 grams.

sodium a mineral present in a variety of 'salty' foods but especially table salt (sodium chloride), cheese and bacon, and which is required for nerve transmission and acid-base balance. The average human requires about 4.4 grams per day and the total body content is about 100 grams.

chlorine as chloride ions, a mineral found with sodium in table salt and in meats, which is required for acid-base balance and for osmoregulation. The average human adult requires 5.2 grams per day and the total body content is about 95 grams.

magnesium a mineral present in most foods, but especially cheese and green vegetables, which is required to activate enzymes in metabolism. The average human adult requires about 0.34 grams per day and the total body content is about 19 grams.

iron a mineral present in liver, eggs, beef and some drinking water, which is an essential constituent of haemoglobin and catalase. The average human adult requires 16 milligrams per day and the total body content is about 4.2 grams.

fluorine as fluoride, a mineral found in sea water and sea foods and sometimes added to drinking water. It is a constituent of bones and teeth and prevents tooth decay. The average human requires 1.8 milligrams per day and the total body content is about 2.6 grams.

zinc a mineral found in most foods, but especially meat and beans, which is required as a constituent of many enzymes. It is also thought to promote healing. The average human adult requires 13 milligrams per day and the total body content is about 2.3 grams.

copper a mineral found in most foods, but especially in liver, peas and beans, which is required for the formation of haemoglobin and certain enzymes. The average human adult requires 3.5 milligrams per day and the total body content is about 0.07 grams.

iodine a mineral found in sea foods and some drinking water and vegetables, which is required as a constituent of thyroxine. The average adult human requires 0.2 milligrams per day and the total body content is only about 0.01 grams.

manganese a mineral found in most foods, but especially tea and cereals, which is required in bones and to activate certain enzymes in amino acid metabolism. The average adult human requires 3.7 milligrams per day and the total body content is only 0.01 grams.

chromium a mineral found in meat and cereals.

cobalt a mineral found in most foods, but especially meat and yeast products, which is an essential constituent of vitamin B_{12}. The average adult human requires 0.3 milligrams per day and the total body content is as little as 0.001 grams.

International System of Units (SI)

PREFIXES

PREFIX	FACTOR	SIGN	PREFIX	FACTOR	SIGN
milli-	$\times 10^{-3}$	m	kilo-	$\times 10^{3}$	k
micro-	$\times 10^{-6}$	μ	mega-	$\times 10^{6}$	M
nano-	$\times 10^{-9}$	n	giga-	$\times 10^{9}$	G
pico-	$\times 10^{-12}$	p	tera-	$\times 10^{12}$	T

BASIC UNITS

UNIT	SYMBOL	MEASUREMENT
metre	m	length
kilogram	kg	mass
second	s	time
ampere	A	electric current
kelvin	K	temperature
mole	mol	amount of substance

DERIVED UNITS

UNIT	SYMBOL	MEASUREMENT
newton	N	force
joule	J	energy, work
hertz	Hz	frequency
pascal	Pa	pressure
coulomb	C	quantity of electric charge
volt	V	electrical potential
ohm	Ω	electrical resistance

Index

THE
COFFEE
AND
WALNUT
TREE

Mel Hanson

Matador
9 Priory Business Park,
Wistow Road, Kibworth Beauchamp,
Leicestershire. LE8 0RX
Tel: 0116 279 2299
Email: books@troubador.co.uk
Web: www.troubador.co.uk/matador
Twitter: @matadorbooks

ISBN 978 1800462 922

British Library Cataloguing in Publication Data.
A catalogue record for this book is available from the British Library.

Printed and bound in Great Britain by 4edge Limited
Typeset in 11.5pt Adobe Jenson Pro by Troubador Publishing Ltd, Leicester, UK

Matador is an imprint of Troubador Publishing Ltd

To Nigel and Kit
and
Jane, for her endless encouragement

1

APRIL

The dawn defied overnight rainstorms, clocking on with just a scattering of picture-book-perfect clouds suspended across blue spring skies. A stream of steadfast workers hurried past her, heading towards offices and shops, but summoning the enthusiasm to match their steely resolve was an exercise beyond April's imagination. Instead, she paused to unbutton her raincoat and grasping the hemline of her blouse she flapped the fabric, allowing wafts of air to cool her fluster. With her composure restored, she trudged onwards, but the sunshine did not cheer her, nor did the prospect of reaching her destination.

Number 31 Westgate was a handsome Georgian residence with grand proportions showcasing perfect symmetry and an impressive stone portico framing the oversized, panelled front door. April owed her love of classical architecture to her historian father, whose collection of beautifully illustrated books on the subject had cluttered her childhood home, and as she approached the end of the street, the sun-drenched façade of this splendid building came into view. Instinctively her mind began to compose a set of glossy particulars to satisfy the mothballed estate agent who lurked within her.

Inside this elegant period home, a wealth of character awaits the family of a discerning buyer. An expanse of monochrome, tessellated hallway tiles offers practicality and the perfect indoor scooter track on a rainy day. The opulent interiors feature high ceilings with spectacular plasterwork, fireplaces with inlaid marble and cast-iron grates where glowing embers would be perfect for toasting crumpets at teatime…

She smiled to herself, chuffed to discover her knack for concocting lavish prose endured, but April knew that modern families searching for a home would not be swayed by an embellished description of a stately pile. It seemed that developers now determined the destiny of many historic properties with their fate being sealed by the dubious process known as repurposing.

Dismissing this calamity from her mind and prompted to press on by the rapidly advancing hands on the face of her watch, April headed off across the street towards the wrought iron railings fronting Number 31. But on reaching the palisade she stalled once more and placing the flat of her hand on her stomach attempted to soothe the queasiness by gently massaging her belly. Her biased reasoning attributed the nausea to a missed breakfast at the start of a calorie-counting day, but her current apprehension was more likely the cause.

'Oh, for goodness' sake, stop being ridiculous… and GET A GRIP!' April's sudden outburst startled an elderly lady passing by, causing her to drop her walking stick with a resounding clatter directly into one of the few remaining puddles on the walkway.

'I am so sorry, please do forgive me.' April retrieved the walking aid for its owner and gave the wet length a cursory dry on the front of her skirt, but she barely glanced at the grey-haired pensioner before returning her focus to the first-floor windows of Number 31. She pictured the reception desk located within and rued two fateful days: the first, which saw this noble dwelling repurposed to provide a workplace for dental practitioners; and the second, which marked her appointment as their newest employee.

*

Undaunted by the laborious process of completing job application forms, April had perfected the art of cutting and pasting from previous documents. The white boxes were swiftly filled with details of her skills and experience, and with each new application came the conviction that this was the one, the role that would be perfect for her, more suitable than all those she'd attempted before.

The vacancy for a part-time receptionist had the lure of employment within the healthcare sector, which she had spuriously concluded would address her need for a more worthwhile occupation. She could leave behind the long hours and stresses at Richmond & Marsh, the upmarket estate agency where three years earlier she'd embarked on what she had then believed was the ultimate career – a thrilling prospect, providing her with an open invitation to check out fine period homes, grand estates and pretty character cottages, all of which could be found in abundance in her home county of Suffolk.

Naively, April had anticipated the pleasures of guiding affluent purchasers as they searched for perfect and forever homes. Sadly, she discovered these fantasy acquisitions are a rarer find than calorific chai seeds – acrimonious divorces, next door's extension plans or a neighbouring boy band being just a few of the pitfalls to rapidly dispel a homeowner's dreams. April had encountered ambitious *must have* lists, unrealistic budgets and an inability to accept any compromise, along with an inexplicable mistrust of all estate agents. A crystal gazer had not been required to foretell the outcome of April's foray into the world of real estate and this profession was inevitably doomed. It was, in fact, just one of the many she had endeavoured to conquer.

*

Removing the strap from her shoulder, April began to rummage in the depths of her cherished Mulberry handbag – a reward purchase for surviving a chocolate-free week – eventually locating her mobile phone beneath a packet of lozenges and a handful of tissues. Having detached a sticky sweet from its screen, she selected Ian's number, hoping he could answer despite his comment at breakfast that he expected to be 'extremely busy the entire day'.

'Hi, can you talk?' she whispered, although uncertain why hushed tones might be necessary.

'Go on.'

Not missing the coolness of his tone, April continued, as directed, 'Oh God, Ian, I feel quite sick. Please say something encouraging. I can't go in, I don't think I can face another day of this.'

'Really, April, you have to give this job a chance. What happened to your three-week pledge?'

'I know, I know… but I'm eight days in and nothing feels right about the combination of me in this job, with these people.'

'Just try and stick with it. You've had more than your share of new jobs, you know what it's like in the early stages when everything is unfamiliar. In three weeks' time you'll be settled in, you always are. It's staying power that you lack.' April was certain she heard him groan. 'I'm sorry but I really should get on. Give those dentists a chance. I'll see you this evening.'

April sighed as she discarded her phone amongst the detritus within her handbag, making a mental note that a clear-out was overdue. Fashion aficionados would no doubt consider it a crime against luxury accessories to abuse such a classy bag – a biohazard had been Ian's description of its contents the last time he had braved his hand inside.

'Right, enough of the dilly-dallying,' April told herself sternly and she took a deep breath before pushing open Number 31's hefty front door. Racing up the steep back staircase, she wondered with each step whether Health & Safety had been consulted regarding the welfare of elderly patients at the time of the building's conversion. On reaching the summit, she averted her eyes to avoid glimpsing the repellent floor covering embellished with brown and yellow swirls which cavorted across the surface marking her route to a partitioned cubicle. In a former life, the lofty galleried landing would have proudly displayed a collection of old masters and coats of arms, but this makeshift office now blighted the ambience.

With her fitness regime at an all-time low, April swayed unsteadily, feeling faint following her rapid ascent of the vertiginous staircase. She gripped the nearby banister to anchor herself and after a moment stepped towards the opened sash window. With the flush on her face slowly cooling, she gulped down a further breath, hoping to neutralise the sharp disinfectant odour of dental compounds which pervaded the atmosphere.

'Hi there, April, you're here. Did you oversleep? What an almighty storm last night. Kept me awake. Who'd have thought it would turn out to be such a glorious morning?' Jessie's greeting rallied April to her post. As she joined her co-worker in the reception office, April saw that Jessie was in a frenzy of filing. This task involved the repeated opening and slamming shut of the drawers of a battered filing cabinet, causing April to flinch with each nerve-shuddering squeak of resistance from its jamming runners.

'Sorry... yeah... absolutely... How are you?' April stuttered out her response.

'I'm fine and delighted that it's my half-day. Just as soon as I've got rid of these reports, I'll go downstairs, put the kettle on and unload the autoclave,' said Jessie. 'The computer is fired up and I'll be back shortly to check you're okay. I'll bring coffee. I'll make yours a strong one, shall I?'

Rotund and engagingly verbose, in April's eyes Jessie was her ally in an otherwise hostile environment. On her first few days in the job, it had been Jessie who had patiently and good-humouredly taken her through her training, helping her grasp the workings of the Smiles software system, which April could only conclude had been named with a large dose of irony.

'Appointments for check-ups are coloured red and allocated twenty-minute slots, whilst the blue represents more extensive treatment, say root canal, or fitting a bridge, and for this we allow forty minutes,' Jessie had explained.

'Okay, but what about all those icons at the bottom of the screen, like that tiny iPhone?'

'Oh yes, that's a really useful one. All you need to do is click on the image and you will be able to send a text directly to the patient, so you could use that to get in touch if, for example, an appointment had to be rearranged.'

April felt there was nothing to smile about whilst striving to understand the myriad of icons and rainbow of codes which decorated the diary and payments processes of this data dashboard. The previous Thursday, whilst attempting to rearrange Mrs Simpson's appointment for a filling, she had been baffled by the sudden appearance of a comic *hand up* symbol which flashed in the bottom corner of the computer screen. In her bemused state, she wanted to imagine this cypher might denote a friendly wave from the dentist in his inner sanctum to the receptionist at her station, but a quick check of her guidance notes – neatly typed by Jessie and placed for her in a pristine clear plastic folder – dispelled all joy at this notion.

The hand icon is a warning – a halt symbol; on no account must this patient be permitted to leave the premises without settling their bill.

April had read this note with unease, wondering whether physical force, even barring the exit, was expected of her in order to fulfil this command.

'I'm back and here's your coffee. Seeing the look on your face earlier I was tempted to add a large teaspoon of sugar to give you an energy boost, but then I remembered you're trying that sugar-free diet.'

'Ha ha… but thank you for the kind thought.'

'Now, let's have a quick run through today's diary. I've already checked in the early arrivals and as soon as you're up to speed I'll leave you to it. I'll only be downstairs typing up clinic letters if you need help with anything, so don't stress, just call me.'

It was a dual-clinic day and Peter Sandford Snr was on duty, along with Maja, the hygienist. Petite and pretty, with short blond hair always pulled back into the tiniest of ponytails, Maja loved nothing more than a good old chitchat with patients. Whilst April considered this commendably friendly and ideal for putting folks at their ease, it inevitably led to clinics running late.

She surveyed the waiting room through the reception window – a rudimentary opening in the partition wall, which she pictured Peter Sandford Jnr fashioning with his handy hacksaw. His slapdash practices gave no indication of dexterity and April had decided never to entrust her teeth to his ministrations. Despite Maja's early start, a backlog of patients was now waiting and, seeking distraction, two ladies were rummaging competitively through the uninspiring pile of aged magazines. It had been whilst tidying up that April had discovered a three-year-old copy of *Cosmopolitan* with dog-eared pages, but on noticing the cover story entitled 'Tips for Pepping up Your Love Life', she'd decided it would be mean-spirited to bin it and deprive the patients of something racy to read.

April was selfishly keen to detain Jessie, hoping to rally her mood with some female banter. The office repartee and gossip, which had been the enjoyable part of her days spent at the estate agency, were proving hard to replicate in this clinical venue.

'Hey, Jess, did you catch last night's episode of *Poldark?*'

'You bet. Never miss it. I'd willingly swap places with Demelza any day of the week.' Jessie smiled as she disappeared down the staircase and with the opportunity lost April had no choice but to get on with her day, minus her soulmate.

A heady waft of sickly scent alerted April to the arrival of Brenda, the practice manager, who was delighted to inform anyone prepared to listen that she had worked for Mr Sandford Snr for thirty-three years. In April's view such inertia revealed a lack of vision and enterprise.

'Please don't hang your coat on the back of the door, April, it looks so untidy. Hide it away under the desk, or better still leave it in the cupboard downstairs.'

Wasn't a door hook specifically asking to have outdoor wear placed upon it?

April was tempted to vocalise this observation, but the sound of the ringing telephone fortuitously interrupted her brain-to-mouth synchronisation. If she'd learnt anything in her first days at this establishment it was never to question Brenda's authority.

'No more than three rings, April. Answer that call quickly, please.'

'Good morning, the Misters Sandford Dental Prac...'

'Dental *Surgery!*' Brenda butted in.

'Err, sorry... Dental Surgery, April speaking, how may I help you?'

'Muft shee dentisht today, the pain ish offal.'

Listening carefully, April did her best to interpret the muffled voice which, once translated, told the woes of Mr Thompson, who had just returned from his holiday in Marseille where he had broken a tooth whilst eating a baguette. Excruciating toothache had left the poor man barely able to eat or sleep – or speak, apparently – since the trauma. After a quick check with her leader, April allocated the remaining emergency appointment slot to Mr Thompson and disconnected the call.

'Really, you must try to remember that you work at a dental surgery, not a practice. We're beyond practising here, the rehearsal is over!' Brenda appeared pleased with her play on words.

April was unimpressed as well as puzzled by the blatant inconsistency. Brenda very clearly displayed a badge identifying herself as the practice manager, didn't she?

With this final rebuke, Brenda deposited a parcel onto the desk and departed, leaving April with a musky taste lingering in her mouth. Dental compounds or old lady scent – there really wasn't much to choose between them. Relieved to see the back of officialdom, April prodded the sealed package apprehensively, suspecting that once opened she'd find it contained sets of dentures. These replacement teeth, no doubt eagerly awaited by the intended recipients, were repulsive to her and she went in search of a pair of protective latex gloves. It was just as she was wrestling one of her hands into a tiny blue glove that April was startled by the sound of Peter Sandford Snr bellowing, shortly followed by Sarah, the trainee dental nurse, rushing out of his room in tears.

April leapt to her feet to go to pacify the girl but was diverted by a commotion in the waiting room.

'Charlie's been sick, all over his trousers and the carpet,' a distraught mother sobbed as April felt her own nausea returning.

Dear God, it really was too much – the false teeth, the gruesome sound of drilling, the chemical fragrances, harassed nurses, bumptious dentists and an uncompromising practice manager. Why hadn't anyone warned her of the perils she'd face in the role of a dental receptionist?

2

ELLEN

'Don't do it, Ellen, it will be mind-numbing – and wreck your career prospects.'

'Absolutely, get yourself a child-minder. Steer clear of the kitchen sink and the jogger bottoms.'

Ellen had ignored the discouraging working mums, choosing instead to wholeheartedly embrace the role of full-time parenting. The rewards were obvious: lots of time with Liam, nurturing him and witnessing his development from baby to toddler. The forfeits were having to cope with boredom and those awful moments of loneliness. Ellen suspected that the status she'd enjoyed amongst her peers whilst working as a marketing manager were now lost to her and four years on she feared she might never be taken seriously enough to warrant re-entry into the workforce. But she had resolved to box up this dilemma for future contemplation; now was the time to bask in the indisputable benefits of her current lifestyle.

Stirring from slumber, Ellen was temporarily blinded by shafts of early morning sunlight streaming through the slats of the shutter at her bedroom window. She closed her eyes and kicked back the duvet to enjoy the sensation of the sun's warmth on her naked body. Relishing this moment of pleasure and the hiatus before her day must get started, she indulged herself in a speed rewind across her thirty-two years.

The life-affirming marker had been the birth of her son, but with the euphoria came all manner of fears, not least that she'd suffer death by sleep deprivation. As the months passed and her baby learnt the necessity for night-times to entail at least six hours of unbroken sleep, Ellen began to relax and enjoy her life as a mother and homemaker. But these mellow days were interrupted – albeit temporarily.

'Do you want the good news first, or the bad news?' her husband had asked on his return from work one evening.

'Well, you look quite pleased, so the bad news can't be too awful,' Ellen had responded hopefully.

'Okay, so… I'm being promoted to marketing director.'

'That's fantastic. Well done, Paul… but what's the bad news?'

'The new position is at our head office in Ipswich… so we'll have to move.'

'Ah, I see… but my friends are here and my family are nearby. That's going to be quite a wrench for me.'

'You'll make new friends and on a positive note, we'll be able to buy a bigger house in the country, with space for visitors.' Paul's enthusiasm was palpable and Ellen resigned herself to accepting the impending changes.

Initially apprehensive on visiting what was to be their new home county and after a number of dismal viewings at uninspiring houses, Ellen had been thrilled when their property search took them to an appealing, if shabby, pink-painted house on the edge of a pretty Suffolk village. Walnut Tree Farmhouse was within easy reach of the market town of Great Aldebridge, the larger town of Ipswich with its rail links to the capital and, joy of joys, the coastline. With neighbouring open countryside and a three-acre plot, Ellen felt sure that if they could own this charming house there might be a silver lining to the relocation cloud. She had been utterly enchanted to discover a namesake tree growing in the orchard garden and excitedly anticipated a bounty of walnuts appearing on its ancient branches.

'At last, I think this move might actually be okay,' Ellen had squealed in delight as she and Paul wandered from room to room. 'I can see our little family being very comfortable in this house, can't you, Paul?'

'Err… it's going to take a bit of work, but if you think you can bear the upheaval of having the builders in, I agree this could become a great investment property.' Ellen was unaware that Paul's only concerns were to keep his wife happy, enabling him to get on with his career and to avoid overstretching the budget.

Their lack of local knowledge might have made the house search an arduous one, but they had been fortunate to encounter the exceptional service of a conscientious estate agent. It had been April who had introduced Ellen and Paul to the idyllic village of Ash Green and the close-to-perfect family home, fittingly described in the shiny brochure as '*a traditional lime-washed Suffolk house, positively oozing pink charm and character, set in beautifully landscaped grounds and with spectacular, far-reaching views across open farmland*'.

Casting her mind back to those first impressions of the sales lady from Richmond & Marsh brought a smile to Ellen's face. Meeting April had not only resulted in their successful purchase of Walnut Tree Farmhouse, it had also initiated a ready-made friendship with a like-minded local. During the settling-in period when she needed advice on finding a reputable builder and a decent hairdresser, Ellen had relied on April's wisdom. April had qualities that Ellen wanted to emulate: her easy way with everyone she met, her sense of humour and her readiness to laugh at her own mistakes (the making of which she seemed rather prone to). Ellen knew herself to be timid and overly stressed by unfamiliar situations. She'd had a privileged upbringing, an expensive education at a private girls' school and an accomplished career, but her lack of self-confidence extended to her belief that it had been her father's phone call to a business associate that had secured her place on the first rung of a professional ladder.

Paul had been a work colleague. He was seven years her senior, fair-haired, ruggedly handsome and an absolute gentleman whilst they were dating. Ellen had felt safe and protected and didn't doubt for a moment that he was the man she was destined to be with. Their wedding took place a year after they met. Eight years later she could no longer squeeze her body into the size 8 wedding dress still hanging in her wardrobe, but now she had a beautiful son and a wonderful life in the country. The perfect setting to bring up a small boy, with wide open spaces, plenty of fresh air and trips to the seaside. Her adjustment to rural living had presented some challenges, like honing her spatial awareness for meetings with tractors whilst negotiating the narrow lanes. This test had seen the appearance of several scratches to the nearside wing of her chunky SUV, but with her recalibration completed, Ellen now considered herself firmly established as a veritable countrywoman.

*

Concluding her reminiscence, Ellen dared to open her eyes to the bright rays of light and having checked the clock on the bedside table she slipped her legs over the edge of the deep mattress, encased in crisp, white Egyptian cotton. She surveyed her domain with complacent satisfaction, the painful memories of the disruptive building work having thankfully dulled. Her redesigned pink farmhouse retained its unique character, but its traditional structure was now complemented by chic, contemporary interiors. As she stepped into the adjoining en-suite bathroom, she revelled in the luxury of the huge glass shower cubicle, where the downpour from the waterfall head created a satisfying tingle across her entire body. She didn't like to consider herself excessively concerned with material possessions, but felt certain others would covet her perfect home and the ideal husband she had bagged to provide this enviable existence.

'Launch the missile... Bang...! Look out!' Ellen heard her son's voice from his bedroom across the landing and guessed that he and Action Man had started the morning on the battlefield. It wouldn't be long before Liam got tired of playing soldiers and wanted his breakfast. She replaced her make-up bag into the drawer of the vanity unit beneath the sink, regarded her reflection in the illuminated mirror and tucking a stray strand of hair behind her ear declared herself ready for the day.

As she made her way to Liam's bedroom, Ellen's mind was focused on leaving home promptly for the journey to the nursery school and the short onward drive into Great Aldebridge. Apart from picking up some provisions and browsing the lovely shops to be found in the historic market town, she was due to meet up with April and she was looking forward to seeing her friend.

'Come on, Liam, time to get a move-on,' she called.

'But, but... I have a tummy ache, Mummy,' he whined.

'Oh dear, sweetie, that's not good, but perhaps you're just a bit hungry. You really should have eaten all your tea last night. I'm sure you'll be fine once you've had some cereal,' she said patiently. 'How about the chocolatey one as a treat?' Ellen didn't want any aggravation and was prepared to relent on a less-than-healthy breakfast option – just for today.

Liam was never too keen to be left at the pre-school group and at times was actively resistant, crying and clinging onto her legs as the staff tried to prise him away.

'Don't worry, Ellen, he always settles once you're out of sight,' Pat reassured her. 'He just likes to protest a little, it's quite normal.' Ellen knew that the group leader was totally unfazed by Liam's heartbreaking performance, although she wondered if being an only child was a disadvantage, causing him to struggle to socialise with other children. But surely having the opportunity to spend time with playmates was a bonus for a child who usually had to play alone?

Ellen really didn't have the answer to this conundrum, if indeed there was one, and there seemed little point in deliberating further as, despite her assumptions that a second pregnancy would naturally follow the first, there had been virtually no discussion on the matter. Paul had been reluctant to share his thoughts with her on this apparently emotive subject and was working such long hours it would have been tricky to find an appropriate moment for a dialogue, let alone an act of intimacy.

'Mummy!'

'Yes, Liam.'

'Mummy… my head hurts.' Liam's whimper interrupted her thoughts.

She placed her hand onto his forehead and felt the abnormal warmth of his skin. Assuming this was a mild childhood fever – easily dealt with by a dose of liquid paracetamol – she went to the medicine cabinet to find the magical cure-all.

'Well, that puts paid to my five hours of freedom.' Ellen vocalised her disappointment. The nursery staff would not thank her for leaving a sickly child in their care. Instead, this would have to be a quiet day at home with her son, one destined to follow the *poorly day* routine of a makeshift bed on the sofa and Liam's current favourite DVD.

Ellen picked up her mobile phone and began tapping a message into the screen.

Liam is poorly, sorry can't meet u today. E x

Oh dear. Shall I call in on my way to town? A x

That would be fab. Thx. C u shortly. E x

Feeling less disheartened, Ellen set about tidying up in anticipation of her visitor's arrival and just as she had finished loading the dishwasher, the sound of a car on the gravel driveway heralded April's arrival. She quickly checked on Liam, now cuddled up under his duvet on the couch engrossed in a *Minions* adventure, then returned to the kitchen to prepare her espresso machine and locate the milk-frother. A loud knocking took her to the back door where Ellen was greeted by a large bunch of pink peonies in April's extended hand.

'They are lovely. Thank you so much.'

'I thought you deserved a treat so I picked these up for you at the village shop on the way over. That shop stocks all sorts of everything. Have you tried their Eccles cakes? And how is the patient?' April said, taking a peep into the sitting room at the dozing child.

'He seems okay... a slightly raised temperature. I'm feeling a little guilty for suspecting that he just didn't fancy a day at nursery. It is Friday, so it's bound to be fish for lunch. He really hates fish.'

'Goodness that is very discerning for a child of such tender years,' said April.

'Coffee? I'm afraid it won't quite be up to the exceptional standards we enjoy at Amarellos' coffee shop, but hopefully a piece of my Bakewell slice alongside will help you over your disappointment.'

'Oh, go on then, I was going to be really good today... but the diet can wait until tomorrow. I can't be expected to resist your delicious Bakewell,' April responded, and Ellen noticed her surreptitiously brushing flecks of pastry from the front of her jumper.

Ellen had always enjoyed cooking but her previous life had allowed her little time to indulge this passion. During recent years she had discovered a culinary talent, especially for cake-making. Whenever she was at home and feeling restless, her favourite distraction was to search through recipes and try out something new. Not every one of these experiments was a total success or worth the time and trouble taken. One evening she was triumphant as she placed a dinner plate in front of her work-weary husband and announced the elegantly presented supper dish.

'A smoked trout and dill tian, with warm potato galette and accompanied by a salad of balsamic tomatoes.'

She probably shouldn't have been surprised when Paul had looked at her quizzically.

'A dill what?'

However, on most occasions her efforts provided tasty suppers – without the need for an internet explanation of the ingredients – and guaranteed that the kitchen cupboards contained cake tins full of her mouth-watering homemade bakes.

'Grab yourself a chair.'

Settled at the kitchen table, Ellen and April both murmured with pleasure as they tasted the moist, almondy sweetness of the tarts. It had been a few weeks since they had seen each other and Ellen was eager to find out how things were working out for April with the infamous father and son dentists.

'Well, I'm afraid I've failed to complete my statutory three weeks.'

'You've left?' Ellen was surprised at the suddenness of April's departure, despite having been certain that it was just a matter of time.

'Well, yes… I sort of left and was sort of encouraged to go,' said April.

'How come?'

April explained that the practice manager had asked at the end of her first week how things were going and replying truthfully she'd admitted she had concerns. Brenda had seemed astonished but asked to be kept informed of how things progressed for her new employee. A few days later Brenda broached the subject again, just as April was recovering from what she considered a moment of her classic ineptitude.

'An elderly patient was attempting to make a card payment for his root canal treatment and it was only after the third failed attempt that I noticed he had his library card, not his credit card, very firmly inserted into the payment machine!' April explained whilst playing out the pantomime of the card being reinserted and pushed more firmly into the slot, then wiggled a bit, then finally, just to make sure, an attempt with the card turned around. 'I'd misplaced my glasses so was none the wiser.'

'Could happen to anyone,' said Ellen.

'Anyhow, I felt I just had to tell Brenda that I wasn't at all sure that the job was right for me and then once I'd got that off my chest I didn't seem able to stop and went on to say that I'd rather like to leave.'

'Oh my goodness, what was her response?' Ellen thought April's honesty had been rather reckless.

'Despite my offer to stay on until a replacement could be found, Brenda quite unexpectedly became rather indignant and said I should gather any personal belongings together and leave at the end of my shift.'

'No way! Just like that, without even offering the compulsory leaving gift to add to your collection!' Ellen couldn't resist this remark.

It seemed that for women employees of a certain age, but repeatedly in April's case, the most popular gift when leaving a job was an exotic blooming plant. You couldn't really go wrong with a potted orchid – could you?

'No, it seems that this time I wasn't there long enough for my exit to be marked by the presentation of a pot plant,' April explained, adding that she felt the whole episode had been pretty tragic and the sudden dismissal a bit of a shocker. She'd had to conclude that Brenda felt the need to take back control, and perhaps she thought April's dislike of the role was a bad reflection on the dental practice and indeed on Brenda herself. Which, of course, April believed it was. It had nothing at all to do with making the wrong choice or failing to persist.

'I'm sure you feel disappointed about how things turned out, but not having to work there anymore must be quite a relief. Don't give those bastards another thought and definitely don't beat yourself up about it,' Ellen said, trying to be reassuring despite the swift demise of another of April's occupations.

April smiled. 'Bless you, you're right. Now I'm a free agent I'm going to take my time and really think about what it is I want to do with my life.'

'You've got loads to offer an employer. Plenty of experience – life experience – as well as the experience gained from your many jobs. Any company would be lucky to have you on board.'

'Thank you, but just now I'd rather not be reminded of my history of bad career choices. I've got a house full of orchids doing just that and as they continue to thrive on neglect their numbers rarely diminish.'

'Muuummmmyyyy,' Liam's wail from the sitting room put an abrupt end to their conversation, 'I feel *really* itchy!'

3

LAURA

When she'd set off from home some forty minutes earlier, Laura's mission had been to get out for an invigorating hike along the seashore with her dog. However, whilst she walked with mindful purpose, the unleashed Retriever romped wildly back and forth over the wet sands, his raucous barking resonating across the wide expanses as he splashed through the incoming waves in pursuit of seagulls. Laura stopped and scanned the area on the lookout for any other visitors to the beach who might be witnessing Max's rowdy behaviour. Inciting the disapproval of a nearby ornithologist was not ideal as her dog refused to respond to her calls and appeared determined to grasp an irate seabird between his jaws. An unruly dog could leave its owner vulnerable to awkward moments and embarrassment, but Laura secretly applauded Max's unrestrained impulsiveness and his zeal for life; she also wasn't particularly averse to him succeeding in the quest to nip the feathery backside of one of those maddening creatures.

Her extreme dislike of gulls dated to an incident involving a granola bar and an enormous herring gull. The hideous beast had imperceptibly swooped down and kicked the snack away from her mouth, up into the air, to claim for himself. Despite the speed of the altercation, Laura had definitely detected its rubbery foot touching her lips and despite rinsing her mouth numerous times, she found the unpleasant sensation difficult to purge. Her carelessness irritated her almost as much as losing the cereal bar; she often walked on the beach and was well-acquainted with these audacious thieves and their ability to move like stealth bombers. They only had themselves to blame as she placed seagulls – along with hyenas and rats – on her *ought to be culled* list. Although this was a sentiment she thought best kept to herself.

The adorable, if wilful, Retriever had been Laura's companion for three years and his loyalty was unswerving, unlike his deportment. Max's presence in her life had prevented any feelings of loneliness she might have experienced having taken early retirement and moved to the country – a departure from her career in the city with its associated entertaining and socialising. These days she partied less and appreciated her leisure time more.

A relative newcomer to Ash Green, Laura had worked hard to ensure her place as a fully immersed member of the community. Joining the gardening society had satisfied a yearning to learn to cultivate, and its members had welcomed her enthusiastically, willingly sharing their knowledge and their precious cuttings with her. Following her successful co-option as a local councillor, Laura was also a member of the parish council and enjoyed the involvement in village matters in a more official capacity. Her previous experience working in finance was greatly appreciated by her fellow councillors, especially when it came to balancing budgets and seeking grant funding for special projects. So far retirement seemed to be going very nicely; she had new friends and plenty to occupy her in this peaceful rural setting. She wanted for nothing more.

But was that really the case? Recently Laura had started to question whether this new lifestyle offered her everything she'd hoped for, or was she just doing an excellent job of convincing herself?

With a council meeting to attend that evening, Laura had been keeping check on the time to make sure they didn't overstay the gallivanting on the sands. But how was she going to convince the dog that it was time to depart? Catching sight of another walker a short distance along the dunes, Laura felt self-conscious as someone was about to observe her inability to control her pet.

'Max... come here, Max... time to go, MAX...' Laura was unsurprised to find him deaf to her calls.

As the rambler approached Laura was relieved to see it was April – a friend rather than a judgemental stranger.

'Hi, there. Max is still doing his own thing I see.' April was laughing. 'I'm pretty certain he is not quite ready to go home.'

'That's true, Max is never ready for home time when we come to the beach. Like all Retrievers his default settings are ravenous, overly excited and with selective hearing,' Laura said. 'My only hope is that when he realises I'm making my way back to the car he'll come looking for me.' Laura turned her back on the rebellious canine and the two women started to walk to the parking area behind the dunes.

'It's so lovely to see you, Laura. I don't suppose you've time for a quick cup of tea at the beach café? I'm pretty sure it will still be open. Having just completed my first attempt at power walking... for... (Laura saw April checking her watch)... close to thirty minutes, I'm feeling decidedly virtuous and planning to enjoy tea and cake as a reward.'

It was Laura's turn to laugh. 'I'm very impressed by your commitment to both your exercise programme and your love of cake! Sadly, I must get home, there's a council meeting this evening. Actually, that's really annoying, it would have been lovely to join you.'

'Not to worry, I'll call you, we'll arrange another time. Say "hello" to the councillors from me. Now I'm no longer the clerk, I'm missing all the village gossip, although it is nice not having to deal with complaints about dog fouling and fly-tipping.'

When Laura had first joined the Ash Green Parish Council, April was in situ as the parish clerk, providing the councillors with six hours a week of her time to carry out admin tasks. April had been keen to report that this position enabled her to '*test my talents and aptitude as a local government employee and provide a service to my community*'. A worthy sentiment, but Laura knew that this position had the added benefit of acquainting April with affluent villagers – most of whom were the owners of prestigious and high-value homes. These folks were clearly ideal contacts for an estate agent and April's day job at that time just happened to be at Richmond & Marsh, selling top-notch homes to those with vast pockets.

'I'm sure you don't miss parish council duties, but it's a lot less fun without you. Those meetings can be wearisome without a few moments of levity and your replacement is a sullen individual. She rarely even smiles.'

'Oh, Laura, that's a bit unkind, Joan is just a bit staid in her ways. The council should be grateful to have her services, especially as that position comes with the burden of having to accommodate piles of local government paperwork in your own home.'

'Well, yes, that was indeed the case, but of course now that the refurbishment of the village hall is almost completed, our humourless clerk will have a smart office space to work from, rather than her own dining table.'

'A much more satisfactory arrangement. So when will those renovations be completed?'

'Just a couple more weeks and the builders will be gone. It's going to be such a fantastic facility and at last we'll have a smooth surface in the dance studio for our Pilates classes, instead of that nasty old wooden flooring we've put up with for so long.'

'Great...' Laura noted the raised eyebrows accompanying April's reply.

'I hope you're not going to continue to excuse yourself from classes when they restart,' Laura chastised her friend. 'It's good fun, especially when we have Ellen's company too, and it gives us a great excuse to have an evening out together.'

Laura knew that April was unlikely to enthusiastically herald the return of Tuesday night Pilates classes. She had been on the receiving end of many of April's imaginative reasons as to why she must decline attending – her most inventive being that she was suffering from chronic bloating and flatulence. On that particular occasion Laura hadn't felt inclined to insist.

'I'm sorry but Pilates is really not fun, you know I find it dreadfully dull. Of course, spending an evening with you and Ellen is an absolute pleasure. Anyway, I do think the new hall is looking pretty amazing and it's doubtless down to your efforts to get the funding. Let me know if there's anything I can help with.'

'Excellent. Thanks, I'll let you know.'

'Looks like Max has finally taken the hint,' April said as the bedraggled, sandy dog appeared at their feet, 'so I'll let you get going and make a dash to the café before last orders. I've a craving for carrot cake. See you soon.'

With Max finally at her side, chewing on a large piece of driftwood – having been unsuccessful in his quest for a seabird snack – Laura prepared for the car journey home. She was sorry that she'd been unable to join her friend for a tea break – she'd seen less of April since her departure from the council and her regular Pilates absenteeism – but more importantly Laura was keen to share news with a confidant.

*

As she drove home, Laura contemplated her pre-Ash Green days spent working predominantly in the company of male counterparts. Over the years she had indulged in the occasional fling with a colleague but had never succeeded in finding a man worthy of her abiding commitment. That was until she met Neil. There was no instant attraction, but the intensity of long hours spent in each other's company whilst working on a company buyout project had given them the opportunity to get to know each other well. They became close companions and their friendship had slowly, but seamlessly, developed into a love affair. But Neil was married to Catherine.

At the time Laura assumed it might be natural to feel some guilt – coming between a man and his wife – but she felt only a longing for Neil's attentions and was prepared to overlook her unsisterly actions. She knew Neil didn't question his digression either; he claimed he had fallen in love with her and that he was struggling in a marriage to the daughter of his father's best friend. Neil had explained that he and Catherine had grown up in each other's company, their families spending their leisure time together. When a few of their mutual friends got engaged and married, Neil and Catherine followed this lead, believing it was their destiny. He told Laura that it was not long into the marriage before he became restless and sensed he may have missed out on something – or perhaps someone – else. And then he'd met Laura.

Neil had sworn that this was the first time he had been unfaithful, that the marriage had been a mistake and claimed he wanted more than anything to be with Laura. Despite being flattered by his declaration, she'd tried not to ignore the realities of the liaison and in moments of sanity was wise to the high probability of an unhappy outcome to this affair. As predicted, ten months later, despite their endeavours for discretion, an after-work drink together at an inn in a remote village would prove to be their downfall. A senior partner at their firm happened to drive past the public house as they were returning to their cars. They tried to fool themselves that he might imagine this to be a work-related meeting, but they realised that their demeanour would make their closeness obvious to an observer. The next day they were summoned separately to the boardroom by the CEO, who claimed that their relationship jeopardised their standing within the firm and that it would be wise for one of them to leave. Laura held a more senior post and it was Neil who was urged to resign. It was not long afterwards that Laura made the decision to depart, believing she had compromised her position of respect within the organisation.

The relationship had ended abruptly and Laura had struggled to adjust to a life without her lover.

But that was all a very long time ago, twenty years in fact, and now quite unexpectedly Laura had heard from Neil. She'd been surprised by the level of her excitement and astounded to experience fluttering sensations in her stomach. Laura thought herself ridiculous – *I'm a fifty-six-year-old woman, behaving like a silly, infatuated teenager.* She was trying without success to remain calm and unaffected to learn that he had invested time and energy in his search to locate her whereabouts.

4

Three weeks after their abandoned outing, Liam's chickenpox spots had healed and he'd returned to his pre-school group. April was making her way to Amarellos' to meet up with Ellen and, quite uncharacteristically, she was early and first to arrive at the café.

'Ciao, April, how are you today?' She heard Gino's exuberant welcome as soon as she stepped across the threshold.

It was cosy and warm inside away from the chilly wind and drizzly rain, and April's senses were bombarded by wonderful aromas and the harsh sound of coffee beans being pulverised into fine grounds. Once the grinding was complete, April opened her mouth to place her order, but voicing her choice of beverage was clearly not required.

'Your usual?' Gino said with a wide smile and before she could affirm or otherwise, he was reaching for the stainless-steel jug of frothing milk to add to the cup.

'Err... lovely, thank you, Gino.' It was the sign of a good barista to remember each customer's coffee habit, but April did wonder, should she get the urge one day to sample a caramel latte, would she actually get the chance to make the request before the cappuccino arrived on the counter in front of her. But it would be discourteous to deny the café owner his moment of personal service, so she just went along with it and, in any case, the cappuccinos served at Amarellos' were the best in town.

The cheerful proprietor and his family had owned Amarellos' café for eight years, bringing their typically Italian ways with them from their Tuscan motherland. Ideally suited to succeed in hospitality, they were hard-working and genial. Their conviviality was shared with all who brought their custom to the café, and consequently Gino and his crew had become welcome additions to their adopted community.

April picked up her cup, admired the skilful coffee art gracing the frothy top and walked over to the squishy leather sofa – the patina of its upholstery revealing that numerous bottoms over many years had rested comfortably there. Seated in the bay window overlooking the pretty cobbled side street, April was able to survey the passers-by, many of whom were struggling with umbrellas against the now blustery rain. The browsers and shoppers in the town appeared to be few that morning, many no doubt having been deterred by the gloomy weather.

A grey sky did not mar the appeal of Great Aldebridge, with its attractive and characterful buildings and at its centre a cobbled marketplace with an ancient clock tower. The town offered three cafés, two restaurants and a selection of independent retailers, whose interesting array of everyday and gift items complemented the convenience of the small supermarket. The town's shopkeepers were known to do their utmost to fulfil every request they received from those without the time or means to get to the larger stores in further afield Ipswich.

April had often heard Ellen extolling their virtues. 'I've never been thwarted in a mission to purchase provisions in

Great Aldebridge. Both the greengrocer and butcher managed to produce all the ingredients I needed when attempting the recipe for *"Aubergine and okra caponata with chargrilled pork belly".*

Goodness, Ellen and her ambitious shopping lists and her constant desire to extend her gourmet skills – April couldn't comprehend this inclination. With this thought, April glimpsed a distorted, but familiar, human vision approaching the café through the raindrops trickling down the windowpane. Moments later Ellen was free of her raincoat and settled on the sofa opposite April, with her favourite vanilla latte and a ricotta cookie.

'Yummyyy… I've been so looking forward to this.' April smiled at Ellen's obvious delight as she savoured her coffee and cake, giving the impression of a deprivation far exceeding the three weeks of Liam's chickenpox confinement she'd actually endured. 'Delicious… I wonder, do you think Gino would let me have the recipe? Remind me to ask him before we leave.'

'Sure, I'll do the best I can, but you'll have to take account of menopausal amnesia. Tell you what, I'll write us a reminder on this napkin.' Satisfied with this solution April retrieved a pen from the depths of her bag and wrote 'CAKE RECIPE' in oversized capitals.

'Anyway, how's it going? What have you been up to since you left the father and son dental practice?'

April was happy to report that she was feeling pretty good – relaxed and much less weary. All her kitchen cupboards had been sorted – out-of-date groceries binned – her oven was now spotlessly clean, the local charity shop was filled to the rafters with discarded items following a major declutter of her wardrobes and her home filing was completely up-to-date.

'But now perhaps it's time I started to think about finding paid employment.'

'Oh… you're not feeling under pressure to get back to earning a living, are you?' Ellen asked. 'I thought Ian was okay about your lack of a job.'

'Ian's been fine. Although constantly worrying about our finances is one of his foibles. Goodness knows why, he's always been so thrifty.' Ian's careful spending annoyed April, but she had to acknowledge that his way was the much-needed antidote to her extravagant tendencies.

'So, any ideas about finding the right career this time around?' said Ellen.

'Well, as you know, I can very easily be swayed by a *good idea* and wholeheartedly inspired when I hear someone talking about their own profession. Ian says one of my favourite phrases is "I've always wanted to be a… dot, dot, dot". Fill in the dots with the occupation of the moment.'

'So, what is presently filling the dots?'

'Sadly, nothing at all. You'll probably laugh, but I even tried completing those online personal profiling questionnaires, in the hope that might throw some light on a suitable career for me.'

'And how did that turn out?'

'Well, I've been presented with a range of options and to be honest some of the suggestions are absolutely not for me – veterinary nurse springs to mind. I've disliked cats ever since I was scratched on the forehead as a child by the neighbour's tabby.' April lifted her fringe to reveal the scar.

'Ouch, you must have thoroughly upset that cat.'

'It was my duty to try and save the field mouse! Anyway, the other careers included things I've already tried and failed dismally at, or tried and got plain bored with, otherwise, just not really me. For instance, can you really see me as a librarian?'

'April, you may love books, but you're far too noisy and accident-prone to work in a library!' Ellen laughed. 'I can just picture it, you reaching up to the highest shelf and, as you pull a book from the tightly packed stack, a cascade of crime fiction comes tumbling down to deck you.'

'Very funny, but I imagine I'd just be trying to share my passion for Lynda La Plante – the duty of any conscientious librarian.'

'Okay, so instead let's focus on what is it that you want from a job – apart from a generous salary, of course.'

'Well, you know me, I love being sociable and meeting new people. But what I really crave is the conviction that I'm giving my time to something that is actually worthwhile. I'm really fed up with working to make money for someone else. Instead, I'd like to think I was making a difference, helping to improve lives – if that doesn't sound far too altruistic.'

'No, it doesn't, that's a very good reason for getting out of bed every day. So, how will you put all your excellent attributes and life skills to good use?'

'Well, I'm done with dull admin jobs and perhaps I'd be better working for myself, being in charge of my own destiny, but that's going to require one of my great ideas and strangely those have deserted me.'

'Well, it's never too late to try something new and in the meantime you can immerse yourself in the latest Ann Cleeves book. Use the free time to lose yourself in a spot of novel sleuthing!' said Ellen.

'Sounds like an excellent interim plan. Now let's forget all about working for a living and just enjoy Gino's excellent coffee.' April sighed with contentment.

'Delicious,' Ellen agreed.

'An exceptionally good cup of coffee, "*made for you with precision and passion by our highly skilled baristas, using the finest Italian coff….*"' As she read from the nearby wall poster April was startled from her narration by the sound of Ellen's tall glass cup being crashed onto its saucer, with an associated coffee spillage.

'Crikey, are you okay?'

'That's it, April! That's the answer we've been looking for and it was staring us in the face... You could have a go at being a barista!'

'Hilarious,' April scoffed, and Ellen dabbed at the coffee overspill with the 'CAKE RECIPE' napkin.

'So, back to the real world, what's new with you, Ellen? You must be pleased that Liam is back at nursery?'

'Oh yes, thank goodness, poor boy, he really was quite poorly, but with my routine back in place, I've even had time to try out a couple of new recipes,' Ellen replied, clearly having reverted to her fall-back position beside the Aga.

'Anything particularly calorific that I could sample for you, but should probably avoid at all costs?' said April.

'A lovely pistachio and almond loaf, which was quite successful, and a new take on lemon drizzle cake. It's a St Clement's drizzle, with lemons, oranges and a sprinkling of lavender in the mix.'

'Really? Should you be messing with your lemon drizzle cake recipe? It's the best I've tasted, as my curvaceous hips will confirm!' April ran her hands symbolically over an exaggerated thigh-line.

'You're probably right, but I couldn't resist just giving it a go. Mostly the tried and tested remain the best and that recipe is a particular favourite of Paul's too.'

'And, how is he?'

'He's enjoying his job, but I am concerned by how busy he is. He seems tired all the time. He rarely gets home before 8.30pm, especially on days when he has to commute into the city for meetings, which is frequently at the moment. By the time he gets back, I'm tired after a day with Liam and neither of us seems to have the energy for conversation.'

'Ah... that's really not ideal.'

'It's not. He comes in, hardly says a word, eats his supper

– reheated in the microwave as I'm usually too hungry to wait for him – then he watches the news on TV and promptly falls asleep. To be honest, I've been feeling quite lonely – bereft of adult company.'

'Don't ever be lonely, you know you can always call me. Now, is this going to be a regular thing – the late evenings and city meetings – because it might be sensible to look at buying a small city crashpad for Paul to use… and it would be a great investment,' April offered.

Ellen sighed. 'Well, funny you should mention that, I'd been thinking the same, but we've not had a chance for a proper chat to discuss the feasibility of buying a second property. I know it would be expensive, especially with London prices, but it would definitely make his life easier and, as you say, a good way to invest for the future.'

'It's not easy, I know. With Liam still so young you must feel like a single parent at times, but Paul is obviously focusing on his career and that ultimately will benefit you all.' April endeavoured to be understanding and supportive, but suspected Paul might be happy for the excuse to opt out of evening bathtime duties. 'Things will change as Liam gets older and with him starting at the village primary soon you might want to consider returning to work yourself.'

'True, it won't be long before I'm out shopping for school uniform.'

'Try and enjoy the time you have whilst Liam is still little, because before you know it, he'll be a sulky teenager. Take it from someone who's been there!' April tried to lighten the mood; the long anticipated get-together was descending into a coffee morning of despondency.

April recalled the tricky times with her daughter, Megan, and was grateful that she and Ian had somehow managed to get through the petulant phase. Now they could enjoy the occasional company of a spirited, ambitious and only slightly exam-stressed university student. These days their biggest concerns were the amount of debt racking up, the quantity of alcohol and, more worryingly, drugs being consumed by students and, in Megan's case, being besotted by a very much older boyfriend.

'You're right and mostly the time at home with Liam has been a joy, but I sometimes wonder if becoming a mother has been a career-ending event for me. I think I'd find it hard to go back into the marketing environment. I feel woefully ill-equipped and anxious at the very thought of it. Having said that, I'm sure I will want to return to work at some stage, only goodness knows what I'll do!' Ellen said.

'Oh, dear God,' exclaimed April, 'not you as well!'

5

The red wine splashed against the sides of April's favourite blue-tinted glasses as she poured their drinks whilst preparing supper. There was a distinct possibility of an unsuccessful outcome to the evening meal, despite Ellen raving over the concoction and the health benefits of lovage as she handed over the jar of homemade fare; this did seem a peculiar variation of the ingredients traditionally found in a pesto sauce.

'So, what's in Ellen's pesto recipe?' Ian asked as he sniffed the contents of the container.

'Apparently it contains fresh lovage leaves, walnuts, garlic, olive oil and parmesan.' His expression revealed that he clearly shared her scepticism, but April was prepared to give it a chance.

He didn't vocalise his thoughts, instead he went back to the chopping board and prepared further quantities of romaine lettuce, cucumber, watercress and grated carrots for the accompanying salad he was assembling. She saw him stand back to admire his work and then throw a handful of croutons on top. April knew Ian's instinct would be to make plenty of the side dish just in case the salad was the only palatable part of the supper they were about share.

The sad fact was that April wasn't a great cook, finding the whole process of preparing meals utterly tedious. This said, she excelled when it came to a full English. Where others would freak out at the prospect of getting all the different component parts of a cooked breakfast ready at the same time, she calmly triumphed. One of her mum's legendary fry-ups – at any time of the day – was a frequent request whenever Megan was home from university.

'So how was your day? How are things with Ellen?' Ian asked. April was never totally convinced that he was particularly interested in what she and her friends got up to. But he had asked and so she felt at liberty to give him the details.

'We met up at Amarellos' this morning. It was lovely to see her and I think she was relieved to be able to get out of the house.' April chatted away sharing Ellen's news. 'Thankfully Liam is now well enough to be back at nursery, with just a few dried-up scabs remaining. Apparently so far he's resisted the urge to pick.'

'Yuk, April, really?' Ian said, grimacing.

'Sorry, darling, inappropriate just before we eat. Anyway, Ellen was saying that Paul is having to work in the city a lot more. I did suggest it might be worth them considering buying an apartment in London for him to use during the week.' April always got excited when property purchase was the topic of conversation.

'That's probably not a bad idea, but can they afford it?' April saw the look on her husband's face and had an inkling that Ian's male ego might be a tiny bit put out that Paul's financial standing enabled him the luxury of purchasing a second property, whereas his own would not.

Moving on with her account of the day, April launched into their deliberations on career options.

'Actually, Ellen thinks she's got the perfect solution to my job dilemma.'

'That sounds promising. What's she come up with?' Ian asked, finally looking a bit more interested.

'She thinks I should become a barista,' April laughed.

'A what! You can't be serious. Is there any money in it?' April was unsurprised that salary was her husband's first consideration and disappointed that he didn't seem open-minded to any role which might have the potential to make her happy.

'I don't suppose I'd make my fortune, but I do appreciate a really excellent cup of coffee, so surely that's a good start. And anyway, no one can have failed to notice the coffee culture phenomenon. There are new coffee shops opening all the time in towns and cities across the country.'

April was miffed by Ian's response and although she hadn't considered Ellen's suggestion to be a realistic one, she was aggrieved that he appeared unable to take her ongoing employment predicament seriously. How did he know that training to become a barista was not the answer to her quandary?

'I really can't see you coping with the complexities of one of those huge coffee machines. The heat and steam would make you perspire and cause your hair to frizz,' he continued to jest, 'and your menopausal memory loss would definitely prevent you from being able to recall all the individual orders as they back up. I foresee an ever-increasing queue of customers waiting impatiently as you endeavour to recall whether the hot chocolate order included marshmallows or cream or a flake, or all three!' He clearly thought this was all highly amusing and grinned as if congratulating himself on his witty remarks.

The pasta water began to rise dangerously close to the rim of the saucepan and April quickly, and defiantly, turned her back on her comedic husband to extinguish the gas. As she did so she resolved to Google '*Training to be a Barista*' later that evening.

With supper ready, they sat together at the kitchen table and ate the surprisingly tasty meal in silence, with the background accompaniment of an uninteresting and barely audible television programme. Ian had insisted they needed a TV in the kitchen and switched it on as a matter of course as he entered the room. April had thought this a terrible idea that would serve only to inhibit conversation. Was it just Ian or were all men incapable of taking part when it came to inconsequential chatting? Thank goodness for women friends, thought April, and by a stroke of serendipity the ringing telephone interrupted the lull.

'Hi, April, you okay? Oh, is this a good time?' Laura could not have known just how welcome her call was at that moment.

'I'm fine, thanks, and we've just this minute finished supper, so a great time for a chat. Please say you want to arrange a coffee date, it's been way too long.'

'Exactly. Look, I know it's short notice, but I'm going to be in Great Aldebridge tomorrow. Don't suppose you could meet me at about 11.30am… Amarellos'?'

'As it happens I am particularly free at the moment, so yes that would be lovely. And…'

'See you then.' Laura disconnected immediately, her swift departure and distracted tone leaving April feeling not only disappointed, but also intrigued.

*

As she walked quickly along the cobbled street towards the café, Laura found haste and cobblestones a tricky manoeuvre, even in her low heels. She was late for the rendezvous, having been stuck behind a milk tanker on her car journey along the lanes to Great Aldebridge. Finally she saw the two wooden coffee crates which marked the entrance to the café – now serving as decorative planters, each containing a small standard olive tree underplanted with pretty blue and white geraniums.

Laura rushed inside and was pleased to see April already seated in a booth at the far corner of the café, nicely tucked away, affording some privacy despite the numerous faithful Amarellos' coffee fans gathered within.

'Goodness, I am so sorry I'm late.'

'Don't worry, I've enjoyed sitting here, people-watching and eavesdropping. Those two at the table by the deli shelves have been engaged in a bit of a domestic about which one of them loads the dishwasher correctly. Despite their hushed tones it's been quite entertaining… Knife blades up or down?' April added controversially.

'Down! Obviously!' they both whispered in unison.

Laura dumped her coat on the seat opposite April and made her way to the counter to order an Americano and a piece of Gino's delicious honey cake.

'That does look like a particularly lovely piece of cake,' April commented on Laura's return, before lifting her gaze away from the plate onto her friend's smiling face. 'Hang on, what's going on? What is your secret, Laura? You're looking particularly fabulous. Have you secretly had a mini-facelift since I last saw you?'

Laura laughed, pleased to receive the compliment. 'Nope, I hope I can leave the facelift – mini or major – for a few years yet.'

'Okay, but I sense that there is something… what's the news? There's an air about you that says it must be more than a little thrilling.'

'Goodness, is it that obvious? Well, there has been an unexpected development in my life – my love life to be more precise.' Laura blushed at the thought of it. *Me, middle-aged Laura, with a love life*; April was right, it was a thrilling notion.

'Tell all.' April was riveted.

'My news involves a man from my past – way back past – and the fact that he wants to see me again,' Laura blurted out in her eagerness to share.

She could see April's astonishment. This was clearly the last thing she had expected to hear. Everyone thought Laura's life was complete, no doubt believing a man would be surplus to requirements for such an independent lady. She was obviously doing a good job of deceiving everyone.

Laura explained that a few weeks previously, she'd been website surfing, reading reviews about the pros and cons of splashing out on a subscription to the Royal Horticultural Society. But boredom had swiftly taken hold and with the rain preventing her from getting into the garden to do some pruning, she'd decided to do a bit of internet-stalking.

'You know how it is, you start wondering what old classmates got up to after leaving school. There was this one girl, Ruth – you know the type, good at everything, from maths to performing arts; her depiction of Toad as in *Toad of Toad Hall* was legendary. I just wondered whether she'd achieved something particularly noteworthy after school – Principal of Cheltenham Ladies' College, perhaps.'

'Sorry, I'm not sure I understand. Was this man from the past in the cast of Toad Hall?'

'No, April, forget Toad Hall, the point is that I decided on a whim to set myself up with a Facebook profile – although I had little expectation of hearing from anyone interesting.'

'So how does that work?' April said. 'Isn't it asking for trouble having your contact details available to any browsing weirdo?'

'No, it's not like that,' Laura said, incredulous at April's social media naivety, 'don't concern yourself. What's important is that this person, this man – possibly the love of my life – has made contact.' Laura's voice rose to a hushed squeal.

Calming herself and lowering her volume several decibels,

Laura swiftly summarised the history of her time spent with Neil, much to the delight of the dishwasher couple, who were now taking their turn at eavesdropping.

'How exciting. But have you spoken with him yet? And why do you think he's got in touch after all this time?' As might be anticipated, April wanted to know everything. 'What went wrong? Why did you lose touch?'

'Err... well, it was complicated and so long ago I can barely remember... but focusing on current news, we have had a brief telephone conversation and arranged to meet up.'

'When?'

'It's this Friday evening, at a pub halfway between his home and Ash Green. I am so looking forward to it. What I do know is that he now lives in an Essex village, having moved to be closer to his brother and sister-in-law.'

'I wonder if he ever married or has any children. Perhaps he's a widower. Gosh, you have so much catching up to do.'

Laura averted her eyes, feigning the imperative need to look across to the counter to check out a new arrival at the café. She was rapidly concluding that despite her desperate need to tell a friend about Neil getting in touch, there should be no supplementing the news with further insights into their past relationship.

'What was he like when you knew him? Tall, dark and handsome?'

'Well, I always found him physically very appealing.' Laura smiled at the thought. 'He was quite tall, with dark hair, blue eyes and beautiful teeth.'

'Beautiful teeth! Well, let's hope they are all still intact.'

'He had lots of appealing attributes, as far as I was concerned, and yes, this did include an excellent set of teeth.'

'It's madly exciting. But I suppose it would be best not to get too carried away. You need to make sure his intentions are completely honourable.'

Having had time to recall those delectable days of passion, Laura was secretly hoping Neil's intentions were utterly dishonourable, but she was touched by the sentiment behind April's benevolence.

6

Laura turned her head to the left and then slowly to the right, scrutinising her facial features, and then she stood on tiptoes in order to thoroughly assess her appearance in the hallway mirror. Placing her fingers in her hair, she ruffled the roots, attempting to give the effect of voluminous locks and, after a spritz of hairspray, she opened her lipstick to apply a touch more of the satin pigment. She smiled at her reflection – elegant, well-groomed, mature but wearing well. It had taken her a while to choose the perfect outfit for the occasion, but she was pleased with the impeccable fit of her navy trousers, teamed with a white linen blouse, which looked both classically smart and comfortably casual and – most essentially – revealed her neat, slim figure to its best advantage. She was proud of her taut and nicely rounded butt, considering herself lucky to have avoided any sign of the onset of sagging. However, she hadn't managed to completely shun the ravages of ageing, and the less-than-perfect condition of her hands and neck did bother her somewhat. A quick search of the small drawer of the hall console table uncovered a tube of hand cream and she applied a large dollop of the lotion. Wrapping her favourite silk scarf around her neck, Laura was finally satisfied that she had done an excellent job and was suitably prepared for the auspicious meeting with Neil.

She was frequently complimented that she looked ten years younger than her chronological age and for this she thanked the exceptional genes she had inherited from her ever-youthful mother. Rose was now in her eighties, but still slender and beautiful, seeing out her days in a care home. Living life with the early stages of dementia was now her burden, but Rose had always been a fun-loving, feisty woman, never suffering fools and never afraid to say what was on her mind. When caring for herself on a day-to-day basis had become more difficult, Laura was delighted to have found The Laurels, an excellent residential home in a tranquil rural location not far from Ash Green, where the staff were always patient and kind, despite her mother's numerous antics.

'Your mother has had a bit of an upset, nothing too serious,' Jane, the care home manager, had told Laura two weeks previously. 'We think it was due to a home makeover programme the residents were watching the evening before.'

'Have you had to turn down a request for her bedroom to be repainted in candy pink, or some other psychedelic colour?' Laura was at a loss to understand how an episode of *DIY SOS* might have sparked an upset.

The explanation, however, was that Rose had set about rearranging the placement of the furniture in her room, heaving objects with a strength a woman half her age would have been proud to muster. Pleased with the new layout, she had failed to register that one of the chests of drawers was now relocated across the doorway. Having no reason to leave the room at that moment – or perhaps secretly delighted to keep others out – Rose had unwittingly left herself a prisoner within the room, unless of course the superhuman strength returned.

'The local fire brigade came to the rescue and we think she'd only been stuck inside for a couple of hours. So, no harm done.'

Mental processes now slightly impaired, Rose was not always aware of the consequences of her actions, although Laura liked to think that an element of mischief remained and could be attributed to the more outlandish of her deeds.

Her mother appeared to be happy and comfortable in her new surroundings and despite the sadness of the ongoing demise of her health, Laura found solace in seeing that Rose was well cared for at The Laurels. When visiting, Laura was delighted when she found that her mother was having one of her better days, able to welcome her daughter with recognition and engage in conversation, animatedly detailing what was happening in her favourite TV soap. The pair occasionally reminisced and Laura loved these moments, taking the opportunity to remind herself of the wonderful, fun times they had shared as a family whilst she was growing up. Her beloved father had died when she was barely a teenager, leaving mother and daughter to cope together despite their disconsolate grief. She knew she was fortunate to have had such loving parents and wanted to guard closely all those special memories. Laura was determined to do everything to ensure her mother's wellbeing until her ultimate departure.

Laura glanced at the clock on the console table – a lovely Art Deco piece which had previously taken a preferential position on the mantel at her mother's house – and noting that she had just enough time to drive to her destination she picked up her bag and keys. Having reached the front door, she hesitated and decided that perhaps a further quick visit to the loo would be advisable before leaving the house.

Arriving at The Beresford Arms – *'a cutting edge gastropub'*; Laura had read its credentials when checking the menu online the previous evening – she parked her car in the furthest corner of the car park, giving herself an excellent vantage point to observe every new arrival, whilst keeping out of clear view. She wanted to ensure she had a chance to check out her date before he caught sight of her. Laura looked at her watch, just five minutes since the last time she had checked, but it was now eight minutes after 7pm. Had he been a poor time-keeper in the past, or should she prepare herself for the chance that he might fail to show up? This sudden thought was distressing and she hoped sincerely that she was panicking unnecessarily.

Since receiving Neil's first message, their subsequent telephone call and the arrangement to meet, Laura had become more eager with each day to see him. She'd even allowed herself to hope there might be a chance they could rekindle their friendship, or perhaps have the opportunity to embark on the life together they had previously been denied. But one step at a time, she reminded herself.

She watched as a car slowed up and then turned into the parking area – a white Range Rover – and her mind immediately questioned the wisdom of this colour choice for a rugged vehicle, designed for off-roading in the mud. Looking more closely she saw a lone male driver and she watched as he manoeuvred the 4x4 into a parking space beside the pub's entrance. As he got out of the car, Laura quickly grabbed the binoculars from the glove compartment to get a better look as the man adjusted his jacket and zapped the door lock with his key fob. Her heart began to pound as she identified the distinguished-looking individual with greying hair – carrying a few more pounds of weight than might be advisable – as her former lover. As she watched him walk towards the building, she chastised herself for being judgemental of both his waistline and his choice of motor vehicle.

Satisfied that he was safely inside the pub, Laura returned the binoculars to their recess in the dashboard, glanced in her vanity mirror one last time and took a deep breath. She got out of her car, brushed a few stray dog hairs off her trousers and made her way to join her date, feeling a mixture of excitement and terror.

Laura saw him standing at the bar, trying to attract the attention of the barman. She approached from behind and with what appeared to be a subliminal sense of her presence, Neil swung around immediately to face her. A smile spread across his handsome face, revealing a perfect set of even, white teeth. She noted too the familiar sparkly blue eyes, now staring at her from a matured face under a receding hairline. Frozen to the spot, Laura was relieved when seconds later he moved forward, greeting her genially, taking her in his arms in a friendly embrace.

'Ah, darling Laura, how wonderful to see you… and looking fantastic.'

'Thank you, Neil, it's good to see you too.' Normally poised and confident, she found herself uncomfortable and self-conscious, a state no doubt caused by the proximity of her past love – the man who had been a significant part of her former life and now rendered a virtual stranger by the passage of time.

'Let's order some drinks and get settled at our table and then we can relax and have a good catch-up,' he offered. He seemed aware of her discomfort, for which she was grateful.

'May I have a glass of white wine, please?' Laura hadn't planned to consume any alcohol, thinking a clear head would be advisable, but decided she would benefit from its calming effects.

They were seated by a waitress in one of the alcoves of the rather quaint and cosy oak-panelled dining area. Laura had read excellent reviews highlighting the quality of the food on offer, but this was her first visit to The Berkeley Arms. She could only hope that her racing heartbeat would return to its normal rhythm, her stomach would cease churning and she could manage the process of swallowing some of the fine cuisine.

Over the years, Laura had thought of Neil often but never dared to dream that she would see him again. Now they were together and her dearest hope was that the feelings they'd once shared might be reawakened. But first, it seemed, they would need to get to know one another all over again.

'So, do tell me, Laura, how has life treated you since I last saw you? What have you been up to and, more importantly, who with?' His opening question was no doubt intended to overcome their initial awkwardness, but she felt it a rather direct starting point.

'Life has been good, actually, and if you're asking did I ever marry, then the answer is no. But I have enjoyed a very successful career in city banking and reaped the benefits of its generous rewards. I've travelled a lot, dated and socialised a lot, and now I'm comfortably retired and enjoying a fulfilling yet quieter lifestyle – despite my excitable dog – in a very pretty village, where I've been made to feel most welcome. I feel very fortunate and contented.' Laura was surprised to hear herself rapidly babbling the précis of her life CV. She was also annoyingly aware of the unwitting attempt to oversell her existence without him.

'Wow, that's a very thorough summing-up, but I am pleased that things turned out so well for you. You deserved to be successful in your career. There's just one thing I would not have expected: I'd never have imagined you as a dog person!'

'Yes, I have Max, the most adorable Golden Retriever. I look forward to introducing you to him.'

'That's a nice thought, Laura, but I have to tell you that I'm allergic to dogs and in particular the hairier varieties! Still, not to worry, as long as I've got my antihistamines and inhalers with me, and I'm not confined in a small space with him for any length of time, I'm sure it will be fine.' Neil's laugh seemed unconcerned, but this piece of news did bother Laura. She couldn't contemplate a future life shared with this asthmatic version of Neil if it would ultimately mean her dog became an outcast. But no, this train of thought must stop; it was important not to overthink the situation or get too far ahead of herself. She searched to recover rational, level-headed Laura.

'What about you, Neil? When we spoke on the phone you mentioned your marriage had ended. I know things were never that great between you and Catherine, even so the final parting must have been difficult for you both.'

'I did hang on in there, but yes, it is over between us. It's been unpleasant, especially sorting out the jointly owned assets, and I do want to ensure I come away from the marriage in good shape financially,' he said. 'Meantime, and whilst the marital home is being sold, it seemed best to move to the small rental property near to my brother's home.'

'I'm assuming you and Catherine didn't have any children together?' Laura was heartened when Neil acknowledged she was correct in this assumption, as another potential complication to any future they might share was removed. 'Well, at least that's one less stress to deal with and you've always got on really well with your brother... it's Ray, isn't it? So you'll have his support whilst you're getting yourself back on track. Are you still working?'

'Sadly, no. I was working for Catherine's father at his construction company in a project management role, but that position became untenable once Catherine and I decided to part. He as good as asked me to leave.' Neil paused and Laura noted his air of defeat. This was surprising; she would never have considered him the type of man likely to find himself in such a vulnerable position. 'Anyhow, enough of that.' Neil's distant look vanished as he focused on the menu. 'Let's get ordering, I'm starving.'

Looking at the food selection together the old familiarity returned, enabling them to recall with relative ease which dish the other would be most likely to choose. They reminisced about the wonderful restaurants where they had dined all those years ago and the evenings at the theatre, which they had both loved. It was not long before any uneasiness between them had been replaced by cordial affection and they were not abashed to speak of nights filled with passion at a variety of hotels and locations. The much-longed-for opportunity to spend an entire night together had been overwhelmingly tempting, even if venue choices were sometimes limited to the short-notice options. They laughed now as they remembered the night they chanced upon the last available room in a hotel in a far from salubrious suburb of Watford. Overlooking the parking area and directly above the kitchens, its location might have brought a night disturbed by slamming car doors and the pungent aroma of curry wafting upwards from the whirring extractor fans below. But these annoyances would never have overshadowed their urgent lust for one another; when alone they were oblivious to everything and everyone around them. This desire had rendered them incapable of acknowledging that their actions would be considered utterly reprehensible. They hadn't had a care for what others might have thought.

Laura found herself aroused by their conversation and felt a longing to revisit the pleasure of this intimacy. Across the table, she sensed the body heat of her companion and the distinct aroma of his maleness. But Laura was anxious not to appear too eager for him, despite his conspicuous virility and the flirty nature of their dialogue, which had her yearning for his touch. Laura wasn't sure how she had managed to mislay her libido, but it occurred to her now that it had been way too long since she had last had sex.

7

Arriving home later than he had planned for the second time that week, Paul drove his black BMW along the lane to Walnut Tree Farmhouse, carefully avoiding the potholes to preserve the suspension of his much-prized company car. Ivy-covered brick pillars marked the entrance to the driveway and as he rounded the corner, he was relieved to see these columns finally coming into sight. The meeting in London had overrun and on arriving back at the train station he discovered his homeward bound train was delayed. Just what you didn't need after a 5.30am start.

Paul was feeling desperately jaded, with no inclination to deal with Ellen's inevitable annoyance at his late arrival. He tried hard to call to mind the welcoming, frown-free face which in former days would have acknowledged his return to their family home. He hadn't stopped loving his wife, but these days his stressful existence and busy work schedules were getting in the way of their relationship.

Ellen didn't understand how important it was for him to succeed and to win. Winning orders, winning respect, winning promotion and bonuses – winning gave him such a buzz, but achieving the buzz and keeping clients happy demanded all of his energy. There was precious little left to give to his wife and child at the end of a long day and that sadly was his current reality.

The arrival of Liam had been a life-changing event and Paul remembered clearly every minute of the day that had involved a gruelling twelve hours of labour for his exhausted wife. But finally, the midwife had presented the bundle of blanket wrapping the tiny form, and he had received this gift with trepidation, with the fear of dropping or holding the child too tightly uppermost in his mind.

Those early days had been testing – interspersed with anxiety and the delirium of fatigue – but he had been so proud of Ellen, who had clearly found her new role challenging. As the weeks passed, he watched her evolving into the capable carer of their child's needs, but he also sensed that he was becoming excluded by the bond that was forming between mother and child. Liam wanted his mum's attention for hurt knees and hunger pangs, and Ellen seemed inclined to push her husband away if he attempted to intervene. Whether this was intentional or not, he didn't know for sure, but he accepted that as she spent the daytime hours and a lot of the nights with Liam, the child inevitably saw her as his go-to grown-up for all crises.

As a married couple, Paul accepted they had new responsibilities; his was that of provider and protector, and to deal with the changed circumstances he turned his focus to his work. He felt in control, respected and effective in the office – something he didn't necessarily feel at home anymore – and on evenings such as this he suspected he would be out of favour even before entering his home. Paul sighed, steeling himself for the lukewarm reception as he got out of his car and made his way to the front door.

'Hi, I'm home,' he called as he entered the hallway.

'Shh, keep your voice down, Liam's finally asleep. He's been such a handful this evening,' Ellen whispered crossly as she hastily descended the staircase.

'I'm sorry, honey,' he whispered back. 'Bad day?'

'No, not especially. How about you?' Paul was relieved to hear her voice soften.

'All fine, just busy as usual and the inevitable issues with train delays – the explanation this evening was that *"we have to wait for a relief driver to arrive"*. I can tell you that after a twenty-minute wait, it was a bloody relief when he finally got there!'

'Might have been a "she". Women drive trains too, you know.'

'Of course they do, darling.'

Paul propped his briefcase by the hall stand, slipped off his shoes and made his way upstairs. 'I'll just get out of this suit, have a quick shower and then we can have a drink together before I eat. I assume you've already had yours.'

'Actually, no I've not eaten yet... In fact, I've made your favourite supper – venison casserole. I was hoping we could relax, eat together, have a few drinks and a proper catch-up. It feels like I haven't seen you for days.'

'That sounds lovely. Just give me a few minutes.'

Paul was now uncertain whether this evening's warmer welcome was a good sign or the precursor to a bollocking. He decided that he must guard against allowing a full stomach and a couple of glasses of wine to cajole him into being unprepared for any potential conflict.

*

Ellen went into the kitchen and reprimanded herself for being grumpy with Paul, especially as she had been determined to make an effort that evening. Her resolve had been weakened by Liam playing up at bath time; irritability had got the better of her and she'd ended up losing the state of serenity with which she had planned to greet her husband on his return.

She made her way into the kitchen in search of a restorative glass of wine and to check on their supper. A wonderful aroma met her and she was certain that despite the delay the casserole had not spoilt, merely improved by the long, slow cook in the Aga. She'd wanted to create a romantic evening: had prepared a meal her husband particularly enjoyed, put on a dress he had previously admired her in – although she wasn't sure that he'd noticed – tidied her hair and make-up. The table was adorned with a vase of garden blossoms and a scented candle, which wafted a bouquet of lavender – ideal for its natural stress-relieving properties. As a special treat, she'd been out earlier and at great expense procured a bottle of Paul's favourite red wine.

'An excellent choice, worth every penny,' the young man in Majestic had claimed. 'Just ensure to allow the wine to breathe before you start drinking. I'd say thirty minutes should do it.'

Ellen wasn't sure whether the extension of breathing time to ninety minutes, which the wine had now endured, would further enhance the flavour. Perhaps sampling it might answer this query, she thought as she filled her glass.

Wanting to give her work-stressed husband some wifely attention, she believed she had achieved the perfect atmosphere in which to pamper him. She did, however, have an ulterior motive. That motive was to woo Paul, lighten his mood and ensure he was totally relaxed – submissive even – in order to establish a scenario suited to a candid conversation.

The sound of water running in the shower prompted Ellen to remove the supper dish from the oven and place it on the table, allowing the meat to rest in readiness for Paul's arrival. Before dropping her weary body into the armchair beside the Aga, she topped up her wine glass and then began to mentally prepare herself for the impending discussion. She'd thought everything through and was now feeling quite confident that she was proposing a good plan – one which Paul should be willing to give his consideration.

Ellen's daydreams now involved the prospect of meeting new people who would take her talents seriously and admire her mastery as a chef and baker. These people might be educators, backers or customers, all playing a part in her return to work – but not just work, more precisely a second career path.

Chatting with April had inspired her and soon after their meeting she began to wonder whether utilising her new-found talents in the kitchen could take her in the direction of the catering profession. At first she'd dismissed the idea as whimsical, then concluded that if it did have any credence she might not have the prowess to make it work. But having mulled the idea over for a few days, accepted that failure was a possibility she must be prepared to bear, Ellen had decided to trust her instincts and allow herself to have some faith.

Internet research had informed her that the first step would be to acquire an appropriate qualification in food hygiene. After all, causing food poisoning to the general public would be seriously frowned upon and unlikely to further a career. But this didn't daunt her; in fact, she was enthusiastic at the thought of studying again and delighted to discover that e-learning at virtual colleges could provide all that she may require without the necessity of attending an actual college – an ideal solution for a mum of a four-year-old. The world had definitely moved on in the fourteen years since she had left full-time education, which was probably encouraging news as her son would soon be embarking on his first days at primary school and she desperately wanted this to be a positive experience for him.

It was just a matter of months before Liam was to start his education at the village school and despite a slight sadness at the passing of his toddler years, Ellen knew this was the next stage in his formative life. She had to let him go and get on with it, whilst she got on with something else. Clearly this something else was not to be the creation of a sibling for her son. Paul had given no indication whatsoever that he was keen to add to their family. She had been dismayed by the dwindling passion in their marriage and their very infrequent lovemaking, but consoled herself with the belief that this was an inescapable outcome for most couples once children joined them.

Ellen had come to realise that having another child may not actually be what she now wanted. The experience of having Liam in her life had been very special, but it had also been incredibly demanding and had left her feeling out of touch with much of the adult population. Added to this was a certain amount of resentment that fatherhood had not resulted in the shelving of her husband's career; he was off successfully pursuing his profession, whilst she was left behind.

With the wisdom imparted by a second glass of wine, Ellen figured that a single-child family would not overly stretch their finances and with the addition of a second salary, should she get the chance to earn one, the pressure would be off Paul. He might be able to scale down his workload and not feel he was the sole provider for his family, or perhaps, with a boosted income they would be in a better position to purchase that crashpad apartment in London for Paul and as an investment for their future.

Ellen was pleased with the logic of her inner dialogue; this was all marvellous. Having fulfilled the very natural yearnings of a woman to experience the wonders of motherhood, she could now move on to a different stage in her life and retrieve a sense of self-worth.

Looking into her glass, Ellen saw just the sludge of wine remained and pulling herself up to standing discovered that sludge seemed also to have impinged on her brain as the room appeared to spin. Inspecting the wine bottle on the kitchen counter, she was surprised to see the diminished quantity which revealed that she'd managed to consume more than half the contents and on a still-empty stomach. Where had Paul got to? Surely he couldn't still be in the shower? At this rate, supper would be spoilt and she'd be too drunk to care.

Stumbling to the bottom of the stairs, she called his name, quietly.

'Paul, supper is on the table. Are you coming down to join me?'

With no sound from upstairs and no response to her call, Ellen made her way up to the bedroom. Despair, along with utter frustration, came to her in a rush as she entered the room and discovered Paul sleeping soundly, lying on top of the duvet, his nakedness barely concealed by an inaptly selected tiny towel.

8

Marvelling at Google's infinite knowledge, April had carried out her research and was now convinced that the first step would be to book herself a place on a three-day City & Guilds course.

'Training and education in barista skills, ranging from the history of the coffee bean to guidance on good customer service.' She read aloud the course strapline.

Clearly there was a great deal more to be learnt than she had initially imagined, including 'acquiring an understanding of the coffee bean, its diverse flavours, the art of blending, and how to expertly grind and brew the precious seeds'. April read on. 'Understanding how to steam and foam milk and the importance of water quality for the best possible coffee experience.' Who knew there was so much to it?

This was looking like the perfect project for her; in fact, the idea was beginning to grow exponentially as she read on further and discovered the course also gave guidance on how to go about opening your own coffee shop. From having fun learning a new skill, she was now considering the opportunities she could explore on completion of this training. Running her own café was certainly something April had never considered, but perhaps she had finally found her vocation; she could be the proprietor of a successful coffee shop, just like Amarellos'.

Gino and his excellent café might be the route to her starting the whole process, as the course she was contemplating required her to have some experience of working in a coffee shop. With a plan to run some errands in Great Aldebridge the next day, April decided that she would call in at Amarellos' and seek out Gino for a chat.

April resolved not to bother Ian with an update on her scheme until she had secured a positive response from Gino and had a firm booking on a course. She feared he might find any number of negatives to her proposals, or put doubts in her mind. Instead she went to find her phone to call Ellen.

'Hi,' she said cheerily when Ellen answered. 'How are things with you?'

'Don't ask.'

'What's wrong?'

'Oh, ignore the grouch, it's just that Paul has really annoyed me. I'm awaiting an apology – flowers and chocolate. Actually, it might even take jewellery...'

'Goodness, sounds serious. How about joining me for one of Gino's fabulous lattes tomorrow morning and you can tell me all about it?' April said, deciding she should keep her news to herself as Ellen didn't sound particularly receptive to chatting.

'It would have to be an early one, as I've got to pick Liam up at 12. How about 9.30am?'

'That sounds fine.'

'Did you call just to make a coffee date, or was there something else? You're okay, aren't you?' Ellen asked.

'I'm better than okay, and it's all down to you. I've decided I'm going to train to be a barista. Who knows, maybe I'll even look at the possibility of opening my own coffee shop.' April couldn't resist sharing after all.

'My goodness, I wasn't entirely serious, but if it's what you really fancy doing, then I'm all for it. What does Ian think?'

'Well, he thought it was a bit of a joke when I first mentioned it, so I've decided to keep things to myself for a while longer until I'm properly organised,' April said.

'How do you mean, properly organised?'

'Well, I want to ask Gino for some work experience, another reason for a trip to Amarellos' tomorrow, and I need to get myself on a training course.'

'It all sounds very promising, although the barista idea seems to have escalated rather swiftly into a whole new enterprise.'

'And why not? I'm up for it!'

*

'I'm sure this would not be a problem,' said Gino, 'so yes, you can join us on our side of the counter for a few days.'

'That's great news, thank you,' April said.

'I am delighted that you love our coffee so much you want to learn more about it.'

'Oh, I do, I really do.'

'Then how about in two weeks' time? Lucy will be away on holiday, so you could be the extra helping hand and at the same time observe how we do things.'

'Perfect, and I really will try to be helpful and not get in the way, I promise.'

'Phase 1 – tick,' April whispered to herself.

April saw Ellen arrive just as she was taking a seat in one of the booths and making a mental note of the interior furnishings and fittings within the café for future reference.

'So, I start my work placement here in two weeks.'

'Goodness, already sorted. Excellent,' Ellen said.

'It is, isn't it? But that news aside, do tell. What has been going on to make you feel so pissed off with Paul?' April asked.

Ellen told the sorry tale of the meticulously planned conversation by candlelit dinner which had been scuppered and although April was disappointed for her, she found herself trying to suppress a grin at the thought of Paul sleeping soundly, his modesty only just intact courtesy of a well-placed towel.

Clearly she had been upset at the time, but April was relieved that Ellen could now appreciate how this image might evoke humour and they both burst out laughing.

'I'm laughing, but I am sorry how things turned out,' April said. 'I'm loving your thoughts on the catering thing, though. Have you had the chance now to talk it through with Paul?'

'Well, when he was apologising profusely for ruining our dinner date, I decided it would be an ideal time to tell him my proposal. I'm not sure he was entirely on my side, but he wasn't in a position to object too strongly,' Ellen said.

'Good timing, and anyway why would he be against it? After all, your plan involves being home-based.'

'He thinks that until Liam is settled in school, it's too soon for me to be planning an alternative career. But you know Paul is such a proud man, he feels he's the family mainstay, and he doesn't really understand my need to move on to something else just yet.'

'Well, yes, Liam will need you around for a while longer, but I don't think that should stop you having a strategy going forward. Anyway, some women do seem to manage to have a career and kids, although I think they must be knackered a lot of the time,' said April.

'I'd like Paul to be supportive and he must know that I wouldn't dream of neglecting Liam's needs. However, I'm not prepared to give up on having a stimulating working future for myself. It's got to be about me, too!' Ellen was adamant.

April was impressed to hear such a willingness to fight from the usually acquiescent Ellen; this was all very encouraging. After a moment of contemplation, April asked, 'By the way, did you get the flowers, or chocolates, you were anticipating?'

'What do you think?'

April shrugged her shoulders in sympathy and then silently picked at the remaining crumbs of coffee and walnut cake on her plate and those scattered down the front of her T-shirt.

9

April slammed shut the car boot, got into the driving seat and, after pulling at the fabric of her just-washed jeans to adjust the uncomfortable crotch, was settled for the short journey from her home to the village hall. This community amenity had the benefit of a picturesque setting at the heart of the village, being adjacent to a green open space. The sward extended all the way to the river's edge, where the bank was punctuated by ancient willow trees, and the limply hanging branches trailed on the water's surface creating mesmerising ripples.

Picnicking in this recreation area was a favourite pastime for families, and as April drove into the parking area she could see a group of mums sitting on their tartan rugs, surrounded by discarded packed lunches, plastic tumblers, rackets and balls. From the back of her car, April removed a garden spade and then wobbled on one leg as she attempted to change her shoes for wellies. The unsteady nature of this manoeuvre was compounded by the startling squawk of a duck, as scraps of unpalatable sandwiches were being hurled in its direction by a pair of tiny hands. April looked up to discover that the accurate shot had unsettled a lone mallard, who flapped in disgust and then, having unruffled his feathers, returned to retrieve the bread.

Seeking out shade on the sunniest days was not a problem for park visitors, thanks to the row of maturing native trees which had been planted to mark the new millennium. As an added benefit their rustling leaves on breezy evenings acted as a welcome buffer between nearby residents and any noisy activities being held in the village hall. The shabby old building, however, had not enticed revellers to book this tired venue for their evening parties and April wondered if the smart refurbishment would change that trend.

Eager to get a closer look at the revamped building and always happy to mess about in mud, she had gladly agreed to join Laura to assist with a tidy-up of the garden area which adjoined the hall. The grounds in the close vicinity were now a churned-up mess following the recent construction works, and Laura had enthusiastically taken it upon herself to re-establish borders and plant them with a selection of shrubs she had managed to acquire at a knock-down price from the local nursery.

April could see that Laura was already busy and surrounded by many pots containing an interesting variety of shrubs. A tad concerned to see that there was rather more to do than she had anticipated, April wondered if she'd have the stamina for this mighty task.

'Hi there,' she called as she approached her green-fingered friend. 'Looks like your negotiations for the shrubs went rather well. Is it just the two of us? Didn't you manage to enlist anyone from the Gardening Society?'

'No luck, I'm afraid it's just you and me. Once we get stuck in it shouldn't take too long. The most difficult part is deciding what plant goes where, but I have my handy plant guidebook with me.'

It wasn't long before they had the new beds marked out and the shrubs placed ready for the more strenuous job of digging holes in earth which had been compressed by the heavy machinery previously on the site.

Taking a short break before attempting to break the surface with her spade, April leant on the handle of her shovel and watched as Laura continued to dig energetically. Overwhelmed by intrigue, April couldn't wait a moment longer to quiz her friend. 'So, don't keep me in suspense, I'm desperate to hear how your first meeting with Neil went?'

'Ahh… well, actually it turned out to be a pretty successful evening,' Laura said, 'although it took me a while to feel comfortable and relaxed in his company again.'

'Oh… why was that? Is he not quite the man you remembered?'

'Really and truthfully… no, not quite the same. Rather more portly and looking a good deal older. I suspect this has not been helped by the strain of his recent marriage breakdown.'

'Aha, so there was a wife. Anything else different?' April asked.

'If you're referring to his teeth, they were still looking wonderful – a perfect line of tombstones in arctic-fox white.'

'I wasn't actually… but how splendid for him. So, it's sad that he's getting divorced, but of course excellent news for you.'

'Well, yes, but it's clearly not been easy for him and to make matters worse he's also lost his job.'

'Bloody hell, no wonder he looked jaded. It's a bit of a watershed moment.'

'I don't think it's all bad. He says they both agreed that the marriage was over.'

'Okay, so apart from his altered appearance – not that we are so shallow as to judge – did you feel any of the magic returning? Could that old flame be reignited?'

'Honestly, yes… Well, in fact, it's already burning brightly.' April was surprised to see Laura blushing as, quite unnecessarily, she tried to explain her impetuous decision to conclude that first evening with Neil in her bed. Overindulgence in red wine, and one taxi being cheaper than two, had all played a part in what had apparently seemed at the time the obvious finale to the much-anticipated reunion, not to mention the frisson that she claimed had sparked between them.

April was taken aback by this news, even a bit shocked, although she didn't consider herself in the least bit prudish. Wasn't it a little incautious for Laura to be rushing into a physical relationship with a man – a relative unknown – who currently appeared to have several personal issues going on?

'So, a few weeks on, how are you feeling about having him back in your life?'

'It's been wonderful. I'm having a great time. There's just one thing, though…'

'Yes?'

'Well, I'm starting to have some niggling anxiety about his seemingly protracted divorce proceedings and his apparent lack of incentive to get started with a job search,' said Laura. 'I'm not sure what that's all about yet.'

Excellent, thought April, *at least she's being judicious concerning her relationship with this well-upholstered ex.*

'But hell, April, I'm not going to worry too much just now because it's so lovely having him back in my life. Just between you and me I'm so enjoying the fact that there's a man sharing my bed again. It's been like an epiphany.'

April felt that her friend was now erring towards sharing too much information, but the return of Neil was obviously having a hugely positive effect on Laura's life. She hoped it wouldn't be too long before she might meet this man who seemed to have bewitched her friend.

'But enough of that, what about you and the barista plans you mentioned? Are you looking forward to spending some time with Gino?'

'Absolutely. It's a great opportunity. I'll get an idea of what it's really like to work in a coffee shop. Even Ian has stopped making fun and is giving me encouragement,' April said.

'And have you booked the course in London yet?'

'Yes, all booked for the end of the month.'

'Sounds like you have everything in place. If nothing else, it will be an interesting experience. And on the subject of coffee, I don't know about you, but I am very much in need of a break and a hot drink. Let's go inside and I'll give you a guided tour on the way to the redesigned kitchen,' Laura suggested as she stopped digging. April willingly threw her spade on the ground and followed Laura's lead.

She was genuinely impressed as Laura led her through the spacious new entrance lobby into the main hall, with its dual-purpose multisports area to accommodate indoor team games as well as providing a large function room for village events. There was a smaller studio room suitable for Pilates and yoga classes, a further meeting room with adjacent office space and at the far end of the corridor, just fitted (yet to be sullied) shower and toilet facilities. As they entered the updated and extended kitchen/canteen area with its impressive vaulted ceiling and glass-panelled wall, they were able to admire their progress in the gardens.

April stood by as Laura took two mugs out of the kitchen cabinet and spooned in the instant coffee. 'I'm afraid this is the extent of the refreshment on offer, but it'll be a welcome reward for our labours. I think we've done a pretty good job so far out there. Once we've got the paving laid and some more benches in place, it's going to make an attractive outside seating area.'

April sincerely hoped that Laura would be engaging the services of a proper contractor to lay the patio slabs – that task would definitely be beyond her – and she stretched backwards, hands on her hips, to relieve her aching spine.

'Yes, it's certainly all coming together now,' she said, spooning an extra sugar into her coffee.

'I'm just in the process of organising a table-top sale to raise some extra funds so we can complete the gardening and buy some benches,' Laura added. 'I don't suppose you'd give a hand with that too?'

'Sounds like fun, but can I request a job where I'll be seated, please?'

10

April riffled through the accumulation of assorted garments hanging in her wardrobe for a fourth time, having already tried on three separate outfits. With all of the clothes now discarded on the bed, she stood in front of the full-length mirror in just her underwear and, despite the lack of time available, carried out an inspection of her thighs for any signs of cellulite. It was her first morning at Amarellos' and as the café opened at eight o'clock, April reminded herself that in addition to arriving suitably attired, it was vital to get there in good time. Finally opting for a pair of black slacks – which didn't cling to her upper legs – and a white sleeveless blouse, to avoid overheating, she was eventually set for the day.

There were no hold-ups on her journey into town and as she hastened to the entrance doorway, April could see that the coffee shop was already open and serving a couple of early customers. She strode in with as much confidence as she could muster and went towards the counter. Thankfully, she was quickly reassured by the warm welcome she received from Gino's staff, who were busy going about their preparations in anticipation of the day ahead.

'Here's our extra pair of hands. Good to see you, April,' Gino said and then led her to the staff room so that she could store her jacket and bag.

'Whilst we know that you are very familiar with what goes on in Amarellos' from a customer's point of view, this morning you can experience how the staff take the orders, prepare the different beverages and respond to the numerous special requests for individual versions of the coffees on our menu. You will discover that everyone likes their coffee made a specific way – extra hot, extra shots, decaffeinated, skinny, cold milk, soya milk, oat milk. The variations are endless and it is our job to make sure the beverage prepared is exactly to the customer's taste.'

April smiled and nodded, desperate to keep up and follow Gino's melodic, accented instructions.

'Yes, of course, essential to get that right.'

'Absolute consumer satisfaction is so important – that's how we keep the clientele coming back. So firstly, you can spend some time with Gemma behind the counter and then by elevenish you'll be ready to take a break, so help yourself to a coffee. After that you can assist Marco clearing tables, loading the dishwasher and replenishing the crockery stocks. It's quiet at the moment, but it will get busy.'

April tied her apron and took up her position beside Gemma in the limited confines of the serving area. Gemma was one of April's favourites amongst the staff; newly graduated with an English degree, this smiley young woman was biding her time in the café until she found the right job. Youth, beauty and brains did not, it seemed, guarantee an instant pathway to career success. This reinforced April's conviction that her own daughter's excellent choice of a physiotherapy degree would lead her directly to a position of paid employment at the end of her final year.

'Morning, Kathy. Latte with an extra shot?' Gemma was adept when it came to remembering what the regulars ordered and was already preparing coffee for the bag-burdened lady as she walked to the counter.

Gemma was a knowledgeable barista and an excellent communicator when it came to sharing her expertise. April couldn't help thinking that this articulate job-seeker should consider a future career in public relations or even newsreading. She made a mental note to offer this helpful suggestion to her new colleague when they had a quiet moment and a further note to herself to put careers adviser on her own list of possible jobs to explore, should the barista thing not work out.

April was conscious that her shoes were starting to pinch her little toes and was relieved when it was time for her coffee break. Taking her cappuccino, she settled into the last remaining seat, close to the door of the café's toilet facility, hoping to have a moment to relax despite the heavy traffic past her table. The early start and frenetic morning had resulted in hunger pangs pre-lunch and April had permitted herself a mini-panettone, selecting the one with dried fruit to ensure some health benefits. She checked her phone as she sipped her coffee and was delighted to see a text message from Ellen.

Hope all goes well. Paul home late – call in on your way home for debrief. x

April was happy to take on the clearing-up duties following her break, certain this would be the simplest task – after all she knew how to do kitchen chores and housework. Being overly confident proved her downfall, and she lost control of an overloaded tray and ended up with coffee dregs slopped down her shirt and onto the floor. Having mopped up quickly, April went to the safety of the kitchen to unload the dishwasher, but she was distracted by a conversation between two of the staff and failed to notice an upturned knife in the cutlery basket.

The rest of the day went by in a blur – her feet and calves ached, her cut finger throbbed – but the basics had been covered and she left the café at six o'clock feeling tired and satisfied by her efforts. A welcome glass of wine with Ellen was her next destination.

*

April took a seat at the kitchen table as Ellen poured the wine. The debris of Liam's tea was still beside the sink and covered most of the countertop – a lot of mess for one child's mealtime, April thought, but it was probably also the remnants of another of Ellen's baking sessions. The little chap was playing happily with Lego in his playroom; the clearing up was obviously a chore for later, as Ellen appeared keen to find out how April had managed her barista initiation.

'It was good, but exhausting and obviously I was just watching and learning and helping with dishwasher and kitchen chores,' April said, holding up a plastered finger by way of a trophy. 'I'm actually quite looking forward to day two, although I might need to choose more comfortable footwear if I'm going to make it through another day.'

'I'm pleased. It's so different to anything else you've ever tried. Well done and here's to the rest of the week.' Ellen raised her glass.

'Whilst I'm here, Ellen, I wanted to run something by you. I've been thinking of inviting Laura and Neil over one evening for a meal.'

'Oh yes, brilliant idea. She talks about him all the time. I'd really like to see what he's like in the flesh.'

'I'd like to meet him, too, but without seeing too much flesh, thank you! I couldn't be happier for her, but for some reason I feel the need to check him out,' April said. 'I was hoping that you and Paul would join us. Actually, I was also hoping that you could help me with the catering. You know I always panic if anything other than eggs and bacon are called for. I'm opting for a straightforward chicken casserole, but one of your lovely desserts would be most welcome.'

'Perhaps we are just being too protective of Laura and to be honest she's more than capable of looking after herself. But she does seem to have rather lost her head over him,' said Ellen. 'How about lemon tart with raspberries?'

April noted that Ellen's thought processes had swiftly returned to the menu. 'Perfect. I'll try and organise something for next week, before I get totally preoccupied by my barista course. I'll let you know dates and I'm assuming Saturday evening would be best for you and Paul, to avoid him not getting home in time. And how is Paul, or perhaps more importantly, how are things between you two?'

'Oh well, things are much the same. I worry that he's still working much too hard and not getting home till all hours. I'm afraid he'll burn out and become ill.'

'Hopefully it won't come to that. But a state of equilibrium is vital, you know, working and living life.'

'Of course, but I know he's really keen to do well and maybe this is just a particularly busy period. On the plus side my repertoire of desserts is growing.' Ellen was baking obsessively to deal with her frustrations.

'It will be good to have an evening together and, if you like, I'll ask Ian to have a word with Paul to see if he can get a guy's perspective on how things are with him,' April offered, whilst also being certain that Ian would not welcome her volunteering him for this task. 'It's probably just that he's thoroughly immersing himself in his job and hopefully this is temporary and in time things won't be quite so stressful for him.'

'I guess so… and thanks.'

April glanced at the wall clock and got up quickly, wincing as she put weight on the burning balls of her feet. 'Well, it's time I got off home. I'll call you once I've spoken to Laura.'

'Yes, it's time I got Liam into his pyjamas and off to bed. Thanks for the invite, it's been a while since Paul and I have had an evening out. I'll look forward to it. Hope all goes well tomorrow.'

'Thanks again for agreeing to help with dessert, Ellen. Now I'm off to locate my peppermint foot soak and my flattest, most sensible shoes for day two.'

*

The next few days at Amarellos' went by at a pace and April no longer dreaded being on her feet all day in the café; in fact, she was enjoying her work experience and had convinced herself that the goal to open a coffee shop was in no way fanciful.

Admittedly, her time at the café hadn't been mishap-free, but April didn't think she'd been too much of a nuisance. Certainly, when the key to the staff cloakroom was missing for two hours, April had been mortified to discover it hidden in the depths of her handbag whilst in search of a tissue to mop the beads of sweat from her brow. She'd surreptitiously returned it to the key safe, rather than admit the oversight which had resulted in Anna having to go home without her rucksack.

Despite repeated practice, the cash register remained her nemesis, although April was certain it was a seriously dodgy device which Gino should consider upgrading. As far as the preparation of the beverages was concerned, well, whilst she had initially been somewhat intimidated by the highly polished, stainless steel brute of a coffee machine, she had learnt to respect its hissing steam arms after just one minor scald.

'Don't swear at it, April, it's a superb instrument – the best quality, designed and manufactured in Veneto,' Gino told her proudly, whilst holding her arm under the cold tap for ten painful minutes.

The week culminated in an early evening session of taste-testing new coffee blends. Gino's favoured supplier, Miles – a very slender and sadly unattractive individual – came to the café to host the proceedings. When it came to tasting the coffee, with Miles's guidance April began to understand his unconventional flavour descriptions – 'chocolatey, nutty and green apple flavours'. No, perhaps that green apple thing was difficult to comprehend.

It was a fun evening, which concluded with everyone getting to vote on their favourite blend of the night and once Miles was despatched, the weary team prepared to go to their respective homes.

As April pulled on her mac in readiness to leave she thanked Gino and his crew of baristas.

'It's been great, but I'm looking forward to being served the next time I come into Amarellos', rather than being the server.'

A small gift for Gemma was clearly deserved – she'd been inordinately patient – and before leaving April handed the offering to her young mentor.

'A young woman with your looks, dulcet tones and communication skills is, quite frankly, wasted in a coffee shop,' she whispered. A carefully considered search on the Amazon Books website had led April to her selection – *A Career Handbook for TV, Radio, Film and Interactive Media.*

'Oh… thank you so much, April.'

April could see that Gemma was touched by the gift and remained totally unaware of her temporary colleague's scepticism regarding a future career on TV.

'You're very welcome. And I look forward to seeing you presenting *Good Morning Britain*, alongside Susanna Reid. Never forget that the world is your oyster…'

11

She felt the heat on her face and as Laura caught sight of her reflection in the cheval mirror standing beside the bed, she also noted the tousled appearance of her hair. The inner glow which follows spontaneous, ardent lovemaking had coursed through her veins and was now radiating its energy onto her complexion. Laura no longer recognised herself; her libido had certainly been recovered, but she wondered if she had now lost all sense of decorum.

They were due to go to April's for dinner and Neil had arrived an hour earlier, looking rather dishy in a new shirt and his best jeans. Laura was now having an influence over his wardrobe and encouraging him to accompany her on long walks with Max. Whilst his dress sense was much improved, any weight loss as a result of the new regime had yet to take effect. Those walks in the open air with Max were doable, but being confined inside the house with her dog resulted in an irritating bout of coughing and the need for several puffs of his inhaler. It seemed that Neil was indeed allergic to her four-legged companion.

The pair were now lying side by side looking dishevelled, the duvet was a crumpled mess on the bedroom floor and there was the distinct possibility that they would arrive late for the dinner date. Laura glanced at the clock on her bedside table and leapt up in a panic.

'Damn it!' she shrieked. 'We're going to be late if we don't get ready and on our way to April's within the next fifteen minutes.'

'You look gorgeous when you're all hot and sweaty,' Neil laughed. Laura was aware of him staring at her naked body and tried desperately to contract her abdominal muscles as she rushed about the bedroom retrieving her underwear.

'And you, you look like a grizzly bear... an old bear. You do know all that greying body hair fails to disguise the size of your paunch.'

'That's a bit blunt, baby.'

'Sorry... sorry, just teasing, but please get up and get dressed, Neil, I really don't want to be late. It's incredibly kind of April to invite us both for dinner and it would be rude not to arrive on time.' Laura rushed into the bathroom to adjust her make-up. On looking into the mirror, she decided that the rosy cheeks were an attractive addition to her appearance, but the smudged mascara definitely needed reparation. From the bathroom she was dismayed to hear Neil farting as he finally hauled himself off the bed; she raised her eyebrows and questioned the niceties of having a guy in residence.

As she brushed her hair in front of the bathroom mirror, she wondered whether she could enquire again about his non-existent search for employment without sounding like a nag. She was a little tired of funding their outings, which seemed to be the expectation in view of his current limited means. He'd also not mentioned the divorce recently and seemed quite contented in his shabby rented cottage. But she could ask him tomorrow; tonight was about having a fun evening with friends.

'Are you ready, Neil?' she said as she left the bathroom.

'Absolutely, ready and prepared for interrogation.'

'April and Ellen are not going to interrogate you,' Laura said as she adjusted the neckline of her navy lace dress. 'They're just keen to get to know you. And it makes a really nice change for me to be part of the third couple at a party, rather than the lone spinster.'

'It's a mystery why a racy lady like you didn't get snapped up years ago,' he said as he squeezed her buttocks. 'Anyway, I promise to be on my best behaviour for the meeting with your girlfriends.'

'Just be yourself, but please do try not to fart in their company,' Laura said, as they made their way to the front door.

*

April slipped on her oven gloves and pulled open the glass door to check on cooking progress. All appeared well: the coq au vin was just at a simmer – her favourite and infallible recipe – the dauphinoise potatoes were bubbling at the edges of the dish and the top was golden. She'd opted for a simple starter of tomato, basil and buffalo mozzarella salad, and the colourful medley was arranged on her square black serving plates ready for delivery to the dining room. Holding dinner parties had the potential to see April's stress levels rising dangerously high; even with close friends she approached these occasions with a degree of trepidation; her desire for perfection was difficult to consistently achieve. Incidents of mistaking teaspoon measurements for tablespoons and incinerated flans caused by distracting TV dramas were just a few of the mishaps which befell her.

'Ian,' she called, 'I think it's about time I allowed myself the first drink of the evening, how about you?' Tempted though she had been, she knew that starting on the wine before the dinner preparation was well under control could prove disastrous; she had dropped a tray full of Tesco Finest canapés the last time she'd been so foolhardy.

'I'm just coming down, pour me a large one, I feel the need.' April knew that Ian was not especially looking forward to the evening; social occasions didn't do a lot for him, although he was fond of Ellen and Paul, and indeed Laura. But this gathering involved bringing Neil into their fold in order to check him out and April was quite certain that her husband wouldn't approve of her strategy. She could rely on him to be a good host, but was anticipating surreptitious clock-watching and the swift clearing of plates by Ian after each course in the hope that the guests would all go home reasonably early and he could get off to bed.

'Now don't forget you promised to have a quiet word with Paul,' April said as Ian arrived in the kitchen. 'Find out what's going on with him.'

'I'll try, but I really wish you wouldn't ask me to do things you know I feel ill-equipped to manage, not to mention totally uncomfortable with,' Ian said.

'He's a friend, surely you can ask him how he is?'

'Okay, but personally I don't think Paul will appreciate me probing. That's your domain and you seem so much more proficient at getting away with it than I do.'

'Just do it, please… Oh, and how does my bum look in these trousers?' April asked anxiously turning around for the necessary inspection. 'I don't seem to have lost any weight since my last weigh-in and I bought these trousers in anticipation of losing a further inch off my arse!'

'You look nice, you always look very nice.'

April really should have known better than to ask, 'Why do you always say "nice", and why do you never spontaneously compliment me on my appearance?'

'You look great, stop worrying. Laura and Ellen are your friends, you don't need to impress them.'

April sighed, but was quickly distracted by headlights shining through the window and the sound of a car turning into the drive. Switching into social-gathering mode, she made her way to welcome her guests.

'Lovely to see you both,' April greeted Ellen and Paul, shepherding them inside out of the cold and taking their jackets to hang on the hat stand by the front door.

'Dessert is safely delivered. I've been holding on to the plate to avoid any mishaps. That BMW seems to incite Paul to drive faster than is absolutely necessary.' Ellen looked aggrieved but then relieved once she'd taken a peek under the foil covering.

'I've brought along some raspberries as well. They should make a lovely accompaniment to the tart… Oh, and some crème fraîche, just in case.' Ellen smiled.

'Oh Lord, I'd forgotten all about cream, bless you.'

*

The two women made their way into the kitchen and Paul followed Ian into the sitting room. 'What can I get you, Paul?'

'A large gin and tonic would be most welcome. Ellen's driving home this evening, but we could always walk if she gets carried away with the wine, it's not much of a hike to ours. Having said that, I feel too knackered this evening to take on a hike to the end of your drive.'

As Ian handed over a generously filled glass, he noticed Paul did look weary and there was a greying of the hair at his temples that Ian didn't recall seeing before. Perhaps there was something untoward going on and April was right to be concerned about his wellbeing and, by association, that of Ellen.

'Take the weight off your feet,' Ian said, rolling his right shoulder to ease the tension, as he directed Paul to one of the precisely aligned armchairs. April's desire for perfection in all things, including the staging of furniture in the rooms of their home, had a tendency to result in Ian wrenching an arm – or some other part of his anatomy – whist lugging items around under her supervision to find the most aesthetically pleasing position for all their household paraphernalia.

Paul sank into an armchair and plonked his feet on the appropriately placed footstool.

'So, how's the job going? I imagine it's a bit of an energy drain.'

'The job's fine, it's all the travelling that's a pain in the arse.'

'April tells me you're having to commute into the city quite a bit,' Ian continued. 'I don't suppose that was part of your plan when you moved out here, was it?'

'I always knew that with certain projects and for some of our clients I'd need to be in London, but at the moment I'm on the train commuting at least three days a week and the long hours are a killer.' Paul sighed, blowing out his cheeks as he did so, before taking a further large gulp of his drink.

Ian pressed on, wondering what exactly he needed to find out in order to satisfy April's curiosity. 'But if you're enjoying the job, isn't it worth hanging on in there? Perhaps in due course there'll be an opportunity to negotiate a slightly different role – more family friendly, and with less travelling?'

'Perhaps… but the truth is, despite the stress I do love it. The problem is dealing with Ellen's displeasure when I get home late.'

'Ah, yes, I see.' Ian didn't really see and wasn't sure he wanted to hear details which might reveal that April's lovely friend was a tetchy, complaining wife. Thankfully he'd managed to hesitate long enough for Paul to continue.

'I really don't appreciate her nagging. What she seems to forget is that I'm working hard for us, for our future.' Paul drained his glass and held it out for a refill. 'Anyhow, she's preoccupied with Liam these days, or more precisely Liam and cake-making…

There's little time for me. Being away from the house for a few extra hours is often quite appealing.'

This didn't sound promising for a harmonious home life and Ian started to wonder how to put a more positive slant on Paul's perception of his current situation. Something upbeat and empathetic seemed to be called for and after a great deal of introspection Ian managed to draw on his own experience to come up with what he hoped was a helpful response.

'Things inevitably change when you have kids and it does take a bit of effort to ensure you make time for yourselves – as a couple. I'm not saying I found it easy when Megan arrived on the scene – the adjustments and the necessity for compromises. Life as parents is a different life. But I'd strongly advise against neglecting your partner.' Ian was pleased, convinced the chat was going well. He was seeing things from Ellen's point of view; April would be impressed.

'I get that, but I'm not prepared to be the one who's making all the effort. It's not a one-way street.'

A loud knock on the front door interrupted their discussion, for which Ian was truly grateful.

'I'd better get that… Here goes, it's time for us all to meet Laura's new man.'

*

Laura and Neil joined the gathering and introductions were made, swiftly followed by the serving of more drinks by Ian. April felt reassured that her husband had slipped comfortably into his welcoming host role; she just hoped he'd been equally efficient delving into Paul's state of mind. She turned her attention to assessing Laura's companion and her initial impression was an unexpectedly positive one. The newcomer seemed unfazed to be the unknown element within the group of longstanding friends and on first inspection Neil appeared to be an affable man – yes, she had to admit it, he was rather charming. Perhaps she'd been uncharitable with her preconceptions and he might indeed be a promising match for Laura.

April started to relax and decided that before another round of drinks was poured, she should seat her guests at the dining table and line their stomachs with some sustenance.

'This looks lovely, I hope you haven't gone to too much trouble… spent the day feeling stressed.' Laura smiled as she sat down opposite Neil and next to Paul.

'Not at all and you'll be pleased to hear that the dessert is provided courtesy of Ellen, so there'll be no need to politely avoid one of my disastrous attempts at a sweet course this evening. The thing is, Neil, I'm not great at making puddings, although I find them very hard to resist. My padded hips will vouch for that fact,' April laughed.

'Now, April, there's nothing wrong with a bit of flesh on the bone. A woman isn't properly womanly without love handles,' Neil said, and seemed to assume this was a flattering remark, although April wasn't entirely sure. 'Us guys like something to grab hold of, don't we, chaps?'

Ian and Paul smiled, but April had never heard her husband express such a desire. With calorie-counting being such a bore, she could only hope that Neil might be correct in his assumption, although she did glimpse Laura's raised eyebrows.

As the evening continued, April was relieved to see that the dishes she'd prepared were well received, with plates being cleared almost as swiftly as wine bottles were emptied. There was a party mood to the get-together, which delighted her, as did Ian's efforts to get to know the new member of the group. She hoped his conversations with Neil were revealing useful information about the interloper. Despite the utterance of one or two rather risqué comments, he was a likeable, not bad-looking man, with lovely blue eyes and, she had to admit, pretty well-arranged dentition. She resolved to make no further judgements and try to get to know him properly because it was so good to see Laura flushed with happiness, or flushed with something – April wasn't sure, but decided not to dwell on that thought.

The social occasion looked set to be rated a resounding success, with palatable food, good company and easy conversations covering all manner of topics, from the serious to the light-hearted: the change in the weather, to worrying episodes of flooding, to the property markets – April's personal favourite – to finding the best supplier of park benches for the village, and not forgetting the most topical – and Ellen's personal favourite – which contestant would be the worthy winner of *Bake Off*.

This was all very jolly, but buoyed by wine April was unrestrained and ready to embark on a more direct line of questioning. After all, she didn't want the evening to come to an end before she'd further satisfied her curiosity regarding Laura's man.

'A toast to Neil, everyone, and welcome to our perfect circle,' April slurred, with everyone duly following her lead and raising their glasses. 'I can't tell you how delighted I am that after decades apart you and Laura have found each other again. It's such a lovely story – heart-warming – and soooo romantic.'

Across the table April saw that Ian was shaking his head, which she thought was an odd gesture. With a clearer head she would have recognised his agitation as she veered towards dangerous territory.

'April, I can't tell you how happy it has made me to find again the woman I always considered to be the true love of my life,' Neil gushed, whilst reaching out across the table to tenderly hold the hand of his woman. 'And despite the passage of time, this gorgeous old girl still knows exactly how to please a horny old man!'

April giggled as she noticed Laura blush and Ellen's mouth open then shut again, seemingly struggling to find a comment that might be appropriate to follow such an admission. Well, Laura's man was certainly very down to earth.

'I hear you're currently renting a rather quaint cottage in Elmdon,' April continued. 'Any plans to buy something? You don't want to be out of the property market too long or you'll find it really difficult getting back on the ladder.'

'I'm in no particular hurry and anyway I've a few things to get sorted before I can start to think about financing a house purchase.'

'Oh, forgive me. I was very sorry to hear that circumstances mean you've ended up without a job. Must be difficult, but I imagine you'll be looking for something else. A fresh start is probably just what you need right now. Time to move on – so long as you continue to keep our lovely friend Laura as happy as you obviously have been doing. She's the best and she deserves the best.'

April heard Ian tut as he walked to the kitchen with a stack of dessert plates which he'd noisily retrieved from the table. *Must be past his bedtime*, she thought.

'Well, one thing is certain, when I do find a job it's not going to involve getting onto the numbingly tedious, eye-wateringly expensive commuting treadmill, Paul. I don't know how you cope with all that travelling every week. Absolute bloody nightmare. Lonely old existence, too,' Neil added.

April looked across the table and saw that Paul, who had started to nod off, was stirred from his slumber on hearing his name. 'Err... well, not necessarily lonely, Neil, my fellow commuters can make the journey a companionable experience. You'd be surprised how camaraderie develops when you travel with the same folks on an almost daily basis.' Paul's words ran one into another, a further victim of the copious amounts of alcohol consumed. 'But it's a fit young man's game for sure, so you'd probably be wise to rule it out.'

Despite the stage of her inebriation, April sensed the party atmosphere deteriorating and received confirmation of this when she spied Ian through the open kitchen door looking her way whilst drawing a horizontal hand across his throat in a 'cut' motion. It was, therefore, somewhat of a relief to see Ellen getting to her feet in readiness to depart, clearly sharing the sentiment.

'We'd better make a move, don't want to keep the babysitter up too late.' Ellen went to retrieve her jacket and the others roused themselves for departure.

'The tart was absolutely fabulous, darling.' April stumbled as she went to open the front door. 'I'll call you. Drive carefully.'

'Thank you for a wonderful evening, April.' Ellen hugged her host. 'Bye, everyone.'

'Bye bye and it's been lovely to meet you all,' Neil said and April was certain she felt his hand brushing her buttocks as he exited past her into the darkness outside.

12

The rolling wheels of the early morning train squealed as the locomotive, followed by five carriages, came to a halt alongside Platform 2 where April was standing. It was the first day of her much-anticipated course and she was excited, but also apprehensive. Her main worry was that she'd find herself in a classroom with a group of adolescents who would look at her with incredulity, wondering what a woman of their mother's age could have been thinking when signing up for a barista course. But surely it wasn't a talent exclusively suited to the young and her challenge would be to prove that advancing years should never be a barrier to mastering new skills.

She walked through the carriage, looking out for a vacant forward-facing seat – in a bid to prevent her motion sickness – and once settled took her book and reading glasses from her bag in preparation for her journey. April was an unaccustomed commuter and hadn't been prepared for the volume of travellers embarking on the same journey as herself. A further two stations along the track and the carriage was at capacity; it would be standing room only until they reached their destination. April found the very close proximity to unknowns disagreeable and being unable to concentrate on the words on the page in front of her, she decided to take in the passing scenery instead.

It was a cold morning with a layer of mist clinging to the fields and resting on the surface of the river as the train made its way through the flat countryside. April thought it an eerily beautiful scene and for a few moments took delight in spotting a buzzard, soaring around the tree tops, whilst young moorhens darted across the water to find sanctuary in the bankside reeds. She took the train into London infrequently and whenever she did get around to organising a ticket for a day's shopping in the city, she always enjoyed the outing, causing her to wonder why she didn't make the effort more often. The downside was that these trips usually involved a very early start and attempting to stir before seven o'clock was always a struggle for April. The previous night she had repeatedly checked to ensure her alarm clock was accurately set. It was such a nuisance that Ian, who rose with the dawn chorus and provided her with the first mug of tea each morning, was away at a two-day conference. Having suffered a restless, wakeful night, April began to fear the effects of an intense day of tuition, followed by the return trek home, on someone of her mature years. Dear God, perhaps she was too old for all this.

With the view through the window now taking in the backs of neglected terraced homes, where gardens were decorated with discarded domestic items rather than flowering plants, April returned to her book. She read for a while, but the character responsible for the murder was too easy to spot and once again she found herself distracted – this time by the behaviours and conversations of those around her.

April noticed the smartly attired young man across the aisle who, clearly preparing for some sort of work presentation, was desperately trying to memorise the contents of his paperwork, his mouth moving silently as he recited his speech to himself. Intrigued to see whether she could lip-read in order to discover the topic of his study, April concentrated her attention on his rather lovely cupid's bow. As a result, when he unexpectedly dropped his head onto the paper-littered surface of the table in front of him, she jumped in fright, losing her grip on the paperback, which disappeared under her seat. The young man groaned, seemingly not from any pain he may have sustained, but from his desperation, as he began to screw up the documents which continued to vex him. April felt sorry for his distress and hoped this didn't mean he was on a pathway to a day of failure; life could be so tough for the young, which reminded her that she must remember to phone Megan that evening to see how her revision was going.

Having rummaged under her feet, she retrieved her battered book and stowed it away in her bag. Minutes later April's attention was drawn to a mobile phone conversation. The obtrusive chatter came from a young woman and April's curiosity was caught up in the rather distressing news she was sharing.

'Poor Mia, it's such a shame... No, really it's been quite painful for her, I don't think she's been able to sit for days!'

Good grief, April was now both shocked and intrigued to learn the outcome to the story.

'I mean, who'd have thought it...', the young woman continued now with slightly hushed tones, 'anal warts... Yes, I know... absolutely, very nasty. Poor darling... although it should clear up in a few days... the vet says the treatment is very effective.'

Thank goodness; this was clearly a canine friend in trouble and although April was initially greatly relieved, she then felt terribly sad for the poor dog's discomfort.

Enough eavesdropping. April decided to sit back, close her eyes and ignore all around her, as the train progressed on its journey. When the announcement came of the imminent arrival at London Liverpool Street station, April gathered her bag and coat and prepared to leave the carriage.

The barista school was a walk away from the station and she was happy for the exercise after sitting on the train for an hour. Fifteen minutes later she arrived at the smartly converted warehouse, easily recognisable as her destination from the website descriptions, and having followed the helpful signage inside the building was soon seated in the classroom with her fellow students – predictably the young crowd she had anticipated, but they seemed a friendly enough bunch.

The morning session got underway with the course tutor, a short and slightly effeminate young man called Guy, enthusiastically in charge of proceedings. The first PowerPoint slide gave an overview of the programme, which explained that the three days would fully cover every detail relating to coffee production, barista skills and opening a coffee shop.

'So, let's go right to the beginning of the coffee story,' said Guy as he moved on to the next colourful slide, featuring maps and plants. 'Where do coffee beans come from? Well, of course, they grow on trees. The coffee tree grows happily between the Tropics of Cancer and Capricorn, being a tropical evergreen shrub. The flowers of these trees, which incidentally have the most amazing aroma, develop into fruits once blossoming is over. These fruits are known as the coffee cherries and each contains one or two seeds, known as green coffee beans.'

Interesting, factual and beautifully illustrated, April felt encouraged by the quality of the tutoring.

Guy continued, 'So green beans must, of course, be roasted if we are to enjoy our favourite beverage. Although, as an aside, you might be curious to learn that green coffee bean extract, from unroasted coffee beans, is thought to be a great aid to weight loss.'

Now that was intriguing and April made a scribbled note to remind her to Google for further information on sourcing this very useful extract.

'There are varying stages of roasting which enable the producer to control the density of flavour, allowing a different roast to be achieved to suit every taste. The very first stage of roasting is called a "City" roast, whereas roasting to a darker stage creates a "French" roast,' said Guy.

April's mind was diverted by the thought of nicely browned French *toast*, drizzled with honey and served with blueberries and crème fraîche. Goodness, breakfast seemed a long time ago; hopefully it wouldn't be a late break for lunch.

'So how many cups of coffee are being drunk across the world each year – anyone care to guess?' The group of students looked to each other for an inspirational answer to offer Guy. Failing miserably, they were amazed to learn that it was in the region of half a trillion cups and that a phenomenal 25 million people were employed in the coffee industry. April tried to imagine how many circuits of the M25 half a trillion cups lined up would cover.

Having discovered more about the growing and harvesting of coffee beans, the morning session came to an end and April was finally permitted to consume her packed lunch of peanut butter and jam sandwiches brought with her from home. The afternoon saw the start of the practical instruction and Guy began by familiarising the group with the workings and different parts of a coffee machine, typical of those to be found in the best coffee shops on the high street. He explained the role of the grinder and the blending of coffee grains to produce the variety of flavours. They were shown how to master dosing, tamping and finally extracting the aromatic brown liquid from the coffee grounds. April's sojourn at Gino's had been time well spent; she was able to follow the instruction with the benefit of some basic knowledge.

'Now that we have an idea of how to create the perfect espresso, we'll finish off the day by covering the techniques for milk-steaming and how to produce micro-foam, the frothy, tiny bubbles which make a beverage sweet and delicious.'

Day one drew swiftly to a close and with a severe case of information overload, April said her goodbyes and prepared to make her way back to the train station. Outside, dusk was making its move towards darkness and the temperature had dropped; April shivered and buttoned up her coat against the cold as she hastened along the pavements.

Once at the station, she read the departures boards and was pleased to see that her train was on time and just seven minutes away. She walked towards the designated platform, along with hordes of fellow passengers, and was delighted to spot Paul ahead of her in the crowd obviously making his way to catch the same train. April was attempting to make her way forward through the throng, thinking how lovely it would be to have some company on the journey home, when she noticed that Paul did not appear to be alone. Holding back now she could see that walking closely beside him was an attractive, smartly dressed and extremely well-shod young woman, with the most glorious shiny, long, dark hair. They were chatting amiably, laughing and apparently heedless to all around them.

Instinctively April kept her distance, whilst also making sure she could see which carriage of the train Paul and his companion entered. She didn't want him to know she was there and had seen him, but wasn't quite sure why she should feel the need to hide away. Trying not to allow curiosity to swamp her sensibility, April settled herself in one of the last available seats in a carriage further along the length of the train.

The train ride home thankfully passed quickly; the scenery now obscured by the darkness, she closed her eyes and allowed the numerous facts of the day to litter her thoughts and in so doing she suppressed any preoccupation as to the identity of Paul's fellow traveller.

*

The following two days of the course were just as exhausting and intensive, but April coped with the tiring commute and the learning process. Inevitably on the journeys to and from the capital she had been tantalised by the prospect of a second sighting of Paul and his friend, but there were none and she had to remain curious as to who the young woman might have been.

April and the youths had experienced, amongst other highlights, the sensory tasting session. 'This is the best bit of all.' Guy clapped his hands to ignite their excitement. 'A technique called cupping will allow you to assess the uniqueness of each coffee,' he explained. April was delighted to discern the 'rich round flavours and dark chocolate aromas with hints of pepper'.

The final day focused on opening and running a coffee shop, including finding a suitable site, sourcing equipment and establishing links with suppliers. April began to search her mind for a potential location, or existing outlet, in her hometown that might be suitable for a new coffee shop enterprise.

'Of the utmost importance to your business is to provide excellent service, as this will ensure your customers are happy and keen to return to your establishment. There are so many coffee shops opening in our high streets, you cannot afford to skimp on offering the best of everything to your clientele.'

April knew Gino was an expert when it came to customer relations and customer satisfaction. She recognised that she didn't quite have Gino's limitless patience; she had witnessed questionable behaviour and bad manners whilst on duty at his café and seen him deal diplomatically with offenders. April's tolerance, however, had been tested by the groups of young mothers who sat gossiping, oblivious to the disruption their unruly toddlers were causing as they raced around the café. But the most annoying were the customers who continued to talk on their mobile phones whilst being served, not taking the time to speak directly to the staff or even make eye contact – which was just plain rude.

Ian had often chastised her for being uncompromising and dared to express his doubts as to her suitability in a customer services role. But if she owned her own café, she would make the rules and therefore bar anyone who did not follow the script, which would definitely solve that problem.

The course was concluded and having successfully completed the prescribed three days, April had gained her accredited qualification. She left the barista school on that final day feeling exhilarated and proud of herself for attempting something totally out of her territory of comfort. It had been a confidence-booster and she was now determined to find an opportunity to utilise her new barista skills. All she needed was the finances to back her enterprise and an empty shop in a prime location. Surely that must be possible?

13

It was Monday morning and Laura and Neil were visiting the local garden centre. Laura was on the hunt for some plants to fill the terracotta planters which adorned the entrance to her cottage. She loved any excuse to browse the rows of garden plants on offer, taking her time to select striking and colour co-ordinated annuals to sit alongside her choices of shrubs designed to provide structure and interest to her displays. She was particularly taken with the graceful, striped leaves of the phormium and in an attempt to claim the finest specimens was removing individual pots from the racks to examine them in isolation. Laura had gained useful knowledge from her gardening friends and knew to check for any signs of disease, dead or misshapen leaves and having dismissed three pots she was satisfied with the two remaining plants, which exhibited vibrant green and yellow foliage.

'Gorgeous. These will be perfect.'

Neil was tolerating this expedition and with hands in pockets had been following closely behind Laura as she systematically trawled each aisle. His lack of interest in anything horticultural was clearly apparent, not least because Laura had seen his unkempt garden. Despite the fact that it was a rental, Laura still couldn't fathom why he made no effort to keep his home environment tidy. On warm, sunny days he seemed happy to sit out in his deckchair amidst ankle-high grass, enjoying a lunchtime beer and sandwich, whilst reading the newspaper, with not a thought for firing up the lawnmower or cutting back the overgrown bushes which surrounded him.

'Great, you've finally made your choices.' Neil was delighted to see Laura place her acquisitions in the trolley. 'I wouldn't mind getting a coffee whilst we're here and I couldn't help noticing that they have some lovely cakes on the counter in the cafeteria.'

'I wasn't actually planning on staying to have a coffee. I'd really like to get these plants home and I'm also planning to visit Mother today. I'm sure I told you.'

'Did you? Can't recall. Oh… won't we even have the chance for a quick lunch out together today?'

'I'll make you a coffee and a sandwich when we get back, if you like, but really I'll have limited time.' Laura was quite sure that she had explained her plans whilst Neil was making his way through the plate of fried eggs and bacon which he'd made for himself in her kitchen whilst she had been taking a shower. She had the distinct impression that he had too much free time on his hands and was clearly looking to her for a distraction.

'I don't know why you bother to spend so much time visiting your mother, Laura, you've said yourself that she often doesn't even remember that you've been,' Neil said. 'Why don't you give it a miss? If we're not having a coffee break here, then perhaps we could try out lunch at The George. I hear the steak pie and chips are to die for.'

'And all that animal fat and cholesterol might well be the death of you.' Getting irritated, Laura just wanted to make her way home, bid farewell to Neil and spend the rest of the day as she had intended, doing the things that suited her without a hanger-on.

They'd been seeing each other for almost four months and Laura had noticed Neil becoming more demanding of her time, no doubt because his days were pretty empty and unstructured. He really needed a job, or even a hobby.

'Did you get your application off for the job I pointed out to you? I think your experience in project management would stand you in good stead for that role and the salary was pretty good.'

'Ah, yes that… Well, I did start filling in the online form, but it was all so long-winded and boring. To be honest, the job sounded dull and I'm not sure it's what I'm looking for,' Neil said dismissively.

Laura was in the wrong mindset to accept this excuse without further probing. 'But you don't really seem to know what you are looking for, Neil, and it can't be doing you any good being at home with a growing gap on your CV,' Laura said. 'I know filling in application forms is tedious, but you've got plenty of free time and I do think it would be great for your self-esteem to get back into the workplace.'

Neil sighed and seemed disinterested, but Laura was determined to make him appreciate just how disappointed she was at his obvious lack of motivation.

'If you really can't see job opportunities that suit your skills, how about doing a bit of volunteering? There are so many charities asking for help. You'd get to meet people, make new contacts and it could even lead to paid employment.'

Neil grimaced. 'That's not really me and anyway I do my bit for charity. I always put a quid into those collection tins people rattle on the streets. Anyway, I'm not going to rush into something I don't want to do, especially as there's no real need. The financial settlement from the divorce will sort things for me. After all, Catherine is part-owner of her father's business and I made a contribution to its success. I don't want the solicitors to think I've landed myself a well-paid job and then try to convince me I'll have to settle for less.'

Laura was disconcerted to hear Neil's comment.

'So, are you saying that your only consideration at the moment is to get as much as you can out of your failed marriage to Catherine?'

'And why shouldn't I? I'm out of work because of Catherine. So yes, as you're asking, my plan most definitely includes screwing her and her old man for every penny that I'm due. Anyway, they've got plenty of money.'

Laura was shocked by Neil's blatant and calculated greed. Surely this was a very different man to the one she had fallen for all those years ago.

'You're out of work because your marriage to Catherine is over. You're not blameless for the relationship breaking down. I'm quite sure our affair was not your only transgression during the time you and Catherine were together.'

Neil smiled coyly and the look on his face made it clear to Laura that their fling was not his only impropriety, something that until that moment she naively hadn't even considered.

The silent drive home was only interrupted by Neil's exaggerated coughing fits. Despite the absence of Max on this particular trip, he claimed the presence of dog hairs in her car – even after Laura had thoroughly vacuumed the upholstery – affected his ability to breathe freely. At the moment she didn't care if he could breathe freely or not and as soon as she had halted the car in her drive, Laura asked him to go. Wisely, he did not question his dismissal and got into his own car and drove away. She was glad to see the back of him and his wheezing.

Left feeling troubled by Neil's disclosures, Laura was jolted into the realisation that the initial euphoria of the relationship was starting to wane and she had even begun to recall her single life with some longing. Neil's neediness was becoming draining – hardly the recipe for a contented and fulfilling relationship – and his comments that morning had been thoughtless, particularly concerning her mother. The realities of how he had landed himself in his current situation were also becoming clearer and she felt annoyed by her uncharacteristic gullibility.

One thing was certain: this predicament needed further contemplation and Laura promised herself that later she would take Max out for a long walk on the beach; the open air and the quiet of the late afternoon would be more conducive to clear thinking.

*

Rose had a pretty pink blush to her sunken cheeks, her eyes were bright and she appeared to be in hearty form when Laura found her in the sitting room of the care home, playing cards with a couple of the other ladies. Her mother smiled as she entered the room, excused herself from the game and led Laura through to the conservatory so that they could sit and chat away from the rather noisy television game show programme, which seemed to be the cause of great amusement for a group of elderly gentlemen in the adjacent day room.

'How are you, Mother?' Laura asked. 'You're certainly looking well today and that blue cardigan suits your colouring perfectly. I don't recall that one, has someone given it to you recently?'

'I'm well, darling, and thank you for the compliment. This is the cardigan you gave me for my birthday, you remember.' Rose smiled at her daughter.

Another mix-up in the laundry room, Laura thought, having never before seen the garment, but it was an excellent accidental acquisition.

'But, how are you?' Rose asked as they settled themselves on cane chairs, in amongst the potted palms, with the comforting warmth of the sunshine streaming through the glass windows. 'You seem a little troubled, is everything okay?'

An impressive moment of mental acuteness, Laura noted with satisfaction, although the last thing she wanted was for her mother to be worried about what was going on in her daughter's life.

'Do you remember that I mentioned I'd recently been seeing an ex-boyfriend, Mother?'

'Did you, darling, aren't you a bit old for a boyfriend?' Rose giggled, which made Laura laugh, because yes, she probably was too old and too set in her ways for a boyfriend, especially one with issues.

'He's an old friend from many years ago and we've been dating recently, but I'm beginning to think he might be past his "best before" date,' Laura mused, although she was perhaps expecting a little too much for her mother to grasp her meaning.

'I remember when your father and I were dating, on summer evenings we used to sneak off to the woods up the hill, behind my parents' house. We'd make love lying on a picnic rug in the dappled sunlight under the canopy of the trees.' Rose looked up to the blue sky and closed her eyes as she was bathed in the sunlight, lost in her memories. 'It was a glorious time and your father was such a gentle and thoughtful lover. He was my first, you know, and intercourse with him was such a blissful experience.' Rose reminisced happily, with a wistful look in her eyes, much to Laura's discomfort. Hearing the details of her parents' love life did not seem at all appropriate and feeling awkward she decided it was time for a diversion.

'Would you like to take a stroll around the garden?' Laura asked. 'It's a lovely sunny afternoon, although we'll need to put our jackets on. There's a chill in the air despite the sunshine.'

'That's an excellent idea, let's go and see the Koi carp in the pond and maybe we could wander over to the aviaries. I want to see how the new penguins are getting on. I do love the gardens here, Laura, they are always so colourful.'

Laura prepared to accompany her spritely, if slightly confused, mother on a tour of the grounds and was happy in the knowledge that she appeared to be in good health and was clearly contented. Rose might truly believe her care home had acquired some penguins (parakeets, perhaps?) – she may not recall every visit that Laura made (as Neil had so needlessly pointed out) – but they were still able to enjoy each other's company. Thankfully she had managed to remove herself from the earlier despondency of a troubled relationship. She was keen to get outside into the fresh air, amongst the aromatic rose gardens, and was quite willing to help her mother seek out a few penguins.

14

The school lunch break was just coming to an end and Ellen was waiting outside the gates for the younger children in the reception class to finish their day. Being a summer baby, Liam was one of the group initially timetabled to half days; they joined their classmates to get used to the routine of lunch at school, and then they departed. Next term he would be in full-time schooling and, despite tears – hers not her child's – on the first few days at the village school, the adjustment had been swift and Ellen was now looking forward to having even more time to herself.

Liam's first term was going well and he seemed happily settled, which was a welcome surprise as far as Ellen was concerned, particularly after the experiences of his pre-school nursery days. He had also made a new friend and Ellen chatted to Caitlin, George's mum, as they both waited for the bell to ring.

'I'm so pleased that Liam and George have struck up a friendship and especially that George has someone to play with at break times,' said Caitlin. 'One of my recurring nightmares before he started school was that I'd be walking past the playground, on the way to the post office or somewhere else in the village, and I'd see him standing all by himself whilst all the other children were playing together.'

'Oh yes, I certainly recognise that concern, so I'm delighted that they seem to be getting on well. I'm sure it's also helped Liam to settle into the routine of school,' Ellen replied. She liked Caitlin and thought George was a nice child – not too rough, not prone to fighting like other little boys in the class. Some people's children were nothing short of savages.

'I'm not looking forward to this afternoon,' Caitlin was saying. 'I've got to make a birthday cake for my sister, Rosanna. She's staying with us at the moment having split up with her boyfriend.'

'Poor girl. Break-ups can be so hard to get over.'

'Yes, and she's also temporarily homeless because she'd been living with him in his flat in Ipswich. Fabulous apartment in the smart development by the marina.'

'I know the ones – lovely position and a nice part of the city. I imagine she's finding things a bit quiet here in Ash Green?' Ellen responded.

'She is a bit down and that's why I'm planning on making the cake, hoping to cheer her up, but to be honest I'm not great when it comes to baking. My relationship with my new oven is very strained at the moment. I put things into it in good faith and it gives me back burnt offerings. The truth is this will be my third attempt to get it right and I so wanted to do something nice for her.'

'It's certainly a lovely thought. Oh, and you've probably got a faulty temperature gauge. You just need to get it checked.'

'I guess you're right, but in the meantime I'm still without a cake and it's her birthday on Friday. That's just two days away!' Caitlin was clearly going to struggle with her task and it occurred to Ellen that she could quite easily step in and help her friend out.

'Tell you what, Caitlin, I love to bake and I've time this afternoon. I'd be more than happy to make the cake. I'll decorate it too, if you like.'

'Seriously, could you really? That would be marvellous. Obviously, I'll pay you for the ingredients and if you could decorate it too that would be a real bonus. I saw the perfect example of a decorated cake on a baking website. I could email you the link.'

'That sounds fine. Leave it with me and I'll do my best.' Having agreed she would bake and put together a three-layered sponge cake, with a butter cream filling, Caitlin's order was placed. Ellen was delighted to feel able to help and it occurred to her that there might be others who would appreciate similar assistance. Baking and decorating celebration cakes for real clients – could this be an opportunity?

'Mummy… Mummy… can we go to the swings on the way home?' Liam called as he ran across the playground towards her. 'Pleeeeease, please.' Ellen took Liam's school bag and turned in the direction of the park.

'I'll email you the link to that cake design as soon as I get home. Thanks again, Ellen.'

'No problem. Right, off we go, Liam, but we mustn't be too long, I've got baking to do this afternoon.'

'Oh goody, can I help you?' Liam was enthusiastic and Ellen was happy to agree, knowing from experience that he would soon tire of helping and leave her in peace to finish the assignment. On the way to the play park, Ellen smiled to herself, feeling optimistic about the possibilities available to her now the routine of school days was finally allowing her some freedom to think about herself. All seemed well with Liam; the village school had an excellent Ofsted report and the teaching staff were friendly and welcoming. This was the education that Ellen had envisaged for her son and Ash Green was delivering again when it came to her aspirations. Even Paul had been in a better mood recently

– still a bit preoccupied and work continued to keep him out late, but thoughtful enough to buy her a bunch of flowers on an almost weekly basis.

Ellen was absorbed in thought, until her daydream was brought to an abrupt end by a high-pitched scream of delight.

'Higher… push me higher… higher… higher!' Liam shouted, as the swing chains squeaked noisily on the A-frame supports.

'Last time, Liam, and then we really must go home.'

*

It was the first Pilates class following the reopening of the new community centre; the old village hall had been renamed to suit its upgraded stature within the neighbourhood. After a successful afternoon spent baking, Ellen was pleased to hear Paul's car on the driveway to the house, as she was due to leave shortly to join Laura and April for the evening. After a very quick hello and brief handover of child-minding duties to her husband, Ellen grabbed her rolled-up yoga mat from the closet in the hallway and blowing a kiss to her son, who was sitting on the stairs, left the house for the short walk into the village.

Her friends were already in the smart dance studio and Ellen was delighted to see that Laura had been able to persuade April to join them for this relaunch session. Past experience dictated that it was unlikely April's attendance would last more than a few classes, but tonight Ellen was looking forward to an evening out with the girls, especially as they had planned a celebratory drink together after the class. An ideal night out for a stay-at-home mum: the conscientious effort of an hour of self-righteous exercise, followed by a decadent mid-week glass of wine at the local pub with her girlfriends.

The class was busy; the participants were all chatting as the enviously skinny Pilates teacher attempted to get the session underway. She explained that she'd be offering a gentle exercise programme to ease them all back into classes after the enforced break of four months. Ellen was looking forward to the satisfying sensation of stretching out her neglected muscles and re-engaging with her core; she was convinced that with persistence she would emerge a toned, flexible woman in the body she'd inhabited pre-pregnancy.

She glanced towards Laura, who was lying on a mat in the space to her left, looking blissfully calm as she took deep and thoughtful breaths – in through her nose and out audibly through her mouth. During a very brief pre-class chat, Laura had mentioned that things with Neil were not great and that she was looking forward to sixty minutes of tension-releasing exercise and tranquillity to restore her equilibrium.

Looking over her other shoulder to the right, Ellen could see April lying quietly staring at the ceiling, already disconnected from the proceedings. She smiled at her encouragingly, but April looked self-absorbed and unwilling to engage with Ellen, or anyone else in the class. Perhaps it was unfair of Laura to continue with the persuasion; it was almost certainly a waste of time, as April clearly gleaned little benefit from attending the classes.

*

Thankful that Ellen was not a mind-reader, April was at that moment trying to think of ways in which she might diplomatically quiz her friend. When she'd mentioned the mystery woman on the train to Ian, he'd dismissed her concerns and tried to convince April that this was most likely a work colleague or travelling companion.

'If you commuted every day, you'd see how unexpected friendships are formed between fellow travellers, all on the platform at the same time every morning, even bagging their favourite seats in the same carriage every day. People bond in situations of adversity, and delays and overcrowding on the trains all equal adversity as far as I'm concerned. You heard Paul say as much when he was here for dinner the other evening.'

April decided that for the moment she would trust Ian's take on the situation, especially as it was the explanation which was least likely to cause Ellen any distress. It was unbearable to imagine that Paul might do anything to hurt Ellen. However, if he gave her any further cause to question his actions, April was determined to follow the example of Jane Tennison and use her female intuition and skilful powers of observation to uncover what others may miss, or dismiss. With this thought, April began to ponder whether she should have considered a career as a private detective – after all, she rarely failed to work out who the culprit was before the reveal.

The gentle music wafted across the room and the dulcet tones of the class leader eventually calmed and distracted April from her thoughts. She decided that having made the effort to get to the class, she should at least try to perfect a few of the less strenuous exercises; the more taxing manoeuvres would have to wait until she was feeling a bit more energetic. April had mastered the art of avoiding the most demanding parts of the workout by choosing instead to close her eyes and lie very still, the pretence being that she was meditating. Everyone surely understood the value of meditating – a most spiritually cleansing phenomenon.

*

'Well, thank goodness that's over,' April said as the exercise class came to an end. 'I've been clock-watching for the last twenty minutes.'

'We know,' Laura and Ellen said in unison.

'For goodness' sake, April,' Laura admonished, 'this is supposed to be an enjoyable, relaxing experience. I feel so much better for it.'

'Good for you. I'm sorry, I just find it boring, my mind starts to wander. I can't be bothered to keep up and then I discover I'm so far behind everyone else it hardly seems worth trying.'

'I loved every minute,' Ellen said, smiling, 'but having said that, I'm ready for a drink now. Let's get going.'

They stashed their mats in the boot of Laura's car and wandered along to The White Swan. The traditional village pub was an attractive timber and lathe building, brimming with period charm, and inside offered a cosy bar area, with a huge open fire. The pub had an excellent menu which ensured that it was a popular venue with villagers and outsiders, who were prepared to make a journey to dine there. The women settled themselves at a table close to the inglenook fireplace to enjoy the benefits of the glowing embers in the grate, each with a large glass of wine and eager for a catch-up.

Their last get-together had been the evening of the dinner party at April's house. Despite the expectation of a debriefing session between April and Ellen, the women had been too preoccupied with everyday life to make time for a post-mortem on Neil's qualities, or lack of them. They were both now keen for Laura's update on the relationship and were disappointed when she offered a limited synopsis of recent events, with nowhere near enough detail – especially when it came to satisfying April's curiosity.

'I'm having to face the fact that Neil might not be the man he was… or perhaps he is and I just didn't realise it when we were together all those years ago.' Laura seemed resigned. 'I'm sorry to say this but I think it's time I called it a day.'

'That's such a shame, Laura, he seemed so charming,' Ellen endeavoured to be diplomatic, 'although I don't feel we've really got to know him that well.'

Without wanting to sound prim, April added, 'Yes indeed, charming, only some of his comments, particularly the ones referring to your... err, bedroom compatibility... Of course, we were all rather sloshed that evening, but as we'd only just met him, it did seem a tad inappropriate.'

Laura blushed and April continued on, sharing her observations.

'That said, you've seemed like a different woman since Neil came back into your life, and Ellen and I so wanted it to work out for you.'

'I recognise that change in me, too. You know, being with Neil reconnected me... how shall I put it...? Yes, reconnected with my femininity. I'd missed that sense of being female and desired by a man. I've liked the intimacy... a lot.'

'And that is great, for sure, but of course there's got to be more to a relationship,' April said.

Her companions nodded in agreement and were unaware that April's mind was now off on a wander, analysing just how their individual sex lives might compare. She liked to think her relationship with Ian was pretty good; their needs were compatible, weren't they? However, she suspected for Ellen and Paul it was a case of too much abstinence to keep them close and that Laura might recently have been overdosing on intimacy.

'I want you both to know that I have no regrets. Neil being back in my life has allowed me to lay to rest an old ghost. I now realise just how often over the years I've thought longingly and fondly of him and wondered "what if?" But I suspect that there's unlikely to be a future for us together.'

'It certainly helps to focus on the positive outcomes of any life experience. If you look hard enough there are always some to be found,' April said, hoping she sounded philosophical and wise.

'Which is true, but being pragmatic when things get really tough and life paths meander off course is a huge challenge,' said Ellen.

'You know something, girls, I think I'm actually going to enjoy a return to independent singledom and just forget about men for a while. I shall busy myself with a new project... and I believe I know just what I'll do,' Laura said.

'Go on, tell us.'

'I'll get myself a greenhouse and learn how to grow veg, perhaps tomatoes and courgettes.'

April was dubious about time spent in a glasshouse being the answer to getting over lost love, but she admired Laura's determination. 'That's a great idea and you could grow all sorts of herbs too, except I wouldn't bother with lovage.'

'Now, that reminds me,' Ellen said, 'just the other day I was reading an article about some traditional and medicinal uses for herbs. There's apparently an old English alcoholic cordial, now let me see if I can remember... It's something like two parts lovage added to one part brandy, which apparently makes a fantastic soothing winter warmer.'

'Lovely though it is to hear your interesting fact about herbs, Ellen, you cannot seriously be considering wasting good brandy on a concoction like that, can you?'

Laura ignored their lovage banter and continued. 'I'll decide on the crops later, but for now, let's raise a glass to new beginnings with one last thought: Neil and Max were ill-suited housemates, so our relationship was probably doomed.' She almost sounded convincing, but April sensed that Laura would not find it easy to put this episode behind her.

'So, to other events... You enjoyed your barista course, April?' Laura asked, determined to take the focus away from her love life.

'It was excellent, I thoroughly enjoyed it and I have the certificate to prove I achieved my goal. I'm going to look out for any vacancies at the Great Aldebridge coffee shops. I'd like to continue with some practical experience to follow on from the training course.'

'That's great news and I look forward to being one of the first customers to sample the delights of your barista talents before too long,' Laura said.

'And what about Liam, is he still enjoying school?'

'Well, I still have my fingers crossed, but it all seems to be going really well. I have to say I'm loving having a bit more time and being able to make plans for myself. Actually, today has been quite interesting.' Ellen explained how the unexpected offer to bake a cake for one of the school mums had sparked the idea for a specific business venture. 'I've already explored the Food Standards Agency website to discover all the legislation around running a food preparation business.'

'Is that what you're planning?' Laura asked. 'Your own home-based, cake-baking company?'

'Well, it's early days… and with Paul working late most evenings I've not yet had a chance to properly talk it through with him, but I'm certainly giving it some serious thought.'

'My goodness, Walnut Tree Farmhouse couldn't be more perfect for accommodating your plan.' April was fired up with enthusiasm. 'Especially that large utility room – too large really just for laundry and wellies. Surely it would make an ideal space for a separate food preparation area. And if things really took off you could even consider converting one of the outbuildings to expand the enterprise. Oh… and what about getting yourself some chickens? Yes, that would be splendid, and fun for Liam, too. You'll be needing a plentiful supply of fresh eggs for your baking…'

'I've been doing lots of research and probably getting rather carried away – just as you are, April.'

'But I really think you could be on to something brilliant. As the self-appointed queen of a good idea, I'd say this one sounds promising… very promising indeed.'

15

Ellen pulled the duvet into place and plumped up the pillows on the bed; it was Saturday morning and the start of the longed-for weekend. She had specific plans, which required sticking to the rules of equal measures of family pursuits and time for relaxation with her husband. The morning had got off to an auspicious start; they had managed to have a lie-in and a somewhat hasty session of lovemaking whilst Liam remained asleep. Opportunities for such pleasures with a small child in the house seemed to be few and had to be grabbed when they presented themselves. The moment of union had left her feeling calmly content and had restored her belief that her husband continued to find her attractive and desirable. It was good to be reassured. As she quickly tidied the bedroom and got herself ready for the day, Ellen smiled as she listened to the joyful chatter downstairs in the kitchen where Paul was now preparing toast for their son.

'Daddy, you know when a pilot is flying a plane…?'

'Yes, Liam.'

'Well, what happens if he needs a wee? Can he take his hands off the steering wheel without the plane crashing?'

'He'll usually have a co-pilot flying the plane with him, so his partner could take over if he was desperate for the loo.'

'What if they both needed a wee at the same time? That would be a problem, wouldn't it?'

'I think that would be unlikely to happen.'

'But it might.'

'Do you want honey on your toast, Liam?'

Ellen chuckled; it was Paul's turn to find an answer to the myriad of questions conjured up by their inquisitive four-year-old son, whose obsession of the moment was with all things aeronautical.

With breakfast consumed and the debris cleared away, the family got ready for an outing to Great Aldebridge. Coffee and cakes together in Amarellos' were on the agenda, with a babyccino for Liam, who always enjoyed this excursion into the world of grown-ups. His afternoon treat would be a trip to the local swimming pool with his dad, allowing Ellen some quiet time at home to prepare a special Saturday evening supper for herself and her husband to enjoy much later. Ellen was hoping that an hour at the pool would tire their lively son and he'd be ready for an early bath and bed, giving them the opportunity for a relaxing evening à *deux*.

Once home following their outing to town, Ellen set to work in her kitchen and having cut a handful of mint leaves from the herb patch in the garden, she was happily chopping and prepping for the supper. Finding herself with the tranquillity that came with her favoured activity, she thought about Paul's reaction when she had told him of her plans for a business. His response had been somewhat non-committal, but equally, having not raised any specific negatives, Ellen decided that non-committal was okay for now and definitely a promising start. She'd spent some time finding out about the necessary requirements to facilitate running a small catering project from her home and was now in the process of obtaining a quote from the local builder to get the minor interior changes organised.

Ellen was certain that her background in marketing was going to prove invaluable when it came to promoting her business, and she'd had fun starting to create a website and thinking about logo options – which she'd decided must include a gnarly old walnut tree. Her pink Suffolk home was well-suited to conjuring up images of artisan baking from the rustic country kitchen of a farmhouse where sweet-scented honeysuckle and rambling roses clambered over an oak-framed porch.

She reflected on April's recent suggestion that she should acquire some chickens and couldn't imagine why such an acquisition hadn't occurred to her sooner. You had to have a few hens if you lived in a farmhouse, didn't you? Especially if you loved to bake. She'd also been doing some research to seek out uses for the glut of walnuts she'd harvested; coffee and walnut cakes, tea loaves and muffins were delicious, but she was looking for something a bit different.

Ellen's thorough internet exploration had revealed limited competition in the area for celebration cake baking. The birthday cake for Caitlin's sister had proved to be a huge success; it appeared that Ellen had created a much-talked-about 'show-stopper', which had delighted the birthday girl and resulted in further commissions at the school gates. She'd been practising her decorating skills, having checked out YouTube videos, and had also armed herself with books full of fabulous and inspiring design ideas for all occasions – fun birthday cakes for children through to wedding cakes for discerning adults. Daring to believe that this could be the beginnings of a creative career, she felt a shiver of excitement at the prospect of a fresh start.

The idea did still require Paul's approval and ideally she should seek this before presenting him with the costings for the upgrade to her workspace. He was bound to be daunted by further works to the house so soon after the original renovations; he had loathed the whole process of having builders in residence.

With her preparation completed for their supper menu of scallops with pancetta and pea and mint puree, followed by fillet steak, and for dessert, chocolate fondants, Ellen removed her apron and with her boys not yet due back from the leisure centre, she made her way upstairs to take a shower and find a change of clothes. As she entered her bedroom she was disappointed to discover a pile of Paul's clothing dumped in the corner of the room behind the laundry basket, which she had failed to spot during her cursory tidy earlier that morning. It was annoying that he seemed unable to make the effort to insert the garments inside the basket or return items to their homes in wardrobes or drawers. The disarray he'd thoughtlessly created jarred with her sense of harmony in the room which should be a haven – their personal sanctuary. But there must be no rows that weekend, so she swiftly set about picking up the jettisoned socks and underpants.

Ellen tutted with dismay to see that amongst this jumble was the jacket to Paul's very expensive suit, which he hadn't taken the trouble to hang up when he'd got home on Friday evening. In his defence, she reminded herself that he had arrived home exhausted after a particularly long day at work, at the end of an exceptionally busy week. He'd appeared overwrought, with his energy so depleted he could do no more than undress, before climbing into bed beside her and within minutes falling into a deep sleep.

On taking a closer look, the suit jacket looked crumpled and creased, bordering on shabby. She couldn't recall when this garment had last been treated to a trip to the dry cleaners, so Ellen decided to do Paul a favour and take jacket and trousers for a freshen-up and press the following week. She wanted to ensure her man always looked his best and continued to impress his contemporaries and clients.

It was on checking the pockets for any stray tissues, pens or coins, that she found the receipt for an extravagantly priced dinner for two in the inside pocket of the jacket.

*

'It's a beautiful day, let's not waste it.' Laura was desperately trying to persuade Neil to venture into the fresh air and join her and Max on a walk. She felt the need for a brisk stroll, but she also wanted to have the outstanding chat with him. As they left the car at the beach car park and approached the exposed shoreline, they were met with a lively onshore breeze, prompting Neil to shiver in an exaggerated fashion and then pointedly tug on the zip of his jacket.

'Bloody hell, Laura, it's cold enough to freeze a bloke's b…'

'Don't dither then, keep walking and we'll soon warm up. We'll need to be swift anyway to keep up with Max this morning,' she pointedly interjected and hastened after the galloping dog.

She heard Neil grumble and glancing over her shoulder saw him making a further attempt to shield himself against the chill by pulling up his collar against the wind and pushing his hands deeper into his pockets.

'Don't you find it bracing?' Laura asked hopefully when Neil finally caught up. 'The sea air smells wonderful and the sight of the sunlight glistening on the water's surface is magical. It makes me feel good to be alive.'

'Just how far are you planning on going? I think I'd feel much more alive with a hot drink and a toasted ham and cheese sandwich inside me.'

Laura sighed and having lobbed the ball an impressive distance allowing Max a good, long run, she felt the moment had arrived.

'I think we need to have a chat about where our relationship is going. I'm beginning to feel that we don't really share the same interests, or have the same outlook, like the importance of making the most of every opportunity and every single moment that we are blessed with.' Laura hadn't intended to add the clout of divine favour to her speech, but felt that it was a dereliction to heedlessly squander precious time. She was sticking to her intention to be diplomatic rather than confrontational, even though she thought a confrontation with Neil might be exactly what was needed.

'What on earth are you talking about?'

'You and I, we're just so different. Perhaps that's always been the case. I was probably too self-obsessed back in the day to recognise it.'

'That's just rubbish. We've been getting on fine, not to mention having lots of fun, particularly between the sheets,' he said as he unceremoniously slapped her behind.

Neil's smirk, along with the stinging slap, brought a grimace to Laura's face. She checked the vicinity to ensure no one else had witnessed his unseemly behaviour; it was totally unacceptable in view of their mature years and, in her case, the cultured woman she endeavoured always to present to the world.

'Of course, I can't deny that it's been good to get reacquainted and yes, we've had fun. There's no doubt that we are compatible bedmates,' she whispered unnecessarily in view of the previously ascertained lack of human proximity. 'But I want so much more from a relationship and I'm starting to doubt your ability to offer that.'

'This is all because I haven't got a job, isn't it? You're fixated on status and so pleased with yourself for having had such a great career, which has provided you with plenty of money for your continued comfort in retirement. And what have you got to worry about anyway, with Rose's substantial home still to be sold and you being the sole beneficiary, once she croaks?'

Remaining as calm as she could, despite provocation, Laura countered, 'Everything I have is down to hard work, careful spending and planning for the future. I've only ever had myself to rely on – no partner to bail me out – and regarding my mother's home, it's likely that most of the proceeds will be required to cover the costs of the care home. Not that it's any of your business.'

'Look, Laura, you just don't seem to get it. I deserve to be compensated for all the wearisome years I spent with Catherine and I'm not giving up without a fight for what's rightfully mine.'

'Ever thought it might have been wearisome for Catherine too?' Laura was livid and yelled to Max, calling him to heel in a shouty voice that startled him into obedience. She then turned to retrace her steps back to the car; her rage brought swift progress, leaving Neil struggling to keep up. As she got ahead his panting was soon more audible than the dog's and for a fleeting moment Laura envisaged the man, his tongue lolling out of the corner of his mouth with associated drool, until she blinked the bizarre and hideous thought away.

Once he was alongside, Laura saw he was ill-advisably about to speak and guessing that he would utter some further justification for his situation, she stopped him in his tracks.

'Enough!' Laura said, holding up her hand for confirmation. 'I don't want to hear another word. I just cannot see any future for us together. What I'm saying, in case there's any doubt, is that it's over.'

'You say that now but you'll change your mind when you're yearning for a man. You'll be desperate to have me back then.'

Having reached the car, Laura opened the tailgate and, with his dog's intuition on high alert, Max, sensing the tension, jumped straight into the boot and crouched down on his blanket.

Laura got into the driver's seat, slammed and locked the doors, imprisoning herself and Max inside the car, and effectively excluding their former companion. She drove off without a backwards glance or a second thought as to how Neil might manage to get himself home.

16

April took her place, seated on a wooden chair behind a small and worryingly unsteady table, which was positioned beside the open doors of the community centre. It was the day of the parish council's fundraiser – the indoor car boot sale – and she'd been enlisted by Laura to collect the entrance fees. She had come prepared, expecting it to be a draughty job – it was, after all, November – and was now feeling smug under the protection of her thickest fleece jacket. To her delight, her allocated task meant she'd be off her feet and with this vantage position, she could also deploy her meet-and-greet skills, making for a jolly few hours of chitchat with village residents. April was very happy to play her part in assisting Laura in her bid to raise the money to purchase the benches for the community garden, where the now established shrubs and perennials displayed lush foliage and pretty late autumn flowers.

It was a chilly Saturday afternoon, but as the sun was shining there was every expectation that the event would attract plenty of sellers and buyers out of their homes. April made herself more comfortable by acquiring a small cushion for her chair on loan from a stall-holder and at the allotted start time a steady stream of locals began to filter in from outside. So eager were they to pounce on the best bargains, however, that there was little opportunity for the catch-up chats April had been anticipating. Ash Green, like other rural villages, had its share of characters and quirky individuals, and she found their eccentricities utterly endearing, adding to the many pleasures of being part of the community.

'Lovely day for it, Derek.' April spotted the village handyman-cum-gardener-cum-pigeon fancier impatiently trying to negotiate a passage past a discarded mobility scooter in the doorway. She was especially fond of Derek, who was always happy to assist with odd jobs around the village for the parish council, from hedge trimming to installing the village sign. She had come to rely on him whilst she was working as the clerk and during this time they had established a firm friendship. April had had the privilege of being invited into Derek's secret world of racing pigeons, which included a tour of the palatial pigeon lofts at the bottom of the garden at his otherwise dilapidated homestead.

'Hello, hope there are still plenty of bargains to be had. Intended to get here earlier but I was waiting for one of my pigeons to get back home. Knew I could expect her any time.'

'And has she got back safely?' Despite numerous discussions with Derek on the subject, April still could not understand how his feathered beauties could navigate their way home from various and far-flung release sites.

'That she did… and her time was good.' April saw Derek's delight, but despite this being his favourite topic of conversation, he excused himself quickly and made his way to peruse the table-tops.

Derek lived in a quaint, old-fashioned farm cottage on the edge of the village, having resided there since he was a child, taking ownership when his mother died. He now shared his home with his grown-up daughter, Milly, but April had no knowledge of a wife and remained curious as to whether Derek had ever married. The pair seemed very contented in their ramshackle abode, which April would always covet. The gardens overlooked a paddock, with the river beyond, but the property would take some updating should it ever become available to a new occupier – which seemed unlikely. Sadly, despite being very capable and obliging when it came to the maintenance of many of the communal buildings and grounds in the village, Derek failed miserably when it came to the upkeep of Holly Cottage. He was also a known hoarder and was no doubt about to purchase all sorts of unnecessary items at this afternoon's sale to add to his stockpile.

An hour and a half had passed, when April spied Laura making her way through the crowds. She was smiling, but looked pale, and April noted the dark circles under her eyes, which she decided were a sign of a demoralised disposition. Despite her efforts at cheerfulness, April was certain that Laura was struggling in the aftermath of Neil and her involvement in this fundraiser was no doubt a welcome diversion following the hostile break-up. When questioned, Laura had revealed that Neil had sent a couple of texts since the day she left him stranded on the beach – a course of action with which April wholeheartedly concurred – but her friend had chosen to ignore his messages.

'I'm starting to appreciate an almost forgotten, but oh so pleasing, freedom away from all of Neil's irritating habits,' Laura claimed when they had spoken on the phone the previous evening.

'Absolutely… men's habits… infuriating.' The ever-curious April would have loved to learn the extent of his many misdemeanours.

'I'm just concentrating on my greenhouse brochures with their evocative photography of fancy structures containing bountiful produce.'

'Right?'

'Don't scoff. I'm certain that the void his departure has created will soon be filled with a project involving organic fruit and veg. Who knows, I could be self-sufficient in a year's time.' Laura sounded upbeat, but April remained unconvinced by her all-consuming enthusiasm for botany.

Laura finally made it through the obstacle course, created by the baby buggy entourage, to reach April and immediately set about counting the proceeds of her cash tin.

'Sixty-six, sixty-seven, sixty-eight pounds… that's not bad.'

'Do you mind if I take a break, Laura? I need the loo and I'd like to get myself a cup of tea. I'm parched.'

'Of course, off you go. You'll find there's helpers' hot drinks in the kitchen.'

With Laura now her stand-in ready to relieve further entrants of their cash, April made her way through the throng of enthusiastic shoppers. It wasn't long before she noticed that she was collecting a small group of followers.

'Can I help you?' April asked curiously.

'Are you off to make some tea?'

'Yes, I'm off in search of a brew.'

'At last,' said Polly, 'me and Cynthia would love a nice cup of tea.' Polly hosted village knit-ins and carried around her knitting bag wherever she went. Seeing the needles poking out of the top of the bag, April made herself a mental note to ask the lovely old lady about some tuition. She'd always wanted to learn the craft, which her mother had been too impatient to teach her.

As April looked around the room, she saw immediately the now obvious problem: when organising the event, someone had overlooked the excellent opportunity it presented to take extra money from a thirsty public, who would inevitably find themselves ready for a beverage midway through their shopping. The kitchen was basic, but adequately equipped, so April swiftly opened up the rolling shutter and set about lining up cups and saucers on the open counter, filling up the urn and the two standby kettles. The instantly prepared teas and coffees were an easy sell, and April was soon having to ask one of the other helpers to run to the village shop for more teabags and milk to fulfil the orders that just kept coming. 'Oh, and get some biscuits too,' she yelled at the runner's back.

April spent a thoroughly enjoyable hour in the kitchen serving a very grateful clientele and was complimented by a member of the organising committee for being so enterprising.

'What an oversight! I can't believe we didn't think about offering refreshments,' said Diana, the bossy leader of the fundraising committee. 'Thank you so much, April, you're an absolute superstar stepping in like that.'

'It wasn't particularly difficult, especially for someone who's a trained barista.'

'A trained what?'

*

'How did it go?' Ian called out from the sitting room as April came crashing through the back door with her hands full of bags, containing irresistible items purchased from the table-top sellers, including the comfy cushion.

'It was okay, actually,' April replied, hoping to find a home for her treasures before Ian came into the kitchen and chastised her for spending good money on other people's 'bits of tat'.

'Really? I'm pleased you enjoyed it because before you left, I seem to recall you complaining about having to spend an afternoon inside the hall when you could have spent it in front of a *Line of Duty* box-set.'

'Ha ha… it was good, actually. There were plenty of people and all the target money has been raised.' April had just enough time to safely stow away her bags out of sight in the pantry as Ian came into the kitchen.

'That's fantastic news, so well worth the effort,' said Ian.

'Absolutely, and I think you will also be pleased to hear that my customer service skills today were exemplary. You'd have been proud of me.' April was smiling as Ian settled at the kitchen table, looking enquiringly, seeking further clarification.

'Really? I thought you were just there to take the entrance money?'

'The members of that events committee are pretty useless. They hadn't considered that the visiting public would be more than happy to spend a few more pounds on teas and coffees in between browsing the table-top goods. So, it was April to the rescue!'

'Okay, what exactly was your rescue strategy?'

'I set up a rather basic refreshment bar from the community centre kitchen, arranged a few spare tables and chairs and served enough customers to raise an additional £43. How about that!' April was pleased with her efforts. 'Just imagine if we'd had time to get Ellen involved and been able to offer some of her scrumptious cakes as well.'

'That's brilliant, April, and I'm guessing by your excited tone that there's more to tell. Could it be that serving teas and coffees in the community centre has got you thinking?'

'Too right. It occurred to me that the centre might be the perfect place to set up some sort of a café facility in Ash Green.'

'I see, the new building is great, but it's a very functional amenity. I'm not sure it would lend itself to an enticing café environment.' April recognised this as a very valid comment, but she was not deterred.

'Okay, I see that, but seriously you have to acknowledge that there might be an opportunity there. I was thinking about it on the way back, and obviously the kitchen and the adjoining small function room would need some additional work, but…'

'Hold on, April, this all needs a lot of thought. I'm not saying it's not one of your better ideas. I've read about these so-called community cafés and how welcome they are in rural locations. Many apparently offer more than a serving of coffee and cake, they can provide a real hub for a community.'

April had known Ian would stress caution when he heard her suggestion, but she could tell he was warming to the possibilities.

'Let's put the kettle on, fire up the laptop for a bit of Googling. We can have a cup of tea and gather some ideas together – we'll have one of those thought-showers,' Ian said.

April looked up at the clock on the kitchen wall and happily claimed that at 5.20pm it was more than okay for them to open a bottle of wine and forget about the kettle.

17

The lure of a *proper* cup of coffee from a coffee shop, and a get-together with Ellen, had April cheerfully making her way to Great Aldebridge. She'd hoped that Laura would also join them and having seen her in the village taking Max for a walk, had extended the invitation. But Laura declined the offer with what sounded to April like a spur-of-the-moment, made-up excuse about Max being under the weather and needing a visit to the vet. To April's untrained eye, the dog looked to be in the peak of good health: his eyes were clear and bright, his coat glossy and at that moment his tail was wagging so enthusiastically that his entire hindquarters were forced to join in on the act.

April had concluded that Laura was happier at the moment taking some time out away from others and was probably absorbed in her lettuce and radish plants. She may also be attempting to avoid any well-meaning opinions from friends on the subject of her ill-fated love life. April couldn't deny that she might be found guilty of this crime, but any advice offered would genuinely be intended with Laura's best interests at heart.

'Hi, Gino,' April called as she entered the café and made her way to the counter.

'And how are you, April?'

'I'm well, and still available if you have any staff vacancies.'

'I wish it were possible, but just now I have plenty of helpers. Don't you worry, I'll be sure to call you if that changes.'

April was talking herself out of a sweet treat as Ellen joined her at the counter and ordered a large vanilla latte and one of Gino's deliciously sweet *cartellate*. They both struggled to resist the Amarellos' delicacies and these crisp pastry spirals, drizzled generously with honey were a particular favourite.

'Well done, you managed to refrain from the pastries today?' Ellen congratulated her friend.

'Nearly didn't, but I'm determined to be good today,' April responded.

The café was busy but they were able to claim their preferred sofa seat by the window. Both women were eager to continue the conversation they had started on the phone the previous evening, as the possibility of a mutually beneficial business arrangement – involving April's proposed café and Ellen's cake-baking enterprise – had sparked their imaginations.

'Really, what could go wrong?' April enthused. 'Coffee with cake, we already know that it's a winning combination.'

'It's such a shame that Laura isn't here. It would be good to have her input,' Ellen added.

'Absolutely, I know I have a tendency to get overly excited.'

'For sure she's the level-headed one of the three of us. She's the person who would recognise any pitfalls and point us in the right direction,' said Ellen. 'I'll be much happier when we've had a chance to run the idea by her.'

'She might welcome the distraction – help her get over Neil. Seems as though events are repeating themselves and she's been knocked off kilter for a second time by that man.'

'Men!' Ellen said, sounding exasperated.

'Oh… what's Paul done now?'

'Nothing as far as I'm aware, although I had a moment of panic just the other day.'

'Why?'

'Well, I discovered a receipt for an evening meal for two people in his jacket pocket.'

April looked up from her coffee cup, horrified to hear this disturbing snippet of information.

'You did ask him who he was dining with, didn't you, Ellen?'

'I most certainly did, because you'll guess what I was imagining and I was desperate for an explanation.'

April's heart was pounding, but the expression on Ellen's face was not that of a woman who had recently discovered her husband was having an affair with an attractive, smartly dressed and extremely well-heeled young woman, with the most glorious shiny, long, dark hair.

'So, what did he say?'

'Well, I felt rather ridiculous for even doubting him, because it was just a business dinner with a potential client. Turns out that at the last minute his MD had insisted that Paul make time for this guy because he was due to fly back to Denmark the following morning. So an evening get-together was the only option.'

'Oh, of course, I see.' April's response was unconvincing even to her own ears. 'And things are alright between you two, aren't they?'

'Honestly, April, Paul is such a conscientious employee he'll readily agree to working late on a Friday night. And yes, things are okay, he even seems enthusiastic about the cake-baking business. Quite willing, in fact, to hand over part of his home to my enterprise. Although, of course, the utility room with all its laundry paraphernalia was never really his domain. He says he just wants me to be happy.'

Happy, occupied and blissfully unaware of what he might be up to, was April's take on this news. She knew she was going to find it very difficult to suppress the urge for some further detective work.

'If it occurred to you that the receipt was a sign Paul might have been enjoying a Friday night dinner in the company of someone else… perhaps a woman… I wonder whether subconsciously the thought was already there in your head. After all, he is often late home and I'm not saying he is, but he could just be claiming that it's work that is keeping him from getting home at a reasonable hour?' April asked cautiously.

'Obviously there was a moment of uncertainty, but I have no reason to believe Paul would stray. He loves me and Liam. I'm sure he wouldn't do anything to spoil things for us. I'll admit that our dwindling sex life bothers me a bit, but Paul is just exhausted all the time.' April noted that Ellen sounded totally convinced, unconcerned even.

'Fatigue from overwork could, of course, be a factor but couples should never allow work to come between them. You must be missing that closeness…'

'Really, April, we're not newlyweds.'

'Sorry, Ellen, I don't mean to pry. It's just I'm a strong believer that you have to find a way to make time for each other to keep the desire alive.'

'Yes, of course, but honestly we're fine. Anyway I've got lots of plans that are keeping me very busy at the moment. I'll soon be too preoccupied running my fledgling business to have time to worry about things like that.'

April was fearful now; this didn't sound good at all; she had to do something. 'Tell you what, I'd be happy to take Liam off your hands for a night so that the two of you can plan an evening away together at a fabulous hotel. There's nothing more conducive to feeling in the mood than a large hotel bed, with freshly laundered sheets, lots of wine and no responsibilities.'

Ellen laughed. 'It's a really lovely offer and definitely one I'll think about, but for now could we please get back to the café and catering projects?'

April had no choice but to leave it; she'd tried, but she suspected she'd also failed. Ellen was now cheerfully reeling off her progress list.

'The builder has almost completed the work to create a second kitchen and my website just needs a little more tweaking in time for the launch. In just a couple of weeks' time all will be in place and The Kernel Kitchen will be ready for business. Oh yes, and we're collecting the hens on Saturday now that Derek has finished the henhouse. Liam is beside himself with excitement.'

'That's great, and did you give all the excess walnuts the candy treatment?'

'A lot of them, yes, I did. I'd never heard of candied walnuts until Google enlightened me, but they're a huge success with Liam and Paul. In fact, Paul asked me to make a further batch for him to take into the office.'

'I don't suppose you've got any to spare?'

'I'd happily make you some, but I thought you were watching the calories at the moment.'

'Spoilsport.'

'And another thing, I'm so delighted with the designs for the business flyers – the depiction of our Suffolk Pink alongside the walnut tree. It's absolutely perfect. I've placed the print order and also for my business cards.'

'I'm afraid I'm lagging behind in advancing the community café plans,' April sighed; she was envious of Ellen's steady advancement towards her goal. Her own proposal was going to take a good while longer to get underway and having researched other similar community initiatives, she'd realised that it might not be as straightforward as she'd imagined. A winning ethos was going to be imperative to gain local support and any sort of funding.

April had made a tentative approach to some of the parish council members to gauge their thoughts on her idea to establish a self-supporting business, run by volunteers, in the community centre. She had outlined her ambition to provide a welcoming environment where locals could meet old acquaintances and make new friends; recent arrivals in Ash Green would find this amenity of huge benefit when getting settled into the village environment. It would be crucial to brighten up the bland canteen to create a coffee shop atmosphere; on April's wish list were lovely tablecloths, pretty cushions for the chairs, little vases for flowers at each table and perhaps some appealing wall hangings. Inevitably she planned to acquire a best-quality coffee machine for this café, appointing herself the principal barista. She wanted to offer coffee on a par with that available at the cafés in the nearby towns, and Ellen's delicious locally home-baked cakes and pastries would be an ideal accompaniment.

'So, as far as the café goes, I'm now waiting for the councillors to have further discussions about the proposal at their next meeting, and I'm hoping that with Laura in their midst, I'll have an insider putting in a good word.'

'When is the next meeting?'

'Thankfully it's next week, so we should know more shortly.'

'If you manage to get the councillors on board, then I suppose the issue will be getting additional funds.'

'Absolutely, we will need some financial support and I'm hoping Laura might be able to help. From what I have learnt, there are funds available for community café projects, you just need to make a good case regarding offering additional value to your community.'

'How do you propose to do that – add value, I mean?'

'Well, my initial thoughts include providing basic hospitality training for local youngsters who are out of work, or having a couple of computers available for local use and maybe getting in tutors to help the more IT-challenged amongst our community.'

'Sounds fabulous and, of course, Laura has previous experience of accessing charity funding,' said Ellen. 'Good luck with the meeting and keep me informed. I'll need time to work out how many tray bakes and cookies I've got to produce to keep your café supplied.'

'I'm sure there will be plenty of local support for a project like this, but first we do need to win over some of the less forward-thinking of the councillors.'

'Argh… and there might lie the problem,' Ellen cautioned.

April chose to ignore this notion because she was convinced Laura would be on her side and that would suffice, wouldn't it?

18

Rosanna had been shocked and distraught when Rob had left her. His explanation was that being tied down at twenty-four was a bad idea and anyway he'd always planned to do some travelling. He had arranged to end the rental agreement on the apartment they were sharing; admittedly it was only his name on the tenancy paperwork. He'd also agreed an extended period of unpaid leave with his employer. His stance was, 'Why on earth would I miss such a fantastic opportunity?' – and of course, he'd have the security of a job to return to. Well, undoubtedly that was an ideal scenario; however, it was only when all of his plans were in place that he deigned to break the news to his girlfriend of one year and one month.

They had been living together for seven months in a fabulous apartment in an upmarket area of Ipswich, convenient for their commute into the city, where he worked in insurance and she for a company involved in letting chalets and luxury villas in France. As far as Rosanna was concerned everything had been just perfect. She couldn't believe that he hadn't considered including her in the arranged trip across Australia and, even worse, had simply ended the relationship in order to fulfil his expedition alone.

When she found herself homeless, her older sister had taken pity on her and that's when she had begun her temporary residence in the village of Ash Green. Pretty though the village was, she missed being in a city, and her daily commute to work was considerably more arduous, especially during the winter months with the shorter daylight hours.

However, despite her enforced rural isolation, things had recently started to look up and her train journeys into work had become much more pleasurable. She had seen Paul at the train station on a number of occasions, noticing the stand-apart, smartly dressed, fair-haired man. Their friendly 'good morning' acknowledgements and cheery smiles quickly became the sharing of a weather forecast for the day, a moan about a train cancellation and before long they were looking out for each other in order to ensure they could have a chat whilst sitting side by side on the train. Paul was not a daily commuter and Rosanna soon realised just how bereft she felt on the days he was not required to be at his city office and her travelling companion was missing.

Paul and Rosanna had shared commiserations: the recent ending of her relationship with Rob and Paul's struggle with a very lopsided work/life balance. Rosanna relished their conversations and quickly came to feel that she was being included in her new friend's deepest thoughts and anxieties. Paul had revealed that he found her non-judgemental and open friendliness the perfect inducement to relaxed sharing – something he had recently found challenging with Ellen.

During the couple of months that Rosanna had been living with her sister, she had become acquainted with a few of the village residents. They hadn't actually been introduced, but she was aware of Paul's wife having heard Caitlin enthusing over

Ellen's exceptional baking skills. She didn't feel awkward about her connection with Paul, why would she? It was merely a mutually appreciated friendship. Consequently, she was flattered but slightly hesitant when Paul had quite unexpectedly suggested they have dinner together one evening.

'I really enjoy your company, Rosanna, and you've been so fabulous putting up with me drivelling on about my personal issues. Your cheerfulness has really pulled me out of my doldrums and I'd like to thank you by inviting you to dine with me.'

'But you've also had to put up with my woes, Paul.'

Rosanna had struggled to resist the temptation to accept his invitation. She questioned whether it was entirely appropriate, especially knowing as she did that Paul and Ellen were not exactly getting on too well.

'It's true, I've listened to your troubles too, but that just reinforces my suggestion that we both deserve to spoil ourselves with a meal out – no strings attached, of course.' Paul appeared to have made up his mind and Rosanna found herself swayed by his suggestion. The evenings in with her sister and brother-in-law were companionable and she adored spending time with her nephew, but her life was lacking variety.

They decided to organise this treat for later that week and Paul said he would put booking a table at the top of his to-do list the moment he arrived at his desk that morning.

He had been true to his word and on the appointed evening the pair made their way to a popular fine dining venue in a picturesque waterside location. They were both in high spirits and looking forward to spending some quality time together away from the noise and prying ears and eyes of fellow travellers. It did feel a little bit naughty to be out on a date, but that just added to the thrill of the occasion and anyway, no one need know.

They were, as always, very much at ease in each other's company and their evening together was filled with conversation, laughter and an indulgently delicious meal – just the remedy they both felt they needed. Rosanna found Paul a witty and captivating companion – more mature than her usual consorts. And she knew that Paul found her a welcome distraction.

'Thank you so much, Paul. I've had such a lovely evening,' Rosanna said as they reached their respective cars in the station car park.

'Well, thank you too for your delightful company,' Paul said as he pecked her on each cheek.

'I'll look out for you on Monday morning.'

'Yes, I'll be there and we really should try to do this again. I'd been wanting to try Brasserie Amélie for months and it certainly lived up to all the hype.'

*

On his journey home, Paul was in high spirits and couldn't control his grin as he recalled the sheer joy of a genial evening in the company of a rather lovely, vibrant young woman. He then thought of Ellen and her current fixation with the new business. Of course, he wanted her to feel she was fulfilling her potential and making the most of her talents, but he saw this as a preoccupation which would leave him excluded still further from his wife's attentions.

It had been the same when Liam had arrived; the proud father had soon discovered that this small child would be responsible for draining all of his wife's energy and take up most of her time. He'd thought the situation would improve once Liam started school, but now Ellen seemed driven to succeed in her own venture and totally absorbed by this goal alone. He knew it was selfish – an outdated concept – but he'd had visions of her as a traditional housewife and mother (just like his own had been) caring for his and Liam's needs, whilst he was the provider, the protector. Through his hard work, he'd been able to provide his family with a beautiful home and had naively anticipated Ellen's contentment and appreciation of his masculine prowess.

In Rosanna's company, Paul's cares and stresses were left behind and for a few hours he'd been allowed a welcome departure from the pressures he placed upon himself, the expectation of others and his daily responsibilities. He'd felt more like his old-self – the younger, carefree self – and more alive than he had done in many months.

*

On her journey home, Rosanna found herself smiling – it was a broad, mischievous grin – responding to thoughts of the evening with her manfriend. She allowed herself to wonder what this unexpected friendship might herald and whether she should even be considering the liaison as a prelude to other things. But it had been such a welcome change – an evening out with an admirer – and he clearly did admire her; all the signs were there. She might be young, but she wasn't totally naive when it came to the male psyche.

The work-free weekend was upon her, but for Rosanna it was Monday morning that she was looking forward to – that and their plans for a second date.

19

April felt apprehensive and linked arms with Ian for reassurance as they walked towards the community centre. She had convinced him that it would be useful if he heard the councillors' comments on her café proposal first hand, which was only a partial truth; she just needed the comfort of him by her side.

It was one of the parish council's monthly evening meetings and April knew that protocol would restrict her involvement in discussions during the order of business, but she hoped that the members would still refer to her with any questions regarding her plans. She was well-versed in the formalities of these gatherings and was intrigued to be present as a member of the public on this occasion, rather than being seated beside the chairman taking shorthand notes for the later transcription into typewritten minutes. April was a speedy stenographer and knew her scribblings were a source of wonder to her colleagues, but they were oblivious to the fact that she struggled to decipher the symbols after the event, relying heavily on memory to produce the necessary official record. She carried on the charade, enjoying the notion that they were impressed by her alien skill.

Her proposal letter to the chairman – which he had requested – set out April's objectives and strategies for the venture. Her aspirations had been put into words by Ian – he being the master of letter-writing and a stickler for the correct use of English grammar. This document would have been circulated to all councillors prior to the meeting, with an agenda item being added under Any Other Business to cover this specific matter for discussion.

With Ian's help, April had also carried out extensive research on the setting-up and running of community cafés and had discovered useful insights, having read the experiences of others. Better to be aware of the hurdles in advance and, worryingly, it seemed a high proportion of working groups failed in their quest to achieve a functioning eatery. Finding a suitable venue and then adequate resources to set up and run the cafés were cited as major stumbling blocks. April was certain she had the venue sorted, but with sources of funding uppermost in her mind, she had called Laura the evening before the meeting.

'Hi, Laura, how's it going? Any spare tomatoes for your friends?'

'Oh hi, April, sorry I'm just about to go outside and water all the veg. Did you need me for something?'

'It was just about funding for the café. We shouldn't have any problems getting the money, should we?'

'Err… I imagine it will be fine… Look, April, I'll catch up with you tomorrow if that's okay, I really do need to get on.'

'Great, thanks, Laura, knew I could depend on you.' But April was talking to herself as Laura was no longer on the line. Those veg must be wilting, close to death, for her to be in such a hurry to hydrate them.

April and Ian made their way into the meeting room and took their places in the front row of seats. Twenty or so chairs had been made ready for an audience, but April knew from past experience that it was unlikely that the meeting would draw such optimistic numbers. Parishioners imagined these events to be dull and only attended when a particularly contentious matter, one likely to affect them personally, was up for discussion; street parking wars never failed to pull in the crowds.

'It's lovely and warm in the meeting room this evening, isn't it?' April commented to Ian, remembering the chill she had endured during the interminably long meetings she had attended in the past. 'The upgrading of the heating system has definitely been a worthwhile part of the refurbishment. But I'm not sure how comfortable these chairs are going to feel after an hour or so.'

'Not very would be my guess, especially for my bony arse, but perhaps your extra padding will be beneficial if we're going to be here for hours.' Ian's laugh was swiftly suppressed by an icy stare from his wife.

'I definitely think tables and chairs more suited to a cosy café environment will have to be included on the shopping list if the café is going to appeal to customers.'

April hushed her chatter as the meeting got underway and led by the chairman, the councillors dealt with the agenda items at an unusually rapid pace. Certain members of the team had a reputation for droning on endlessly. As a former minute-taker, these long-winded discussions were a nightmare when it came to maintaining concentration and April's mind had had a tendency to drift to infinitely more interesting and pressing matters – like what to have for supper when she got home.

April recalled a never-ending discussion about narrow pavements and the difficulty this presented to buggy-pushers. This subject had been raised by Isabel, a recently recruited councillor and a relative newcomer to motherhood, who appeared to think she was the first person ever to navigate an unwieldy pushchair in a tricky situation. Whilst April fully appreciated the difficulty this posed, Ash Green was a rural village location and the Highways Department could not – even if they wanted to – widen all the footpaths in this historic village in order to fit the width of Isabel's buggy. Honestly, why choose rural living when you want urban amenities?

Ian's elbow making sudden contact with her ribs returned April's attention to the present; he had obviously noticed her lack of concentration and disinterest in the recent dialogue. The unsmiling, but clearly efficient, clerk confirmed she had noted the appointment of a new village dog warden, leaving the councillors free to move on and discuss the community café proposal.

'I have to say I was rather excited at the prospect of a much-needed facility to serve the isolated residents of Ash Green,' Isabel commented. *A good and positive start to the debate*, thought April and she immediately took back her reservations about Isabel's suitability to reside in her beloved village.

'But is there really a need?' said Graham, a long-standing councillor with a mind resolutely closed to new initiatives which he believed would destroy the rustic aesthetic of the location. 'Surely very few of our residents can be described as isolated. I mean, we have a bus service into Great Aldebridge, don't we?'

We do indeed, thought April, *one bus in the morning and the last bus passing through at 7.10pm*. Graham clearly had no personal experience of using public transport and had no idea what life might be like without the use of his Jaguar.

'I think we must consider the benefits that such a facility could provide for our village. Community cafés, as April's proposal so concisely explains, offer a much wider range of services than the obvious provision of refreshments. I, for one, am most open to this idea,' said the chairman. 'I also believe we would be failing in our duty to our parishioners if we did not explore every avenue to ensure that this fabulous new building is utilised to its full potential.'

Retired now, Jeffrey had in a previous life been an executive director at John Lewis and as such was a much-admired and well-respected village resident. April was delighted to have his vote of confidence, knowing it was likely that the other councillors would follow his lead; many were in total awe of this distinguished silver fox whose presence in a room of dignitaries never went unnoticed.

'I understand that April has started to look at possible funding options and Laura has very kindly offered her help. Any grant application will, no doubt, have to be made in the name of Ash Green Parish Council, as we are the caretakers of the intended café premises. I propose we give our conditional support to the proposal, and I would like Laura and April to set up a café sub-committee reporting back to the council. I trust you would all be in agreement.' Jeffrey noted the nods of assent and April dared to imagine her first victory.

'As April has outlined, the costs for setting up the café should not be excessive, as we already have a splendid venue and the operation will be run by volunteers,' Jeffrey continued. 'However, I must request that April undertakes an in-depth local survey to find out exactly what our parishioners think of the idea. The community centre is their village facility and they should have a say in how it is used. Also, funding must be sought and the resources available not only for the project's inception, but for its ongoing success. This may prove to be a tall order, but Laura does seem to have a knack for this sort of thing.' Jeffrey concluded by once again acknowledging Laura's previous efforts.

'Thank you, Jeffrey. But before we close this discussion, may I just mention something that has occurred to me. There is a distinct possibility that this parish council may not warrant further Lottery funding, bearing in mind how recently we received a large sum towards the community centre's refurbishment programme.'

All of April's joy was gone in an instant. Why hadn't Laura thought to mention this when they were speaking on the phone the previous evening?

'Well, of course, you may be correct, Laura,' said Jeffrey, 'and may we leave this with you to ascertain? But without finances being forthcoming from one or other funding body, I don't believe the council could give its full endorsement to the café project.' The members nodded in agreement and Joan recorded the minute which April feared had the potential to scupper her café.

Out of courtesy, April and Ian stayed in their seats to the end of the meeting, even though they were more than ready to ease their aching buttocks off the uncomfortable wooden chairs. At the meeting's conclusion April made her way across the room to Jeffrey to express her thanks to the council for taking the time to consider her proposal. She wanted to assure him of her commitment to the project and determination to find the resources.

'Whilst I believe this idea is worthy of further exploration, as I have said, you'll have to convince all of us that it is a viable proposition and that village residents are properly canvassed and supportive of the venture. I look forward to receiving an update from you in due course.'

Ian had collected together their coats and was chatting to Laura when April returned.

'I'm really sorry, April, I only remembered after we spoke that I'd seen something in the small print of the guidance documentation about repeat pay-outs to the same organisations within a specified timeframe.'

'I'm not giving up, we'll have to think of something, and anyway you might be wrong.' April didn't want to listen.

'I'm not usually wrong on things like this,' Laura responded, rather curtly April felt, as she waved a swift farewell.

'Don't be dispirited, there are probably still options to take the project forward and I know you'll do everything you can to find the way.' Ian tried to reignite her spirited approach as they made their way home, but April wasn't in the mood to be heartened by his positive take on her ability to succeed.

'What the hell's the matter with Laura? How could she let me down like that? And she didn't even give any real sign of enthusiasm or support for the project during the meeting.'

'Okay, you are obviously annoyed, but help is at hand. We're almost home and I'll pour you a gin and Slimline tonic as soon as we get back. Less calories than a glass of wine, didn't you tell me that?'

'Can't you see that I'm too upset to worry about calories?'

20

She was starting to feel worn down by his persistence and Laura had to concede that she was also missing male company. Neil's call came on a dreary, damp afternoon four weeks after their break-up; her anger had subsided and everyday humdrum had returned – despite the beginnings of a promising vegetable crop. His previous voicemails had sounded sincerely contrite and although she had been determined to ignore him, put the episode behind her and move on, she just couldn't quite come to terms with the finality of that decision. After all, she had waited nearly half her lifetime for an opportunity to revisit her relationship with Neil.

'Darling Laura, thank you so much for speaking to me. I'm so sorry about how I behaved and what I said. Well, it was entirely thoughtless and hurtful. I can see that now. Please accept my apology and believe me when I say that I was behaving out of character – frustrated by my current situation. I miss you terribly. Could we perhaps meet up and talk?'

'I appreciate your apology, but I still have concerns…'

'I know, I know. I've been selfish, but I've started to look for a job. I've also asked my solicitor to conclude negotiations with Catherine's lawyer as swiftly as possible. I want to move on and I want that moving on to be with you, my darling,' Neil pleaded. 'I'd allowed myself to fall into a self-pitying rut and you were right to be very annoyed with me.'

'I'm pleased to hear it. Unless I'm looking back with rose-coloured specs, the Neil I used to know would definitely grasp control of a difficult situation and sort things out.'

'I am that person, I really am, and I'm hoping you'll give me a second chance to prove it to you. To begin with, please may I invite you to join me for a meal – my treat – somewhere special, your choice. We could go into Ipswich for a change, stay over if you fancy… No, of course, that's presumptuous of me, we'll get a taxi back?' Neil was falling over his words, clearly desperate to make amends.

Laura felt herself being drawn in, warming to his offer, warming to the man again, despite herself.

'Okay, let's go on a date in a week or so, that'll give you enough time to start turning things around. And maybe we can try again, no promises, we'll see how it goes. Oh, and as you're asking, I'd really like to go to that smart French restaurant, the one on the marina.'

Laura said her farewell and disconnected the call. She wandered distractedly from room to room, deep in thought, questioning her sanity. Was it a mistake to go back, having distanced herself from him? But could she let him go without being absolutely sure there was no hope for them together?

"Oh my goodness, Max, have I lost my mind?'

Max was following closely at the heel of his mistress; his hairy coat brushed up against her leg, his intention being to convey reassurance. Instinct and loyalty had alerted him to her discombobulation and in his desperation to affirm his faithful presence, he darted across her

148

path just as she was stepping into the kitchen. His strategy worked; she had stopped roaming, but she was very vocal as she tumbled.

'For goodness' sake, Max, you silly old dog, what are you playing at?' Laura recovered her composure, but stayed sitting on the cold tiled floor for a moment in order to assess the damage.

The dog instinctively saw this as the opportunity for play fighting and he leapt on her excitedly, pushing her to the ground, with a further thud.

'Enough, dog, let me be,' Laura said, desperately trying to push the animal away.

Max was confused; his mistress seemed to be a bit cross now, so he resorted to giving Laura a good licking with his enormous tongue to rally her and remind her that he needed to eat. It was way past his breakfast time and he might actually starve to death.

Laura rubbed her arm and her thigh, both of which had made contact with the door frame, and once upright she hobbled across to the freezer to look for a bag of frozen peas to put on her cheek.

'Ouch, ouch... dear God, that's sore.'

The good news was that no bones appeared broken and although the pain was alarming, it had in an instant put an end to her pondering.

'Right, I'm not going to give it another thought, he's getting a second chance,' Laura said to no one in particular and certainly not Max, who was bound to see things differently when it came to Neil. Sharing Max's scepticism would be April and Ellen, although they wouldn't say as much; she knew them both well enough to sense their not-so-favourable opinion of Neil. Clearly they cared about her wellbeing, but they both had partners to share their lives and could have no idea of living life without a soulmate. Anyway, it had absolutely nothing to do with them.

The telephone rang again and Laura limped her way to answer Jeffrey's call.

'Ah, Laura, just wanted to check if you've received the quote for the new signage planned at the community centre, you know, the sign we need to erect to acknowledge the Lottery funding?'

'Yes, I've received a couple of estimates. Can I email them to you later?'

'I'd really love paper copies, if possible please, my aged eyesight struggles to read from a computer screen. Could you drop them in to me when you and Max are out walking?'

'Will do, but bear with me, I'm not really feeling up to a walk today.'

Max whined; he knew he'd completely lost her attention now and he pottered off for a sulk in his basket by the kitchen range.

*

Neil lay back down on his sofa and breathed a sigh of relief. His persistence had reaped the reward he was after; he'd inveigled himself back into Laura's favour and now that he had secured that second chance he could not afford to mess up.

He'd been lonely and bored since their split; he really missed female company. A future with Laura was likely to provide him with all his requirements: a home, financial security, the lifestyle he craved and a woman most amenable to his carnal needs. Things looked to be moving back onto the right track, but he knew that Laura was unlikely to release him from his pledge to find work and get on with the divorce proceedings. He could hardly bear the thought, but he'd have to at least try to find a job; she clearly had no idea how difficult that would be at his age. Perhaps he could just fire off a few CVs to companies that were not currently hiring, then he wouldn't need to lie too much when she quizzed him on his efforts.

But firstly, he would give Donald, his solicitor, a call. Much as he'd wanted to be obstructive and hang out for the most generous settlement, it was time to agree to the meeting with their respective lawyers that his estranged wife had been requesting. He hadn't been keen to see Catherine again; she was probably only slightly less disgruntled despite the cooling-off and calming-down period which the months apart should have provided. Clearly, the face-to-face discussion was the only way forward to get any sort of a deal agreed. Lowering his sights was likely to be the answer and that wouldn't be a problem as long as Laura was lined up to be his future partner. His mission was to ensure that this former mistress remained captivated by her aged lover.

Neil checked his watch to ensure he was avoiding his solicitor's customary long lunch break; it was 2.15pm, so Donald had probably returned to his office, feeling comfortably replete. He reached into his pyjama pocket to retrieve his mobile phone.

*

Laura ran the hairbrush through her silver-blond bob and checked her appearance in the mirror. Leaving home without making an effort to look her best was never an option as far as Laura was concerned, even for a trip to the village stores, and this evening she had applied her make-up carefully in preparation for her date with Neil. As the day had drawn closer, her eagerness to see him had increased, but she was determined that this inclination should not be too obvious to him. This time he had to prove himself worthy of her allegiance and she needed him to have made progress with his assignments to be free of a wife and free of idleness. Following her first instinct, she hadn't confided in April or Ellen regarding her decision to see Neil again, believing it would be prudent to wait and see how the reunion – second time around – turned out.

They had agreed to meet outside the restaurant and when Laura's taxi pulled into the parking area, she could see him waiting for her. She felt encouraged that he had ensured to arrive before her, as a gentleman surely must, especially one with an agenda to impress. Neil was wearing the jacket she had encouraged him to purchase, with a crisp white shirt. His attention to detail made her smile and feel hopeful.

'Darling, you look divine,' he said as he kissed her tenderly on each cheek. He then took her gently by the arm and led her towards the entrance of Brasserie Amélie. Laura noticed immediately the smear-free glass doors with their polished steel fittings; she envisaged an underling being permanently on hand to wipe away any unwelcome fingerprints threatening to ruin the pristine appearance. Beyond this immaculate entranceway was an atrium where three tall white planters were symmetrically aligned – so perfectly positioned that Laura could only assume that a tape measure had been employed for the task. Each contained an exotic agave plant, creating a stunning architectural display to usher the diners onwards to the *maître d'* at his table.

Laura's eyes were instantly drawn to the array of astonishingly dramatic light fittings adorning the ceiling of the restaurant.

'Oh, Neil, look at those striking chandeliers. They are fabulous.'

'Err… bit too quirky for my tastes.' Neil took a brief look upwards.

'Really? I'd say rather than quirky they are… amusingly ingenious,' Laura said, whilst scrutinising the fittings which were constructed from an eclectic selection of wine glasses. 'I wonder if I could purchase – in mini-version, of course – something similar for my dining room.' She thought this was an intriguing proposition, but Neil did not respond; he was already out of earshot en route to their reserved table.

Laura's initial impressions were that she had entered a fashionably chic restaurant and she was looking forward to sampling the cuisine, which she had heard was exceptional. The waiter had seated them at a table positioned beside the window – a large expanse of glass which ran the length of the dining room – providing wonderful views over the marina. Lights twinkled in the darkness outside and the gentle sound of rigging clanging in the evening breeze created a romantic, atmospheric setting.

'This is all amazing, Neil, thank you so much for inviting me.'

'I am humbled and delighted that you agreed to join me and this evening is entirely on me, so let's get started and order something to drink. How about we push the boat out – sorry, that's a bad joke in view of the location (he laughed anyway) – and order a bottle of champagne?' Neil was effusive in his generosity and seemed in excellent humour. Laura began to feel hopeful that he had taken to heart all of her helpful comments.

'Well, if you're sure, that would be an absolute treat.'

'Does it still hurt, your face?' Neil reached over to gently stroke the bruising.

'Just a bit. That will teach me to be more careful,' Laura responded and in order to deter Neil from an accusation aimed at Max, she swiftly moved on to the crucial question. 'So, have you managed to sort things with Catherine?'

'I'm assured by Donald that we're almost there now and with a mutually acceptable resolution. Fear not, Laura, I am just a matter of weeks away from being free for you and, most importantly, solvent. So, let's get on with our celebration.' His response was, of course, welcome and he certainly looked pleased with himself, but Laura was slightly bothered that his answer sounded pre-prepared and too conclusive to allow for any further questioning.

Chastising herself for being overbearing, Laura let the subject be, after all what she wanted primarily was to have a wonderful evening with her manfriend. Eager to enjoy the fine dining experience, they began to peruse the enticing menu together. They took it in turns to read out the descriptions of mouth-watering dishes and slipped easily into their convivial custom of guessing what the other would choose with proficient accuracy.

'So, it will have to be the scallops for you, Laura.'

'And the chicken liver parfait for you, Neil.' They laughed and relaxed in the comforting familiarity of their close friendship.

Starters and mains were ordered and another glass of champagne poured. As they toasted each other for a second time, Laura took a moment for a further appraisal of the luxurious interior of the restaurant – clearly a popular venue, with only a few of the tables left unoccupied. She didn't have her glasses on, but as she looked across the room she was almost certain that she could see Paul at a table in the far corner of the dining area where a large parlour palm obstructed a clear view.

'Is that Paul over there?'

'Where?' Neil started to swing his head from side to side, scanning the dining room.

'Don't be so obvious, Neil, you're embarrassing me.'

'Sorry, darling, I couldn't make out where you meant, but I've spotted him now and yes, it is Paul. Oh, and he's with the most gorgeous dining companion. Lucky bastard!'

21

Ellen's professional kitchen – her starkly functional and squeaky-clean den – was now officially open and ready for the preparation of the first tart, cupcake, or anything else in fact that took a cake-lover's fancy. The Kernel Kitchen website was 'live' for all to see, posters had been posted and all of the advertising flyers had been delivered.

Along with a selection of celebration cakes and desserts, Ellen had dared herself to be bold and was extending her services to offer dinner party menus for those with a passion for entertaining, but no time or talent to do-it-themselves. She didn't want to overlook any opportunities and quietly commended herself on her courage and hard work which had seen the project get underway in a relatively short space of time. The new enterprise was ready to be put to the test, and Ellen was primed for the phone to start ringing and her email inbox to start filling up with order requests.

Paul had been an amazing help when it came to the final push towards the designated launch date. He'd even managed to get home reasonably early most evenings to take charge of bath and bedtime duties for Liam. Despite his initial resistance, Ellen was delighted when Paul appeared to join her team and was certain this was due to him finally understanding what the endeavour was all about – her need to find a fresh challenge and to regain her self-esteem. He hadn't said much, but his attitude towards her seemed to have changed in the last few weeks; this was all the reassurance she needed.

'Liam, come on now, it's time we left for school.'

'Oh no, not yet.'

'Tidy the Lego pieces away and get your coat on,' Ellen called from the kitchen, 'and don't forget your PE bag.'

'But I haven't finished building barns for all the farm animals. The greedy chickens will get mixed up with the pigs and they might fight over the corn whilst I'm at school. It's very important!' Liam babbled on insistently.

'Enough, Liam, just get ready. If we're late Miss Jackson will be very cross with us and anyway, I need to get back home quickly today, I'm going to be busy.'

With Liam's eventual co-operation, mother and son were in their coats, out of the door and on their way to the school. It wasn't a particularly long walk, no more than fifteen minutes on a good day, and having commenced their route march, Ellen sensed Liam's new-found eagerness to get to the playground to see his friend.

'You'll be able to tell George how you are getting on with the farmyard construction. Perhaps when it's finished, he could come for tea and you can show it to him.'

'Yeah… and, Mummy, you must make sure not to mess it all up with the hoover when you're doing your housework today.'

'It's all change at Walnut Tree Farmhouse,' said Ellen. 'Mummy isn't going to be doing any housework today. From now on I'm going to be doing a lot more cooking – it's my new job.'

Liam was bemused. 'But cooking is not work, Mummy, it's Daddy who goes to work.'

Raising her eyebrows, Ellen made a mental note to find time to address her son's gender preconceptions.

Caitlin was loitering outside the school and waved as Ellen and Liam approached. Liam ran off to find George without the merest sign of a backward glance and although Ellen was delighted that he was continuing to cope well with life at school, she would have liked a chance to kiss her son goodbye.

'That's us free for the day then,' Caitlin said, smiling at Ellen. 'Don't suppose you've time for a coffee before I tackle the laundry?'

'Yes, that would be nice, but come to the farmhouse with me and be prepared for my new obsession with checking emails and phone messages. It's Day One and you never know, someone might urgently require a Victoria sandwich cake or a dozen vanilla cupcakes.'

'Of course. No one could fail to have noticed your attention-grabbing poster in the village shop window. Well done, Ellen, I'm sure you will be very successful. There's bound to be lots of other hopeless bakers – just like me – who will be keen for your assistance.'

With a few drops of rain appearing in the air as they walked, both were pleased to reach the house and get inside, ensconced in the kitchen by the Aga – although Ellen had eagerly opened her laptop to check her emails prior to settling herself.

The two mums chatted companionably about how the boys were getting on at school and shared their concerns regarding what they considered the disproportionate amount of homework being set for children. They agreed that, of course, they wanted their boys to make good progress with their reading and writing skills, but the regime known as 'Read, Write, Draw' set each evening did seem an overly arduous task.

'Of course I'm happy to read with Liam, but getting him to practise writing the sentences and then having to portray the story with a little drawing is a painfully long process when he's already tired after a day in the classroom,' Ellen complained.

'I agree, it's a lot to expect of them and I have to say I've taken to doing the sketches myself in order to complete the tasks before there are tantrums,' Caitlin confessed. 'I was particularly pleased with my depiction of a helicopter!'

'Practising their dexterity is clearly important, but it does seem a bit much. Dare we say as much to Miss Jackson, or does she scare you, too?' Ellen laughed and they both agreed to keep quiet.

'So, how is your sister?' Ellen asked. 'Any sign of her moving out, or are you still enjoying having her staying in the house with you?'

'I enjoy her company, but I think Stephen gets a bit fed up having another soap fan in the house – he's lost his domination of the remote control.'

'Every man's worst fear!'

'But I've noticed that she seems happier and has been socialising a bit more recently.'

'That's progress. It will be good for rebuilding her confidence after the Rob fiasco.'

'She's being quite secretive – which is slightly out of character – but I suspect she might be dating. I hope this does mean she's finally over Rob.'

'Has she heard anything from him?'

'I don't think so, but she might not say, and I sincerely hope that when he returns from his gallivanting abroad that he doesn't try and get back together with her.'

'She's young and carefree. She's unlikely to be single for long and as you say, she may already have an admirer.'

'I hope you're right. I really want her to find a nice man and settle down. Actually it would be good to have the spare room back, a tidier bathroom and less laundry to do… Our mother was remiss when it came to training her youngest daughter in preparation for married life.'

'I'm shocked, Caitlin, that is such an archaic take on division of labour when it comes to household chores. I was only thinking this morning that I have a duty to Liam's future wife to make sure he understands that there's no such thing as women's work.'

'Good luck with that,' Caitlin replied, resigned to the daily clear-up after the other occupants of her home had all departed for work and school.

'Oh my God, that was my email notification ping. I've an incoming message!'

*

With a bulging folder under her arm, April was making her way to Laura's cottage. She was thankful that the early morning rain was easing as, despite a hasty search through her belongings whilst outside the village stores, she seemed to have mislaid the useful plastic carrier bag which normally accompanied her on every outing. With nowhere to deposit the Chelsea bun she'd purchased from the shop – intended as a treat for later – April had no choice but to munch on the soft, delicious dough as she walked. A bag would also have protected her precious paperwork – a dossier of all things related to community cafés – from a soaking if the rain returned.

April was very proud of the fact that the information gleaned had doubled in size since she had recovered her determination to get the project moving forward and she had even found time to make a start on the village survey. She hadn't seen Laura since the incident at the parish council meeting, but two days previously had been pleased to receive her text arranging the get-together to discuss the café. Laura had suggested they make a day of it and had offered a lunch of homemade soup and bread to follow their labours and an early afternoon walk on the beach with Max. Such a lovely invitation, April decided, was a sign that Laura acknowledged she had let her friend down, although she knew Ian had a different view. He thought April had overreacted and urged her to draw breath and review the situation before careering towards the scraping of one of her better ideas. Why on earth would he come to that conclusion?

'It's not Laura's fault, and anyway you'd be wise not to alienate her when she's the one person that you might need to rely on.'

'Oh, I suppose you're right. Her lack of enthusiasm was just such a shock.'

'You know that's not fair. She's just being cautious and it's her level-headed approach to business that you value.'

'Okay, but do you have to be so reasonable all the time? It draws way too much attention to my tendency towards hysteria.'

Despite her revived gusto, April now found herself dreading the prospect of preparing the funding documentation, which would undoubtedly be the tedious part. However, she felt bolstered to take on this chore, as the agenda for her day with Laura sounded pretty perfect.

'Hi, April, come in.' Laura opened the front door. 'I've cleared the dining table so we can spread out our paperwork and the laptop is fired up ready to go. Dump all your files and we'll get a coffee first. Max, leave April alone, she doesn't want you slobbering all over her.'

'Hi to you too, Max, you gorgeous boy!' April made a fuss of the hairy beast, who was barring her way to the kitchen to ensure he got her full attention. 'My goodness, is everything okay with you, Laura?'

'I'm fine. Oh, you mean how did I come by the very unattractive bruising on my face?'

'Err, yes… What's happened to you?'

'A silly tumble courtesy of a soppy beast.'

'I'm sorry, who?'

'Well, Max, of course, who'd you think?'

'No, absolutely, I knew you meant the dog beast,' April swiftly responded with a forced laugh; she didn't want any misunderstanding or offence caused that might spoil the day. 'Right, where should we start? Oh, and thanks so much for agreeing to help with all this.'

'I'm keen for us to try to get the funding applications sorted. We'll have to scour the internet to find which trusts are currently funding community projects. I'm afraid we're not going to be eligible to go back to the funder of the hall refurbishments.'

'That is a shame. But, there's no point dwelling, we'll just have to try others. I know we can approach the district council to see what's available in their pot and there must be some local businesses willing to sponsor the venture.' April was determined to be upbeat and positive. 'Ian's been brilliant and has helped me with a business plan and costings. I'm sure the fact that we have suitable rent-free premises is going to be a major advantage.'

'Have you had many responses to your residents' survey?' Laura asked.

'I've had one or two and there does appear to be some interest in our plans. I've also heard from a couple of potential volunteers, which is fantastic.'

'Okay, I know you only circulated the village consultation document last week, but if you don't mind me saying that sounds like a bit of a lukewarm reaction.'

'Early days, Laura. We need to give villagers time to consider the proposals and to get their completed questionnaires back to me.' April refused to be discouraged. 'I'm doing my best to engage with residents when I see them, as I'm sure the personal approach helps. I had a chat with Derek recently and I think I managed to get across to him all the plus points of having a café.

Derek knows everyone in Ash Green and does love a chinwag with those he meets when he's out and about – he'll be a great help to have on side.'

'Strangely, I wouldn't have thought a community café would be Derek's "cup of tea" – excuse the pun. But anyway, as you say, getting local residents chatting about a café is a start. I just hope they are not going to take forever to respond, because we could be doing all this work for nothing,' Laura said, failing to disguise her reservations.

'Just give them time, it's a new concept. They just need to understand how wonderful it would be to have a coffee shop in the village,' April continued to enthuse.

'We'll see. On that note let's take our coffees into the dining room and make a start,' Laura said whilst looking out of the window. 'The sunshine looks set to win over the clouds, so let's get this task out of the way as early as possible so that we can enjoy the rest of the day.'

It was the monotonous process April had envisaged; however, after a few hours they had found some possible sources for the finances they would need. The only issue was that each organisation was offering no more than a few hundred pounds, so it looked as though they were going to have to make multiple applications. On discovering this stumbling block, they both decided they'd had enough for one day; it was definitely time to take a break.

Laura put the saucepan of stilton and celery soup on the hob to heat through and the bread rolls in the oven to warm.

'Mmm, that smells good,' said April, salivating. 'So, what else is new, Laura? Tell me to mind my own business, if you like, but I can't help but wonder whether you've been in touch with Neil.'

'Was it that obvious that I'd eventually relent? I suppose you think I'm mad to see him again,' Laura said, whilst also taking the opportunity to apologise for being distracted, if not utterly crabby, over recent weeks.

'I wasn't convinced that you were convinced it was over. But it's not up to anyone else. Only you can know whether he's the companion you want in your life.' April was determined to sound non-judgemental and it was, of course, entirely up to Laura who she dated. She looked again at the black eye. Good Lord, was there any possibility that Laura was hiding a worrying truth? But no, she swiftly reprimanded herself; he might not be an entirely ideal partner, from April's viewpoint, but it seemed highly unlikely that he was a bully. And anyway, Laura wasn't the sort of woman who'd suffer a tyrant… was she?

Laura spooned the steaming soup into bowls and placed them on the kitchen table. Before sitting, she held a dog chew in front of Max's nose and led him away to his basket so that they could eat their lunch undisturbed.

'Now it's no longer my secret, you might as well know that Neil and I went out for a meal together at the end of last week. I thought it would be good to have an opportunity to make peace. We hadn't parted on particularly good terms.'

'No, it was a rather dramatic dumping. I wonder how long it took him to find a way back home?'

'Yes, well… Anyway he does finally seem to be getting out of his rut and I've decided to give him the benefit of the doubt. I'd hate it if you thought I sounded weak, but the truth is, I really missed having a man in my life.'

'Not at all, as long as Neil makes you happy,' April replied.

They concentrated on the contents of their soup bowls and April was delighted to find she had plenty of appetite, despite the unplanned bun calories consumed earlier. She noticed that Laura was lost in thought and seemed to be troubled by her contemplations. 'Okay, there is something wrong, isn't there?' April asked, praying it was anything other than black eye-related.

'Have you seen Ellen recently?' Laura asked in a rush, almost choking on her bread in the process.

'Not since last week when I dropped in to Walnut Tree Farmhouse to admire the new cake kitchen and sample some of her candied walnuts. Goodness they are delicious.'

'I'm looking forward to trying them. They can't be entirely off-limits. I read somewhere that walnuts are rich in antioxidants.'

'An excellent excuse to eat more the next time I'm at the farmhouse,' April said, smiling. 'The kitchen looks fabulous. Ellen has done such a good job there... but, why do you ask?'

'I've been in two minds whether to say anything, or not. But I just can't stop being bothered by what I witnessed when I was at the restaurant with Neil the other evening.'

'What was that?'

'I saw Paul. He was at Brasserie Amélie dining with a young woman.'

April was momentarily rendered mute and wide-eyed, clearly indicating to Laura that her news had made a worrying impression.

'What is he playing at?' April finally spoke out. 'I was concerned when I saw him at the train station with a young woman – attractive, slender, long dark hair?'

'That's the one – young being the key word. So you've seen them together, too?'

'When I told Ian, he said I was reading too much into it and that she was probably just a colleague, or travelling companion.' April was annoyed now that she had allowed Ian to dispel her fears so readily.

'She could be either of those things, but they did seem a bit... err, friendly. Does Ellen have any idea, do you think?' Laura asked.

'After the train incident, I did have a chat with her, trying to gauge how things were between them. I know she has been thoroughly fed up with him working all hours, but I'm certain she hasn't any knowledge of this friendship.'

'Just what I thought.'

'And, another thing, she did confront him,' April added.

'About what?'

'Well… she discovered a dinner receipt in his jacket pocket. He told her he was entertaining a work client. I sense she assumed this was of the male gender.'

'Oh no, she actually found the Brasserie Amélie receipt?'

'Yes, but that was a while ago, rather than just last week, so from that we can deduce these dinner dates are a regular occurrence.'

'This is dreadful.'

'But what the hell are we going to do now we both have reason to believe he may be up to something?' April blurted.

*

Laura saw that April was looking to her for guidance and could only be thankful her friend was unable to read her thoughts.

She was both horrified and overwhelmed with anguish. Having been eager to talk to April about her fears, her hope had been to receive reassurance that there was nothing to worry about. April was especially close to Ellen and was likely to know if there was anything really wrong between the couple. Now it appeared that April too had suspicions. Could someone they both cared about be sharing her life with a lying cheat?

The reality was stark and unbearable; thoughtless infidelity could have destructive repercussions, the sort of repercussions that you didn't want a dear friend to suffer. Laura felt her cheeks starting to burn with shame. She had never bothered to consider the indisputable facts that real lives got wrecked and people got hurt; poor, dear Ellen, how would she get through this?

'I tell you what we are *not* going to do and that is jump to any conclusions!' Laura said with emphasis. 'The last thing Ellen will want is us prying and interfering in her personal life. The rules of friendship have boundaries. This is one that should not be crossed.'

'Yes, but she's our best friend. Shouldn't we do something?'

'Actually no, I don't think we should do anything at the moment, except just continue to be available if she needs to talk and perhaps be on our guard in case...'

'In case?'

'Look, I've always thought Ellen and Paul were really well-suited – and with little Liam, they are the perfect family unit. We're probably barking up the wrong tree – or should I say, wrong walnut tree.' Laura laughed almost hysterically, hoping to lighten the mood. 'And talking of barking, let's get this dog out in the fresh air. Like us I think he has been cooped up inside for way too long.'

22

'You asked for my opinion, April, and I say again, I think you and Laura should tread very carefully.' Ian was getting irritated; April had launched into her tirade the moment she had crossed the threshold and before removing her sandy walking boots. 'And talking of treading carefully, watch it, you're trailing the contents of the beach across the carpet now.'

'Calm down, I couldn't get the laces undone and was desperate to tell you what Laura told me.' April was now hopping back to the boot room, hiking footwear still stubbornly attached to her left foot.

'You really don't know for certain that Paul is having an affair and if you wade in with your suspicions it could be disastrous – for your friendship and in the worst case for Ellen's marriage. Think about Liam, please leave it alone.'

'But that's just it, I am thinking of Liam and…' April's attempt to continue was abruptly halted.

'Enough, April.' Ian held up his 'stop talking now' hand.

With any further discussion clearly off-limits, April shut

her mouth, but was resolute in her concern. She thought there was a real possibility that Paul was being unfaithful and the ramifications, if Ellen were to discover this, well, it was just too distressing to even consider.

To April's surprise, Laura's opinion had been similar to Ian's. After sharing her fears, Laura had surprisingly quickly set them aside, forcefully expressing her opinion that they should not approach Ellen with their concerns. Whilst they had been out with Max, April had made one final attempt to scrutinise the suspicions they both clearly shared, but Laura had been adamant.

'It's conjecture, April, we have no real proof. For sure, I felt uneasy when I saw Paul, but equally I wouldn't be comfortable having this discussion with Ellen. I just don't feel it's appropriate for *us* to involve ourselves in the marital affairs of others.' Laura had stressed *us* but April knew that the reference was directed at her.

April didn't want to agree with her husband, or her friend, and she wasn't prepared to ignore her hunches. Privately, she made the decision that, regardless of their views, she would keep a close eye on Paul and continue to check that all was well with Ellen. This was clearly her duty as a good friend.

'Did you and Laura finally get the grant application sorted?' Ian asked April, ensuring that the conversation was rechannelled to a more pressing and tangible matter.

'Oh, that… Well, because the parish council has already received Lottery funds for the village hall, we've had to find alternative grant providers, just as Laura suspected, but we did manage to unearth a couple of other organisations that might look favourably on our project. I left it with Laura, having input as much of the information as I could.'

'Told you that you needed her on side.'

'Yes, thank you, you were right, and I am grateful that she is willing to assist.'

'Do you really think you've got a chance of getting any money?' Ian had the air of someone leaning towards scepticism.

'Let's stay optimistic, but in the meantime I'm going to approach some of the local businesses for sponsorship and Laura's going to see what help the district council might offer.' April had to acknowledge that urgent project-café challenges had to be overcome, and so for the moment she might have to put worries about Paul and Ellen aside and focus on her coffee shop.

'I think it all hinges on how you sell the scheme. You need to stress your intention to create a café primarily as a centre for community cohesion and inclusion, run by volunteers and not relying on making a profit. In those circumstances you must be able to attract support.'

'Hey, Ian, you don't fancy helping me with canvassing for funds, do you? I think people will warm to your eloquent interpretation of the cause.'

'I'll write all the begging letters you'd like me to, but don't ask me to do anything that involves socialising. No barn dances, pub quizzes or karaoke events to raise funds. I know my limitations.'

'Any help you are willing to give will be most welcome, but right now I must make arrangements to meet up with the volunteers. They are going to be vital to the success of the café.' April wandered off to make some phone calls and update her spreadsheet.

'What plans are there for our dinner this evening, April?' Ian was concerned that his wife would become engrossed in her mission and he was famished.

'Oh yes… dinner. Well, I'm not sure what there is and I had some rather delicious soup for lunch, so I'm not really hungry just yet. Take a look in the fridge and see if you can find some inspiration in there.'

'Okay, I get the picture… Looks like a fish finger sandwich for me.'

'Yum, that sounds nice. Put three in the oven for me too!'
April called from the study.

*

'Look forward to seeing you next week, and thanks for
putting yourself forward, I really appreciate it.' April completed
her round of calls to the village ladies; she hadn't been surprised
that none of the local men had put themselves forward. Everyone
was happy to commit to the e-training to gain their Basic Food
Hygiene certificates and now they just needed the group to form
the official sub-committee reporting to the parish council.

One of the ladies, Janet Stewart, an up-and-coming artist,
had an intriguing proposition which April had found most
compelling. She suggested that the community café could double
as a gallery for arts and crafts.

'The large expanses of wall space in the community centre
would provide the perfect backdrop for exhibiting artwork,' she'd
said.

Janet's paintings and drawings, many illustrating the glorious
Suffolk landscape, could be offered for sale, with a percentage
of the sale price being returned to the café's coffers. April had
seen some of the artist's stunning work when she had held an
open studio event. She knew that these would not only provide a
lovely environment for café users to enjoy, but would also support
a local artist, whilst providing some welcome additional income
to keep the café afloat.

April envisaged that other gifted and talented local artists
and craftspeople would appreciate the same opportunity, and she
reopened her spreadsheet to add yet another item to the list of
things to do.

She'd also had a conversation with a lovely young mum
called Caitlin, who apparently was a friend of Ellen. Caitlin was

happy to take on the supervision of training and any mentoring, explaining that her involvement in tutoring in a previous life would stand her in good stead for this role. It was becoming apparent that there was a wealth of currently untapped expertise in their small community. April saw herself on a crusade to ensure that the talents of all these wonderful women should not be squandered.

The enterprise seemed finally to have gained some momentum and April foresaw a busy few weeks ahead. It was clear she would need to use her free time wisely and was very doubtful that she'd have time to attend any Pilates classes. Ian had been right – although she wouldn't tell him so – when he said she didn't have the time to delve into the possible failing of Ellen and Paul's marriage. Her covert plan of a spell of surveillance on Paul's movements would just have to be put on hold – but just for a while.

23

'Bloody hell, it's freezing.' Neil winced as he lay in bed shivering. He wrapped his arms tightly around his upper body and ran his cold feet up and down the surface of the bed sheet briskly, hoping the friction would generate a modicum of heat. The winter weather was being spiteful and he was finding it a struggle to keep warm in his rented cottage; it was quaint, with old-world charm (April's description), but so draughty and in desperate need of a boiler update. Having woken to find that yet again the heating had failed to click into action at the appointed hour, he was grumpy and dreading having to remove himself from underneath the duvet to brave the chilly climate in the bathroom. This barely adequate facility was located in a tiny extension to the rear of the property, on the ground floor and off the kitchen, offering function over form. It was a far cry from the home comforts he had enjoyed in his former marital home.

He had kept his promise to Laura; the terms of the divorce settlement had been all but agreed with Catherine and her lawyer. This left his lack of employment to address and he had reluctantly forwarded his CV in response to a couple of job vacancies Laura had drawn to his attention. Being in his mid-fifties – well, in reality, speeding towards sixty – he was confident that he wouldn't get a look in when these companies were sifting through the numerous applications that would no doubt land in their in-boxes.

Going back to work had not been an option that registered on Neil's radar; he'd hoped there would be no need for a job following his divorce, but Laura clearly had a different point of view. It was alright for her, she obviously had plenty of money to see her through a very agreeable retirement. He couldn't help being slightly resentful that her life was stress-free; she had time to enjoy her pastimes and was in good shape – fit and well-conditioned.

'Mmmm…' Neil smiled at the thought of Laura's womanly shape. He was happy to be back in her favour, which ultimately meant he had been invited back into her bed, where she never failed to satisfy him. He might currently be enduring the polar environment of his grotty accommodation, but he was envisaging a way out.

The sound of his mobile phone stirred Neil from his thoughts and required him to reach out across to the bedside table from under the covering provided by the bedding. An icy blast of air rushed along his arm and down his body, having found an access route as he moved.

'What the…' He grabbed the phone as quickly as he could, returning beneath the duvet to take the call, which by happy coincidence was from the lady herself.

'Morning, luscious lady, how are you?' Already aroused by previous thoughts of her naked body, this enjoyable sensation was further sustained by the sound of her voice.

'I'm fine, but where are you? Your voice sounds distant.'

'Err, out for a brisk walk, darling, talking to you through my scarf!'

'I'll keep this brief then and anyway I must dash as I have a job to do for April. I was just wondering if you fancy coming over this evening for supper. I've a couple of rather substantial fillet steaks that risk exceeding their use-by date if you don't come over and help me eat them.' Laura sounded happy, provocative even.

'That sounds like an offer I'd find hard to refuse. May I ask whether your lovely offer might extend to a sleepover?' Neil was hopeful and had visions of finally being warm, perhaps even warm enough to raise a sweat.

'Let's just say, bring your toothbrush and we'll see how we feel. See you at six o'clock.'

'It will be my pleasure.' Neil's day was definitely looking more auspicious. It was just a shame that Laura didn't see fit for his toothbrush to be a permanent fixture in her en-suite bathroom, rather than an occasional visitor.

*

After finishing her call, Laura bent down and wrapped her arms around her dog for a hug. She was trying to be cautious, but had persuaded herself that the reunion with Neil was the right decision and with a bit of gentle moulding there were signs he could become her ideal partner. He was clearly listening to her, taking heed of her suggestions, and perhaps he might soon be rejoining the workforce. Laura felt for Neil's own sake it was important that he kept himself occupied, not only to boost his income by his own efforts, but good for self-esteem and dignity. He'd been through a difficult time, and in fairness it wasn't surprising that he'd succumbed to inactivity and self-absorption.

'Well, dog, you might not be too delighted, but I'm going to give Neil the benefit of my stewardship and take advantage of the benefits of having a love interest back in my life.'

With her plans for the evening in place, Laura settled herself in front of her laptop determined to get some of the dreaded funding applications emailed off before the end of the day. She wanted to help April, but felt her enthusiasm for the café proposal waning somewhat. Laura knew that if the grants were forthcoming and the café did open, then her services would also be required further along the process. But she was happy to assist as a volunteer – a role that would fit well alongside being the parish council's representative overseeing this project. For April's sake, she hoped her current pessimism was misplaced.

'Okay, Max, let's try and get this boring job out of the way and then we can get outside for a walk. Not to the beach today, we'll go across the fields and over to Jenny's Hill. You can chase rabbits instead of seagulls for a change.' Laura liked to keep her dog informed of her objectives for the day, and he seemed happy and in agreement with her suggestion, wagging his tail in response.

The screen of her laptop was showing her the spinning beach ball, as it whirled slowly into action, and before she was properly logged into its technology, her telephone startled her with its shrill ring. She hesitated and contemplated leaving the answering machine to take a message, but reconsidered not wanting the task of returning a call later.

'Yes, hello!'

'Hi, it's me, Ellen. I hope I'm not calling at an inconvenient moment.' Ellen sounded apologetic and Laura hoped she hadn't sounded unnecessarily brusque.

'Not at all, it's lovely to hear from you. Oh gosh, I've just remembered it's Pilates this evening. Were you calling to arrange to go together, because I'm really sorry but I forgot all about it and have other plans now.' It was Laura's turn to be contrite.

'Actually, no I wasn't and I'm not able to go either because Paul has an evening meeting so won't be back in time. It's a little too late now to try and find a babysitter for Liam,' Ellen said, adding that she wasn't too bothered about missing the class as she was feeling weary having spent two days baking and then decorating a seventieth-birthday cake for a new client.

'That's a shame, we must make sure we go next week and drag April along as well.'

'Yes, I'll see... Anyway, I wanted to ask you whether you'd recommend that restaurant overlooking Ipswich Marina – I can't recall its name – the one you went to recently with Neil. Hope that's all going well, by the way,' Ellen added, then hastened on, 'You see, I'm planning on surprising Paul with an evening out to celebrate our wedding anniversary and I thought it might be just the place. It appears to be everyone's favourite at the moment.'

'Ah yes, that's... err... that's... Brasserie Amélie. The phone number should be easy to find.' Laura suddenly felt hot; a trickle of sweat ran between her breasts – the physical manifestation of her horrid discomfort. 'The food there is very good and it's a romantic setting – perfect for a wedding anniversary celebration.' Oh God, why had she said that?

'Great, I'll see if I can book a table. I want to make it special and show my appreciation for his help getting my business off the ground.'

'And how are things? Are the orders flooding in?' Laura was relieved to change the subject.

'It's going okay, thanks. I'm actually just back from delivering another completed cake order. Home-baked and bespoke-designed cakes are very much in demand, as, it seems, am I.'

'That's great news, really fantastic. Good for you.' Laura was keen to get back to work. 'I'll have to get on now, Ellen, I've promised April I'll get her funding applications off.'

'Of course, thanks for the info about the restaurant and I promise to try to get to Pilates next week.'

Laura went back to her desk and stared blankly at the screen of the laptop. The irony of Ellen making plans for a surprise dinner for Paul at Brasserie Amélie – the venue of choice for his love trysts – was not lost on her.

24

Paul's rush hour journey had been fraught and on reaching the train station he could feel his blood pressure rising as he struggled to find that illusive of all goals: a free car parking space.

'What the fuck!' In desperation, he had completed two speedy circuits of the aisles of parked vehicles, before divine intervention engineered a miracle. He spotted the reversing lights of an Audi about to vacate its space and quickly and aggressively positioned his own car so that he was poised to take possession.

Sprint-walking the distance from the parked car to the station entrance, he refused to allow his briefcase – although heavy with paperwork – to hinder his pace. He needed to move quickly or he was in danger of missing the train and therefore the opportunity of spending some time with his favourite travelling companion.

The pair had managed a few exceedingly pleasurable clandestine dinner dates over the preceding month. They had also found the opportunity to meet up after work on a couple of occasions for a drink in one of Ipswich's popular gastropubs, keeping company with the young professionals of the town. Paul particularly enjoyed these excursions which made him feel vibrant and part of the energetic crowd. It had, however, been a while since Paul had spent any time with Rosanna; instead he had listened to his conscience which told him it would be judicious to be on hand to offer Ellen support whilst she was at the crucial final stages of getting the cake venture underway.

He was proud of his wife; that went without saying, didn't it? She'd actually surprised him by showing a previously unrevealed tenacity to see her idea through from its inception to the launch of the business. This was all fine, but he still resented being sidelined as Ellen's vision and focus narrowed to a sole purpose. He had doubts as to whether she would see the deluge of cake orders that she seemed to anticipate and although failure would be gutting for his wife, it would mean there was a chance that life at home might return to the normal he preferred. He'd be understanding when the pursuit ran its course, ready to offer commiseration and comfort to his wife. This would obviously be expected and he would oblige.

Until that time, Paul was happy to bask in the success of his recent game plan. Ellen had clearly appreciated his efforts and it looked likely that he might now reap the rewards of her gratitude. He'd been pleasantly surprised when she'd initiated lovemaking the previous night, reaching down to his recently dormant manhood as he'd lain beside her. It was a very rare occurrence for her to make the first move. Perhaps the adrenaline buzz of her recent accomplishments had sparked this welcome development, or she saw this as his treat for all the support. Whatever her motives, he wasn't about to complain; Ellen seemed happier and he had played his part as the accommodating husband and, apparently, he had played it particularly well.

With the atmosphere at home now free from any hostility, occupying his thoughts on his journey that morning – apart from the anticipation of seeing Rosanna – were some untimely feelings of insecurity at his workplace. His company hadn't fared well during the financial year and there were whispers of a clean sweep of management and possible redundancies. He didn't think his own position was particularly vulnerable – he was doing the necessary business to hit his personal targets – but there was an uneasy vibe in the office which wasn't conducive to good staff morale.

Determined not to spoil the moment, Paul tried to put all work thoughts aside; there was no benefit to pondering employment concerns until he was at his desk. Instead, he had the train journey and the bliss that came with having Rosanna beside him on his commute to the capital.

With less than a minute to spare he barged his way through the crowded station entrance and onto the platform, desperately scanning the sea of faces to locate his prize.

*

Rosanna was waiting beside an open train door, standing back very slightly to allow passengers to alight, but close enough to get into the carriage at the first opportunity and secure a seat or, better still, two seats. She'd had her eyes trained on the entrance to the platforms, but there was still no sign of Paul. It would be just like him to overlook the need to allow sufficient time to accommodate the roadworks on the main road and she felt irritated by his habitual lack of time-keeping. She hadn't seen him for a couple of weeks and it was beginning to look doubtful that she would see him that morning, which would be a pretty rubbish outcome to what had promised to be an exhilarating start to her day.

Their relationship to this point had remained chaste; however, Paul had slightly overstepped that mark at their last meeting and she shivered with pleasure at the thought of the advancement. They had been sitting together enjoying a glass of wine at the George & Dragon; she had been telling him about a gorgeous villa in the French Alps that had recently come onto her company's books.

'The panoramic views are unimaginably stunning and the interiors are wood-clad, as you might expect, with designer furniture and fittings that are straight out of *House & Garden* magazine. I'm sure we'll have clients fighting over dates to book it for their holidays,' she had said and then gasped as she felt his hand on her inner thigh under the table. He had smiled and gazed into her eyes whilst sliding his hidden palm upwards towards her crotch. This act, fleeting though it was, was the first clear signal of his ultimate intent.

Rosanna liked Paul – no, it was more than that, she really fancied him, he was a good-looking guy – but she was wary about getting involved with a married man. She'd heard plenty of desperate tales from her girlfriends of the disastrous sequels to embarking on affairs. A liaison with Paul could be way too complicated, not at all appropriate and likely to end in tears. Her tears. She'd shed enough of those when Rob had dumped her and it had taken a long while to get over that hurt. But there was no denying his appeal – the startlingly blue eyes and his mature masculinity, which had a surprisingly powerful draw. Perhaps a connection with this man might be worth a bit of risk-taking and anyway didn't she deserve to indulge herself in some harmless fun?

*

'Rosanna! Rosanna, save me a seat.' Paul's yell, as he sprinted along the platform, clearly startled her but he saw she was quick to smile with pleasure at the sight of him.

The train was busy, but not quite at standing room only, and they were relieved when they spotted two free seats having walked the length of three carriages in their search. Succeeding in this quest had looked exceedingly unlikely, which added to the joy of the elusive opportunity for a precious hour sitting together.

'So, how are things with you?' Paul settled with his outer arm pressed against his companion's, allowing him to feel her warmth.

'I'm pretty good, although work has been hectic.'

Paul couldn't help noticing Rosanna's full pink lips, which looked invitingly kissable on this serendipitous morning. 'You're looking really well, I must say... and it's great to see you again.'

He saw a blush appearing on her exquisitely formed cheekbones. 'Yes, I've missed seeing you, but I hope Ellen's new business is now fully operational.' She seemed to be trying to divert the conversation, perhaps feeling uneasy under the intensity of his affectionate gaze, although he was convinced she was flattered by his attention.

However, he followed her lead. 'Yes, baking activity is well underway and with a consistently low failure rate it's unlikely that my waistline is going to suffer as a result of having to eat the rejects.' Paul held in his stomach and laughed. It was so good to feel this relaxed.

'She's such a good cook, I can't see how she could fail. I really hope it all works out for her.' Rosanna sounded so sincere; such a sweet girl, he thought.

'It's early days and who knows, perhaps she has discovered a winning formula. Best of all she seems so much more contented now that she has this new distraction,' he said.

'That's really nice for her.'

'Anyhow, enough of the baking news, what plans have you this week? Any chance that we could meet up after work? I've gained a few brownie points recently, so feel justified in having some "me time" – with you, of course.'

Rosanna grinned. 'I'd like that. What about the end of this week? Things at work usually wind up swiftly on a Friday afternoon and I'll be more than ready for a well-earned glass of wine by then.'

Fridays were not great for Paul; he knew Ellen liked him back home promptly in anticipation of spending as much time as possible enjoying a family weekend after the madness of a busy week. But after the merest moment of hesitation he agreed to Rosanna's suggestion and their next date was firmly diarised.

They chatted amiably for the rest of the journey, catching up on the events of the past couple of weeks. Rosanna revealed that she was concerned Caitlin might be getting a bit fed up having her little sister living-in and so she'd started to look at the rental market with a view to moving on. However, working within the limitations of her budget, her search had been a rather depressing one so far.

'My options seem to be a shoebox in the city – and in a not especially desirable location – or something a bit better in Ipswich, but with commuting costs to be added into the equation. I think on balance my preference now would be to rent a city apartment.'

Paul felt saddened by her obvious dilemma and also at the prospect of her moving into the city, putting an end to their most congenial train journeys together.

'Are you sure Caitlin wants you to go? Surely she can see that it's not going to be that easy for you to find something affordable.'

'It's probably Stephen more than Caitlin, to be honest, but actually I do need to get my own place. It's a bit restrictive living with family – I can't really invite any friends over – and my sister likes to be kept informed of my movements. It's like living with your mother all over again.' Rosanna grimaced.

The train slowed as its arrival at Liverpool Street Station was announced, signalling the end of their journey – an anti-climax for them both. As always, the simple joy of being together, with easy, flirtatious conversation at the start of another demanding workday, was a scenario they both delighted in securing.

They parted company outside the entrance to the tube station; Rosanna's office was just a saunter away, but Paul still had to brave the underground to reach his final destination. He kissed her cheek as they parted, his hand on her arm, their bodies closer than entirely necessary for the platonic kiss.

Paul watched as Rosanna walked away; she handled her four-inch heels with well-practised technique and a provocative sway of her hips. Her shapely arse and giraffe long legs had him mesmerised. Yes, he was enjoying her friendship and her company, but ultimately he was yearning to bed her and he was certain she shared his longing.

25

Gino had glimpsed the first rays of sunshine illuminating the street outside the glass frontage of his café. This was the sign he was looking for to determine that it was now an acceptable hour to make his phone call to April. He didn't want to assume that because he was up and ready for his day that other mortals would be keeping similar hours. His trusted employee, Gemma, had sent him a text the previous evening letting him know that she needed to take some time off work at short notice. In other words, she wasn't coming into work at Amarellos' today and Gino knew he'd struggle to cope unless he found a replacement for her. He couldn't delay a moment longer; surely by now even the unemployed would have risen from their beds.

He was delighted to hear April's chirpiness. 'Hello, Gino, this is an unexpected pleasure.'

Gino rushed in with his request. 'I was wondering whether you might be able to come into the café, April. I'm so sorry that it's such short notice, but Gemma has gone to visit her mother in Frinton-on-Sea. Seems as though the poor dear lady may have

broken her wrist, and Gemma wanted to go and help her out.'
Gino was speaking quickly – his urgency apparent – and with
his Italian intonation April had difficulty comprehending the
gist of the conservation.

'You'd like me to come in to work with you today?'

'Oh, thank you. Thank you. If you could come in this
morning, I'd be so very pleased. Just as soon as you can.' Gino
disconnected the call and dashed off to empty the dishwasher.

*

April had other plans for her day but she didn't feel she could,
or should, pass up the chance of working at Amarellos', not that
Gino had given her that opportunity. It had been a couple of
months since her training and she felt a bit apprehensive, but
then again this would provide her with a refresher course. She'd
drunk plenty of coffee since her barista training, and with a new
appreciation of its skilful preparation, but she hadn't had the
occasion to make use of the tutelage.

Having already consumed two slices of toast, with lashings
of butter and marmalade and two mugs of tea, April had only to
touch up her make-up and find suitable clothing for her day at
the café. Her black skirt and white linen shirt would be fine –
imperative to stay cool – and most importantly her comfy, flat
black shoes. It took her just a further fifteen minutes to be ready
to leave her home, and with keys in hand she closed the front
door and got into the car.

'Damn and blast,' April swore as she joined mums on school
runs and employees on their way to work, all participating in
the queuing traffic. It hadn't been that long since she'd become
jobless, so how could she have already forgotten the frustration
of rush hour, even in this rural idyll? The stop-start journey into
Great Aldebridge was a wholly frustrating one.

With her car finally parked in the town's main parking area on its outer perimeter, she leapt from the vehicle, eager to get on her way to Amarellos'. As she dashed between other parked cars, she became aware of someone calling out her name – a loud call, one of apparent urgency. She looked around her wildly, unable to see anyone she recognised.

'April, April… it's me…'

April was searching for a name to give to the woman, who she could only assume she had met somewhere previously.

'We've spoken on the phone. It's Caitlin… I'm so pleased to have caught you.'

'Oh hi, yes, of course… Caitlin. I'm afraid I'm in a bit of a hurry – but feel free to walk with me, I'm heading for the high street.' April didn't really have time to chat, but as a potential café volunteer Caitlin did deserve her attention, even if briefly.

'It's about our community café meeting on Friday evening.' Caitlin blasted April's eardrums, continuing to speak at the same level despite the fact that she was now walking alongside, matching April's strides towards the town centre. 'I was wondering whether you might consider the merits of my sister.'

'Well, yes, of course, all volunteers are most welcome.'

'But she has talents that might prove especially useful.'

'More talents… Excellent, we seem fortunate to have an endless supply of talented women, Caitlin.' April couldn't figure out where this conversation was going and glancing at her watch picked up speed in the hope of losing the foot passenger who was delaying her progress.

'If I could just explain. She's spent a couple of years working in France – firstly as an au pair and then at one of the ski resorts – and is a fluent French-speaker. She's now got a fabulous job in the city, which requires her to use her linguistic skill, so she's never lost the ability.'

'I'm not sure how useful that's going to be, unless you

anticipate French tourists will be dropping into our café during their tours of Suffolk – although I suppose it's not impossible.' With her mind preoccupied with thoughts of the mechanics of coffee bean grinders, April was struggling to follow the sequence of Caitlin's conversation.

'You are funny,' Caitlin laughed, but April didn't think she was being remotely funny, she just wanted to get to work; she hadn't been pressed to get to a place of work for a long time and she was beginning to feel stressed by the necessity this morning.

'It occurred to me the other day,' Caitlin continued, 'maybe we could offer some French conversation sessions. There are quite a few folks in the village who holiday regularly in France. One or two are even lucky enough to have second homes there. Perhaps we could arrange a few evenings at the café for some informal conversation? Oh, and we could even offer French-inspired fare.' Caitlin was obviously very pleased with her suggestion and was looking to April to check that she was equally impressed.

The fast march had succeeded in conveying April to the side street where Amarellos' was located, now she just had to lose her companion. But she did pause a moment and reflected that Caitlin probably was offering up a unique proposal for drawing customers to the café.

'I think you may be on to something rather promising, Caitlin. Unfortunately, I really must go. Can I call you later, or perhaps just get your sister to join us at the volunteers' meeting?' April waved and dashed off.

*

'Thank you so much, and welcome.' Gino was clearly delighted to see April. She hoped to goodness he would be equally thankful for the help she could in reality offer. But she mustn't put herself down; he had asked her because he believed she was capable, so she put her bag and coat in the staff room, found herself an apron and went to find out what her first task would be.

'I think you may need a brief time working alongside me just to familiarise yourself with the workings of our splendid coffee machine – it has been a little while since you were last here. We've managed to get through the early rush this morning – a bit stressful with just two of us – so we need to get you up to speed before the next burst of activity at mid-morning.' And it didn't take her long to orientate herself behind Gino's counter with merely a few minor spillages, a barely suppressed shriek as the hissing steam arm took her by surprise and a couple of hot flushes.

The café was busy and April didn't have much time to chat, or be sociable, but she was surprised by how much she was enjoying the hectic morning serving a seemingly endless line of customers. Despite the fact that she'd enjoyed the luxury of her unemployed freedom, April was reminded that a day with purpose made you feel so much better about yourself. As time progressed, she felt more in control and was finding moments to converse with those around her, customers and colleagues alike, and everyone appeared to value her efforts.

Ian texted her during her lunch break with words of support and a teasing reminder to maintain her patience with all the customers, but for once his motivational message was not required. April frowned at his certainty of her losing her cool at some point during service, making her determined to prove him wrong.

It was a good day, but when her watch told her it was 5.30pm, April sighed with relief and went to remove her apron. She was exhausted, but delighted when Gino asked her to return the next day. Obviously, she hoped that Gemma's mother would make a good recovery, but she was happy for Gemma to stay in Frinton-on-Sea for a little while longer.

'Well done, April, your training prepared you well. We'll make a very proficient barista of you in no time at all. See you early tomorrow morning.'

April waved to her new colleagues as she exited the café feeling chuffed.

'Well, who knew. This could really turn out to be my new future,' she informed a nearby seagull, who wasn't paying attention as his beak and most of his head was inside a discarded crisp packet.

26

April plumped up the sofa upholstery and then stroked her favourite teal-coloured cushion, taking pleasure in the texture of the soft, velvety fabric. A sudden thought then consumed her and she crossed the room to the open fireplace in order to run her finger along the mantel shelf above to check for sooty residues that had a tendency to loiter there. Using the tissue which she found scrunched up in one of her pockets, she removed the offending deposits.

It was Friday evening and she was preparing for the gathering of café volunteers. Despite the anticipation, she felt jaded and if given any choice she would have opted for putting her feet up, with a large glass of wine and the secret pleasure of an episode of *Love Island*. It had been an exceptionally busy day at Amarellos' and having now completed three full days at the café, April was ready for a rest. She tried to rally, drawing on her belief that she was an excellent multitasker and would find a way to muster the energy. Anyway, the meeting would be over in at most a couple of hours, surely?

She retrieved a couple of extra chairs from her dining room and then laid out wine glasses in readiness – for herself and her fellow imbibers – and coffee cups for the more restrained of the group.

Ian had made plans to be absent that evening and although Paul had turned down his invitation to join him for a pint at the pub, he assured April that he was happy to take a chance on finding some interesting company for a chat at the local hostelry. April noticed him tuck *The Telegraph* newspaper under his arm just before he'd left and with a pen in the top pocket of his jacket, the crossword might instead be keeping him occupied. She knew he had no desire to remain in the house that evening.

'A bunch of women all together – with wine – it will definitely get rowdy,' Ian had claimed.

Arriving at the designated hour – a good omen for their time-keeping skills, April thought – the women were swiftly settled with refreshment of choice, ready to work their way through the items for discussion. Some introductions were required, but most of the ladies were already acquainted and the room hummed with the sound of their chatter.

They numbered six and included Laura and Caitlin – minus her sister. Also missing was Ellen, who was at home with Liam having been unable to find a babysitter when Paul had to let her down at the last minute. Laura was representing the parish council and her first task was to confirm to the new committee that she had managed to send off four applications for funding and expected to hear whether any were successful within eight weeks. This timeframe would allow them a sufficient interval to seek out any other sponsorships, identify any additional equipment they might require and work out how the café would operate.

'I will start looking into sourcing the all-important espresso machine, which I envisage will take centre stage of our new enterprise.' April was just ever so slightly biased in this respect.

'I'd like to put myself forward as the person responsible for seeking out some extra-café activities, perhaps craft demonstrations and, as I've already mentioned to April, gatherings for those wishing to practise their French dialogue,' said Caitlin.

'That sounds good to me, as long as it's okay with the rest of the group?' April noted the nods of agreement – others no doubt grateful to allow someone to put themselves forward. 'As you say, we've had a brief chat about offering French conversation group sessions, which might indeed prove to be appealing, and it's marvellous that your sister is willing to get involved. Such a shame she couldn't join us this evening and meet everyone.'

'It would seem she was easily swayed by a better offer, I'm afraid,' Caitlin explained, 'and I suspect that offer might have come from a man.'

There were raised eyebrows from one of the older ladies, but the rest of the group nodded their understanding that a young woman would obviously choose a date over attendance at a gathering such as this. Was she really up for offering language instruction, or had Caitlin, in her excitement, offered her services prior to seeking her sister's accord?

'We still haven't talked about opening days and times,' said April. 'Your thoughts on this would be most welcome.'

The audience looked to Laura, being confident in her qualities as the voice of reason and good sense, and she was happy to contribute. 'I have given this some thought and propose we initially open three days a week, perhaps between the hours of 9am and 4pm. We should have enough volunteers to cover the three days without it being an arduous burden on our willing – not forgetting unpaid – helpers, and the hours would enable us to catch trade just after school drop-off in the mornings and around pick-up time in the afternoons.' Again, a murmur of assent came from the committee members. 'Our elderly villagers might like to come for mid-morning coffee, or afternoon tea, and with any luck we'd also be recognised as the village venue for a light lunch. I don't believe we'd be taking any business away from the pub, but I will have a chat with the landlord to offer some reassurance.'

'That's great, Laura, and I'm sure I speak for all the volunteers when I express our thanks for all the assistance you can give,' April said, 'and is there anything else that you want to add before we move on?'

'Actually, yes... I also think we should be prepared to offer some of the additional activities we are considering at weekends, and perhaps some evenings to make them accessible to all, just so long as we don't feel this would be burdensome in terms of personal commitment,' Laura concluded.

Whispers of uncertainty were just audible, but general agreement was noted and the decision made that progress could be reviewed again once the café was up and running and they had the benefit of some experience of their venue's popularity.

April glanced at the wall clock; it was nine o'clock and the ladies were now onto their third bottle of wine; the coffee cups had remained untouched. She was tired and keen for the meeting to be over and mindful that she had more than enough new points to add to the ever-growing spreadsheet.

'Coffee before you all leave?' April asked politely whilst making her way into the kitchen with the spent glasses. But thankfully the team, especially those with small children and an inevitable early morning start, said their goodbyes and departed, leaving just April and Laura to reflect.

'How do you think it went, Laura?' April asked as she placed a cup of coffee in front of her friend.

'Not too bad. It's great to have the volunteers' interest and new ideas, but I think we will eventually need to have more people on board. I suppose that once things get going, more of the locals might be inspired to get involved.'

'Thank you so much for your input this evening and for being the link between the café committee and the parish council. I imagine you plan to give them an update on how things are progressing at the next meeting?' April asked.

'Absolutely, they just need to be kept informed, but once we know about funding the councillors will want a more formal meeting with your committee to agree the way forward,' Laura said.

'Such a shame Ellen wasn't able to join us. I'm still hoping that she will be able to supply the café with cakes and other baked delights. Trust Paul to let her down. Another late evening meeting, no doubt?' April was sceptical and looked quizzically at Laura. But Laura refused to be drawn.

'Has Gino finished with your services yet?'

'Thankfully I'm not needed tomorrow, as he has a couple of students who work weekends, but it looks as though Gemma won't be able to return to work for a while yet. Her mother's wrist injury turned out to be rather serious − a nasty break − and living alone she's not going to be able to manage without Gemma's help.'

'Poor lady. What happened to her?'

'Apparently she tripped over her dog − how careless!' April scoffed, but Laura thought this was merely one of the minor hazards of co-habiting with a dog.

'Could happen to the best of us,' said Laura, rubbing her thigh. 'So, you must be delighted to have the weekend off. Have you and Ian any plans?'

'No doubt he will want to take a drive out to the coast on Sunday in his recently acquired mid-life crisis purchase.' April had entertained Laura with the tale of her husband's malaise and his self-indulgent purchase of a sports car.

'The weather hasn't been conducive recently for trips out with a "top off", but he's been watching the TV forecast avidly and predictions for this weekend have given him the incentive to plan an outing.' April was apathetic about being asked to accompany him; she still didn't quite get his passion for a convertible.

'April, where's your enthusiasm? It will be exhilarating – the warmth of the sun on your face, the fresh air filling your lungs and the breeze blowing away a week's worth of cobwebs,' Laura effused.

'It's the howling gale whipping my hair about my face that's the absolute off-put, not to mention unsuspecting flies getting stuck to my lip balm, but if there's the offer of a pub lunch at the end of it I might be tempted. Have you tried the beef and ale pie with triple-cooked, hand-cut chips at The Wheatsheaf?' April's mouth watered at the prospect.

'Diet's going well then?' Laura asked – rather cuttingly, April felt.

27

Paul's last-minute evasion of babysitting duty on Friday evening, followed by his lukewarm response to her surprise that morning, had left Ellen feeling thoroughly exasperated. As a consequence, her instructions were barked abruptly as father and son were leaving the house en route to the swimming pool.

'Please make sure to shower Liam properly and wash and blow-dry his hair after his lesson. Oh, make sure you have a 20p piece for the hairdryer. I want him returned to me in time for an early tea.'

'Right, will do.'

'Once you're back, it will be your job to get him into his PJs and ready for bed before Izzy arrives.' Ellen gave Paul no room to doubt what was required of him. He had agreed to her plan that he and Liam would go to the local sports centre, allowing Ellen the time and uninterrupted peace to put the finishing touches to a rather unique design on a wedding cake, which had to be delivered to the clients on Monday morning.

When Ellen had received the enquiry about this particular

wedding cake the clients were most specific about the cake they had in mind. She was more than happy to put herself to the test and produce the two-tiered cake; the bottom layer was to look like a black tuxedo, the second layer to include the first initials of the couple and the cake was to be topped by two little black top hats. Her first gay-wedding cake – its photograph would be a great addition to her website.

Since the launch of The Kernel Kitchen, Ellen had received what she considered to be a good number of orders, mainly for celebration cakes, and she was thrilled that the venture looked set to endure. Developing the business beyond cakes and sweet treats into fine dining services was still her long-term goal, although she knew she had to start somewhere and her instincts so far had proven to be correct. There was most definitely a market for her goods and services, and she now felt confident that she had the ability to make this work. It was early days, but this wife and mother had added a new role to her repertoire and she wasn't prepared to allow either husband or child to hold her back on her path to success.

Buoyed by her conviction and some extra cash in her personal bank account, Ellen had been eager to arrange an evening out for herself and her husband to celebrate their forthcoming wedding anniversary. She hadn't missed the fact – not least because of comments he'd made – that Paul was feeling ignored and neglected. She knew he'd tried to help her out within the confines of his own work commitments, but it irked her that she felt an innate necessity to pander to the needs of her man.

It was, therefore, hugely disappointing to be met with Paul's lukewarm response to her surprise announcement that morning. She'd expected him to be touched by her thoughtful gesture, but he'd seemed anything but delighted to hear of the preparations she had made – table reserved, babysitter at the ready, taxi booked

– so that they could spend a romantic evening together. Was it really too much to have anticipated an appreciative reaction?

'It's a good surprise, isn't it?' Ellen asked him again a little later in the morning.

'Of course, ignore me. Sorry, just a bit tired.'

*

Paul had been taken aback to be told that morning that he'd be dining at Brasserie Amélie for the second night in succession. This time in the company of his wife. He hoped his indifference to Ellen's news that she'd organised a surprise celebration hadn't been too obnoxious. But he convinced himself that she'd accepted his apology and the explanation of a manic week at work resulting in exhaustion and his apparent lack of enthusiasm.

He and Rosanna had initially planned to meet for a drink the previous evening, but in an attempt to impress her, he'd booked a table at their favourite restaurant, which had become their special place. He knew he'd made the right decision when she'd clapped her hands together and squealed with delight.

'Oh, Paul, how lovely! That's a fabulous treat at the end of a hectic working week. You're spoiling me,' she'd said, her red-coated lips gently brushing his cheek.

He'd been happy to spoil her; in fact, he was now at the stage where he was prepared to do almost anything to please this nubile young woman, who seemed to find him equally interesting and, if he was reading the signs correctly, sexually attractive.

Reposing poolside, Paul pondered their blossoming friendship as he dozed in his galleried seat overlooking the blue quadrant of water, its surface choppy with children's frantic attempts to master the front crawl. The warm atmosphere made him impossibly drowsy and, closing his eyes, he allowed himself a moment's respite and the opportunity to relive the delicious details of his date with the bewitching female. He switched his mind away from the sound of the noisy children and the shouted supervision of the instructor and focused on Rosanna.

'Another glass of wine, Rosie?' Paul knew she enjoyed hearing him use this pet name.

'Much as I'd love to, I shouldn't. I've still got to drive home later.'

'Oh, go on… It's Friday and I'd be happy to give you a lift back to the village.'

He'd seen her hesitation, but it was momentary and she'd soon ordered another Pinot Grigio Blush.

As with all their assignations, Paul had relished every moment, but this time their conversation was definitely more flirtatious than on previous occasions. Rosanna clearly shared his enthusiasm for their liaison and had expressed her disappointment when Paul declared it was time he was getting back home. She wasn't intoxicated, but, having consumed two further glasses of wine, was clearly incapable of driving herself back to Ash Green. Paul had been delighted when she'd happily accepted his offer to drive her home, rather than take a taxi.

'You can leave your car in the car park overnight. I'm sure your sister won't mind driving you back tomorrow morning to retrieve it.'

'Thank you for a great evening, Paul.' He noted her flushed cheeks and it was clear that she was ever so slightly tipsy. It was endearing; she looked hot, smouldering with uninhibited desire. Was that just the effects of the alcohol? he wondered.

Once in the quiet confines of his car, they both sensed the thrill of the intimate closeness, away from prying eyes. Whenever they were out together, Paul understandably felt the necessity to frequently scan all in proximity for any familiar faces. He knew this annoyed Rosanna, but of course she'd had to accept that this was the disadvantage of spending time with someone else's man.

'Will you be travelling on the train next week?' Rosanna had asked, leaning towards him and placing her hand on his knee.

'I've meetings in my diary for Tuesday and Wednesday, but I'm not sure if I will need to be in London any other days.' Paul was aroused by her gentle touch and very keen to plan their next encounter. 'You know, when I've meetings on consecutive days like that and they are likely to be late ones, it does reinforce the practical sense of getting myself a small studio apartment in the city.'

'You couldn't make that a two-bedroom flat, could you, then I'd be able to rent your spare room?' Rosanna had said flippantly. 'That would solve my current accommodation problems.'

He laughed at her remark as he pulled the car into a side street, parking up just around the corner from Caitlin's house. As he'd switched off the car engine, he'd turned to his passenger.

'Thank you for your company, I so enjoy our evenings out. I would find it difficult to imagine you not being part of my life now. The truth is I've become rather too fond of you over the last few weeks, Rosie.'

He'd been delighted when Rosanna's response had been to place her hands either side of his face and pull him towards her. It was a passionate and unmistakeably eager embrace.

These reminiscences had caused a stirring in his boxers, but his eyes shot open and the tantalising arousal was abruptly interrupted by the sound of his son yelling, 'DADDY... Daddy, did you see me...? I swam the whole width, without my armbands.'

28

Neil's lack of employment continued to bother Laura and she had started to check out the job vacancy websites on his behalf. She couldn't understand why Neil was not being invited to attend any interviews for the roles that she had been so sure would be perfect for him. It was probably time to check out his CV to make sure he was selling himself properly. He may not understand that it was all about overplaying your strengths and skills these days – not something that Laura's generation would necessarily feel comfortable doing, but if that made the difference between getting noticed or having your application consigned to a waste bin, then so be it.

Laura was finding the fine-tuning task she had set herself – the conscious but caring remoulding of Neil – a slow-paced endeavour. He did appear to be making an effort to find a job and following a number of accompanied shopping trips his wardrobe was looking in better shape, even if his waistline continued to resist her attempts to assist with its reduction. However, the most important development was the news that Neil's divorce from Catherine was now one step closer.

'Let's mark the occasion properly,' Laura said. She hoped her enthusiasm would be a further boost for Neil. 'I feel as though we can consider ourselves a proper couple now.'

'Sounds like an absolutely excellent plan,' he said, grinning broadly. Whilst she was encouraged by his delighted response, she knew that Neil's vision for this celebration would probably look very different to her own. His would entail the pair of them spending a night at a luxury hotel (away from Max), gourmet food, plenty of Merlot and a sumptuous four-poster bed.

'I was thinking... we could make it a double celebration. Let's invite April and Ian, and Ellen and Paul to dinner,' she said. 'We can congratulate Ellen on the promising start she has made with her new business. I've heard so many lovely comments from people who've commissioned cakes.'

'So will the party be about me – I mean us – or about Ellen's cakes?' Neil failed to disguise his disappointment, doubtless foreseeing Ellen's accomplishments hijacking the festivities.

'For goodness' sake, Neil, it's about fresh starts, leaving behind the old and moving forward with the new, events to be applauded. I'll give April and Ellen a call and get a date in the diary.'

Acting immediately on this thought, Laura went towards her snug room, where she could sit comfortably with her phone and enjoy a chat with her friends at the same time.

'Come on, Max, come with me and leave Neil in peace to watch the cricket on TV.' Laura encouraged her dog away as Neil's wheezing was starting to get on her nerves. 'Did you remember to take your inhaler this morning, because there's no point having an asthma remedy if you don't take it conscientiously?'

Laura heard him mumbling as she left the room – dog at her side – but the words were indistinguishable.

*

'Blast it!' Ellen carefully placed her rolling pin and icing cutters on the kitchen surface to prevent them escaping onto the floor before brushing her hands on her apron and making her way to the telephone.

'Hello, Ellen speaking.'

Laura noted Ellen's breathy voice; she sounded agitated. 'It's me, Laura, sorry have I dragged you away from another work of art?'

'It's fine, I am working but this cake is for a child's birthday party and it's quite a straightforward design, thankfully.' Ellen had completed the sugar paste train and was about to start on the Kit Kat train track around the base of the cake.

'Still getting plenty of enquiries?' Laura asked.

'Yes, I really can't complain, and all the extra decorating practice does mean that I feel so much more confident in my skills. There's a creative side to me that's finally getting a long-awaited airing, and I'm loving it.'

'That's so brilliant, Ellen, well done you. And that actually leads me to the reason for my call. I'm planning a dinner party – for the six of us. It's primarily to mark the official demise of Neil's marriage – a moving-on celebration really – but it would also be a great excuse for your most avid supporters to salute you and your creativity.'

'My goodness, thank you. Although celebrating a divorce does sound a bit insensitive. But, of course, I understand the sentiment and we do indeed appear to have new beginnings to acknowledge and toast.'

'So, I need to check April's diary, but I was thinking of suggesting a week on Saturday. Does that give you enough time to find a babysitter for Liam?'

'That should be fine. Izzy has become a bit of a regular helping us out with Liam and he's grown quite attached to her. She's great with him. Such a sweet girl and she lives in the village.

It's so convenient,' Ellen said, making a mental note to call said godsend straightaway to check her availability.

'Excellent, let me know if she's not free. We can always rearrange, but in the meantime, I'll get in touch with April.'

'Would you like me to assist with any of the catering? I mean as a friend, not in an employed capacity.'

'No way, it's an evening where you can relax and let someone else prepare the meal. But thanks for the offer.' Laura was adamant that Ellen was to be treated.

'Well, I do feel I am having an indulgent time at the moment, especially having just enjoyed a rather splendid meal at Brasserie Amélie only last Saturday evening. Thank you for the info and the recommendation. Paul and I definitely marked our wedding anniversary in style. The service at that restaurant is exceptionally good. The staff were so attentive.'

'That's good to hear,' Laura said, now on her guard as Ellen continued.

'Once we got through the door, Pierre, that rather handsome *maître d'*, was so warmly welcoming. We were treated like regulars with all the benefits – moved to the best table, overlooking the marina and presented with a complimentary bottle of Prosecco.'

'How marvellous. It was obviously a very successful surprise then?' Laura managed to get out the words, despite alarming thoughts running rampantly through her head.

'To be honest, when I first sprung the surprise on Paul, he didn't seem that pleased – probably just a little overtired and not relishing the thought of a late night – but once we were there and had polished off that first bottle he seemed to relax and enjoy the evening.'

Ellen remembered the train, still without its track. 'Must dash now, Laura, the cake is calling, or is that whistling?' she laughed. 'But thanks again.'

Having felt so cheerful and upbeat, Laura now felt inexplicably saddened.

She sighed as she dialled April's number and stroked Max's soft head.

'Well, Max, what is going on there? But we must not say anything to April. She'll just get anxious and we are supposed to be planning a celebration.' Max wagged his tail in agreement to keeping quiet.

Laura was adamant, under no circumstance would she be drawn to delve intrusively into the personal affairs of others. Her mind bizarrely chose to liken the encroachment to someone removing the lid of her precious wormery and rummaging around in the dirt inside without invitation, or proper cause, to do so. She swiftly returned from her revelry as April answered.

'Hi, April, how are you?' Laura forced herself to sound cheery.

'I'm fine, thanks. Have you got funding news?'

'No. Not yet, it's too soon.'

'Oh, I know, but I'm just keen to find out if we can get our café plan underway. Oh, and I think I may have located a second-hand coffee machine. It was a contact given to me by Gino, a friend of his who is closing his café in Bedford and returning to his home village in Italy. His premises are being taken over by a local florist who obviously doesn't want to keep any of the equipment, so it will be worth a visit to see what else is on offer.' Laura could sense April's eager enthusiasm.

'That all sounds very positive – although perhaps not for Gino's friend – but my call is of a social nature, nothing to do with the community café,' Laura said, feeling relieved to be temporarily free of the wearisome demands of that project. 'I'm arranging a dinner party and I'd love you and Ian to join us, along with Ellen and Paul, for a small celebration.'

'You know me, I'm always up for a party. Whatever we're celebrating, count me in. Oh yes, and Ian!'

29

All her preparations were complete; Laura was ready for the arrival of her guests. She had decided to keep the dinner menu simple – asparagus to start, salmon fillet with new potatoes and salad as a main course, and pears in red wine to finish the meal. Well-balanced and light – devised to keep the calorie count down, which of course was a huge disappointment to Neil.

'Not steak and chips? I thought this get-together was prompted by the impending end of my marriage, so surely the meal should involve my favourite things to eat?'

'It's a tasty combination of fresh flavours… and it will also assist April. She's dieting again,' Laura said, whilst prodding Neil's fleshy girdle; no words were required.

'It's a special occasion. Surely even April is allowed a day off dieting.'

'You know April – absolutely no will power if confronted by a calorific option. Just last week when she and I were doing our monthly tidy-up of the community garden, she was incessantly munching Maltesers, claiming they were the dieter's treat of choice at just ten calories each.'

'So, what's your point? She'd clearly selected a low-calorie confection?'

'As I said to her, that's all fine, but not if you eat the entire contents of a family-sized bag. I still think she'd benefit from a bit more exercise. It's such a shame she won't commit to Pilates sessions.' Laura smiled smugly as she placed the flat of her hand on her stomach, enjoying the comparative tautness of her fifty-six-year-old abs.

Not wanting to be left out, Neil took the opportunity to sidle up and grab her from behind, pressing his groin into her buttocks. 'Whereas you, foxy lady, have a gloriously trim and toned body, which is much appreciated by your man.'

'I do, indeed… and I also walk miles with Max,' Laura said, attempting to push him away.

'Mmm… yes, very fit…' Neil's hiss at her lobe was swiftly followed by the tip of his tongue probing her ear.

'Stop it, Neil, now is not the time. Behave yourself and please promise me you'll be mindful of what you say in front of my friends this evening,' Laura said as she managed to free herself.

'I'll only promise if there's a reward for me later when your guests have gone home and left us in peace.'

'Enough, our guests will be here any minute,' Laura said, her voice firmer this time.

'Shall I put Max in the utility room, out of the way?'

'Not just yet, he'll want to say "hello" to everyone when they arrive.'

'Honestly, Laura, you treat that dog as if he's almost human, no wonder you have no proper control over him.' Neil backed off as the dog approached, teeth very slightly bared; the mutual dislike had not abated.

Laura's look told Neil all he needed to know and he crept

away into the sitting room to pour himself a whisky. 'And don't drink too much this evening,' Laura called over her shoulder.

*

'Absolutely delicious, thank you so much,' Ellen praised her host as she savoured the salmon, which was cooked to perfection, with its accompaniment of Laura's classic French hollandaise – courtesy of the local deli. The other guests murmured their appreciation in unison; they were all enjoying a convivial evening and the free-flowing wine.

'It's my pleasure, honestly. I'm just so happy for you, Ellen. Setting up The Kernel Kitchen was such a brave and inspired choice of career change. I've huge admiration for how you've got the business up and running. Paul, you must be so proud of your talented wife.'

'Well, of course. It's certainly been keeping her occupied. Liam and I are just about coping by ourselves. So yes, all good at the moment,' Paul said, draining another glass of wine.

Laura looked across the table and shared a frown with April.

'There's no reason why Ellen shouldn't have the chance of a fulfilling career, Paul, and I've no doubt that she's ensuring Liam doesn't miss out on having his mum around. I imagine you are giving him as much of your time as you can, too.' April wasn't at all sure, as her sarcasm made apparent. She was irritated by Paul and found his comments insensitive and egocentric. How on earth did Ellen put up with him? Arrogant arse.

'Paul does work really hard, and it's long hours, April, but he does try to do his bit when it comes to childcare. It has been a bit of a juggling act recently, but now that I've got Izzy to call on it's become much easier.' Ellen had sensed April's tacit disapproval of Paul. His comments had annoyed her too, but she wasn't happy for others to condemn her husband.

Finding the conversation a little on the dull side, Neil was determined to play a more centre-stage role, after all, this evening was also about him and his new life post-Catherine.

'I have to tell you, guys, it feels great to be a – very close to – free man again.' Neil launched into his speech, without thought or discretion. 'Not sure I'd get married again in a hurry. And I've no regrets about not having had kids, either. What an encumbrance they turn out to be, especially when a marriage breaks down!'

The chiming of warning bells rang out for Laura as the well-oiled Neil relapsed into his predilection for speaking without caution. A change of subject might save the moment, but before she could utter a word Neil unwisely continued.

'Pity I didn't have the balls to give Catherine the heave-ho all those years ago when I knew Laura before. But better late than never and now it's definitely time Laura took me on a romantic break so we can really celebrate my untying of that knot. A bit of sun, sea and plenty of the other should do the trick nicely,' Neil slurred.

Her host's wince did not go unnoticed by April, whose brain tried desperately – despite alcohol saturation – to determine whether there was something underlying in Neil's remark that she ought to be questioning.

Laura swiftly turned to Ian and attempted to draw him in with a new topic of conversation. Ian was being his usual well-mannered and courteous self, but had also failed to contribute much by way of conversation.

'So, Ian, I hear you've treated yourself to a rather racy little sports car. How are you enjoying it?' Laura enquired. 'It's a Porsche Boxster, isn't it?'

'It is, and I love it, Laura.' Ian's eyes lit up for the first time that evening. 'Driving along the country lanes, with the roof down, feeling the sun on your face and a connection with the environment. What could be better?'

'Ending up with a sunburnt forehead the other day was not so great, was it?' April laughed and the haze of her previous thoughts instantly evaporated.

'I was just a little unprepared on that occasion,' Ian said, blushing slightly. 'As you can probably gather, Laura, April is not exactly a fan of the convertible.'

'Honestly, ladies, let me tell you, my husband now spends more time stroking the sleek curves of that car than he does stroking my own womanly form.'

'I saw April trying to get out of your car the other day, Ian,' Neil sniggered. 'You were dropping her off at the community centre. She virtually crawled out onto the verge. It was hilarious.'

'Actually, Neil, my slightly arthritic hips were giving me a bit of bother – I was just a little stiff that particular day,' April was indignant, 'and don't say it, Laura.'

'Don't say what?'

'That attending Pilates classes would help!'

30

Her mouth was as parched as a gritty desert and a throbbing pain rested just over her left eyebrow. With a concerted effort, April managed to open her eyes and, having done so, peered across at her husband in the bed beside her. Ian was sitting upright, propped up on a hillock of pillows, dunking ginger-nut biscuits into a large mug of tea. The Sunday newspaper and its associated magazines were scattered around him on top of the duvet.

'So, finally awake, are you?' Ian smiled at his dishevelled bedmate.

'Dear God, I feel awful. How come you look so bright and, by the way, where's my tea?'

'As is my practice, I managed to be so much more disciplined than you. I stopped after two glasses precisely.'

'Of course you did, you're utterly virtuous. It's enough to freak out even the most moderate of drinkers.'

'You only have yourself to blame, April.' Ian was unsympathetic. 'Your tea is on the bedside table, but it's probably lukewarm now.'

Hauling herself to a slumped sitting position, April reached over to the mug and took a grateful sip. 'Yucky,' she said with a grimace, but continued to consume the tea.

'Bit of a strange evening, wasn't it?' April said, after a few minutes of recall.

'It might have been better if everyone had drunk less. Unrestrained drinking does have a tendency to lead to unrestrained talking. Unfortunately, that in turn leads to details being disclosed which might be better left undisclosed.'

'It wasn't that bad... was it?' April was hopeful.

'It was as if you'd all taken a truth serum. But no, you weren't too badly behaved – quite reserved, in fact. But Neil didn't do himself any favours, which he's probably regretting now because I'd got the impression that he'd made up some ground with Laura recently.'

April was intrigued. 'I don't remember him saying anything especially controversial. With Neil the comments are just inappropriate.'

'Still not a fan, I gather,' Ian said.

'I hope for Laura's sake that I'm wrong about Neil, but I still don't think he's genuine. And he's such a lecherous old git.'

'Really? What's he ever done to you?'

'Nothing... I don't imagine for a moment that I am his type. Anyway, what did he do, or say, that I appear to have missed?'

'You did look a bit sleepy at one point. Perhaps you dozed off for a moment and missed the clanger.'

'A clanger – excellent. Will it result in him being cast out again? Do tell.'

'You are awful, April.'

'Just tell me!'

'Well, we were talking about his rented cottage and he said that the landlord doesn't want to renew his tenancy, which is due to end in a couple of months. Laura apparently wasn't aware of his impending eviction – Neil should have drunk less wine, for sure – and she immediately quizzed him on his plans. Of course, this is Neil we're talking about and it's pretty obvious he doesn't make plans. He appears to just bumble through life until Laura takes him in hand and forces him to make a decision,' Ian said.

'I bet Laura wasn't too pleased.'

'Things got worse, because he implied it wasn't really a problem as he could always move in with Laura.'

'Now why doesn't that surprise me? That was probably his plan all along.'

'But then he went on to make the situation even worse.' Ian was grinning now.

'How so?'

'He said the only stumbling block to this arrangement would be his allergy to Max, but that in his opinion dogs should be kept in kennels in the yard and not permanent residents inside human homes.'

April laughed. 'Anyone would think he doesn't actually know Laura at all. That has to be one of the biggest *faux pas* he could have made. Do tell, what was Laura's response?'

'In accordance with decorum, particularly at one's own dinner party, she said nothing. Always a lady, our Laura. But the look she gave him was quite terrifying. If he'd imagined he'd been on a promise last night, I'm afraid all was lost in that moment. I suspect he was relegated to the sofa, at best.'

'Better still, he'll be relegated to the kennel he has planned for the yard.' April was feeling so much brighter. She just felt it was a pity she couldn't recall more of the evening's entertainment, as it seemed Neil had opened up a tidal floodgate with his careless talk, and with any luck this time he might drown in the deluge.

*

After a decadent lie-in and leisurely brunch, April was cajoled by Ian's suggestion that they get out for a walk by the sea.

'Yes, okay, some fresh air might do me good,' April said, 'and the weather forecast does look promising.'

As they arrived at the beachside car park and got out of the car, they were immediately blasted by a lively on-shore breeze, which was accompanying the sunshine.

'This will sort out your fuzzy head, if nothing else,' said Ian.

'I'm feeling a lot better, actually, so don't make fun of me. And by the way I've been thinking... and I've decided to give up wine for a couple of weeks. It's just too calorific and my liver probably deserves a bit of a break, too.'

'If you say so.' Ian seemed unconvinced by his wife's pledge.

Well equipped with their sturdy walking shoes and fleeces – along with Ian's essential backpack containing camera and mint humbugs for emergencies – they linked arms and set off, planning to hike to the next beachside village further along the shoreline.

'With any luck, the Beach Hut café will be open and we can get a cup of coffee. Oh, and they usually have that delicious red velvet cake,' April said. 'You know, Ian, it's red because of the beetroot... so, do you reckon...'

'Don't tell me, you imagine that will be one of your five a day taken care of.'

'Absolutely.'

'That is so not the case, April. Your reasoning is faulty, as I'm sure you are only too aware.'

April ignored her husband's comment and they fell into a companionable silence as they walked. Ian was enjoying the peace and the invigorating exercise, unaware of the thoughts whirling through April's mind as she reflected on the seemingly troubled relationships of her two dearest friends. She couldn't really understand Laura's ongoing interest in Neil and could only assume she must have been really lonely for male company and had never shared these feelings with anyone. April wondered if Laura had just chosen to live up to everyone's preconception that she was perfectly happy living alone. And what was really going on with Ellen and Paul? Their marriage pact showed no sign of being strengthened, despite his recent support of Ellen and her fledgling business. If anything, her growing confidence appeared to have given her a self-sufficiency that permitted limited time for others. It was good to see Ellen forging ahead successfully, but it might be wise if she also kept an eye on Paul. April was in no doubt that he had been overly friendly with his travelling companion; anyone who saw them together would surely draw the same conclusion. She knew that Laura also had suspicions, even though she'd shrugged off April's concerns. And what was that all about?

'I do really love you, Ian, ever so much, and I always will.'

'I love you too… but what's with the treacly sentiment?'

'Oh, I was just thinking… I am lucky to be sharing my life with you. Despite the fact that you are infuriating at times, we are good together, aren't we? We enjoy doing the same things, like this – walking –and visits to the cinema… Oh, yes and meals out.'

'I suspect there's a compliment lurking somewhere,' Ian laughed.

'I love Ellen and Laura, but you're my best friend really. You're so patient and pragmatic. Over the years we've been together, you appear to have mastered an almost infallible method of dealing with my, shall we say, foibles and my tendency to the odd mood swing.'

'Have I?'

'Yes, you have. Like when I'm having a rant about something, you just let me get whatever is bothering me off my chest and I suspect you sometimes pretend to agree with me, even if you don't, and when I cry you always give me a hug, even though you might not be sure of the cause, because, as you know, a whole raft of things bring on my tears.'

'They certainly do, like that animal advert, with the sad donkeys.' Hoping to join in, Ian was willingly searching for examples, but April wasn't quite finished.

'And when I laugh – which is frequently because I find so many things hilarious, things you probably don't even get – you laugh with me. Our relationship works like a dream, doesn't it?'

'Are you still drunk?' Ian laughed, but April sensed he was touched by her impromptu outpouring of affection.

'We're quite different, aren't we?' April continued on her mission to prove their marriage was better than most and Ian decided there was nothing to be gained by trying to stop her. 'You see, I'm a social butterfly and you're a bit of a homebird. But we make it work because you recognise there are times when I really need you to accompany me, but equally you are happy for me to venture out without you.'

'Relieved, actually, to get out of some of those outings.'

'But we have trust and that's so important.'

'Okay, I see what this is all about, you are worried about your two dear friends, whose relationships, even to my untrained eye, appear rocky.'

'You see, just as I said, you might not say a lot, but when you do express your views you are always subtly diplomatic and insightful.'

'I am a saint and you are indeed very lucky. I suspect that a lesser man might not cope with you.' Ian's laugh was checked by a jolt as April's elbow made sharp contact with his body.

'I love this place. We are fortunate to have easy access to such a great beach.' April appeared happy, which made Ian happy.

'It is glorious, especially on a day like today.' Ian was enjoying the tranquillity; there were other people sharing the beach, but they were few and the expanse of sands were so large, it was inconsequential.

'I am feeling less anxious now I'm no longer concerned about not having a permanent job. I'm enjoying helping out at Amarellos' and being involved in the café project. I hope it comes to something, but in the meantime I've loved spending a bit more time doing the barista thing. Who'd have thought I'd so enjoy making coffee all day for other people!' April laughed.

'I'm pleased for you and it shouldn't be too long now before you hear one way or another about funding for your café. If it doesn't work out at least you'll have the satisfaction of knowing that you had a great idea – one with the enjoyment of the community in mind – and that you did all you could to make it work.'

*

The Beach Hut café was finally in sight, much to April's relief. The wind had strengthened, its potency stirring up the sandy surface of the beach, and the sun was obscured by the increasing cloud cover.

'Thank goodness, it's open. Let's get inside for a warm-up,' April said, 'then we'll have to think about making tracks to get home before dusk.'

But the muffled sound of a ringing mobile phone distracted them from their mission and instinctively they both began to search their pockets to locate the offending device.

'We must select more identifiable ring tones, April. I never know whose phone it is.'

'Mine this time,' April said, and was just in time to answer before voicemail prevented her.

'Oh, Mum,' April heard Megan whimper.

'What's up, darling? Tell me.'

'It's all such a mess. What will I do?'

Their daughter was barely managing to speak and a brief silence was then followed by the sound of sobbing. The few words she did utter were lost amongst the tears and the sounds of seagulls screeching overhead.

31

Ellen stirred first, sensing movement at the foot of the bed, and Paul roused moments later as his wife reluctantly received the incomer. Paul's was an agonised sigh; he was not ready to be awake, but then neither was the child's mother. It had been a late night, one of overindulgence followed by restless napping, with the result that the quietest possible wake-up call was essential. But it was not to be; they were the parents of a four-year-old live wire, who was ready at six in the morning to face the day at full speed.

'It's morning time… and I'm starving.'

'Dear God, sort him out please, I'm knackered,' Paul said, turning over in denial that this awful inconvenience had anything to do with him.

With some effort, Ellen rolled off the edge of the bed and led Liam downstairs, out of Paul's way. 'Come on, sweetie, let's ignore Daddy, he's just a grumpy-head this morning.'

Liam laughed and went off with his mum, chanting, 'Daddy's a grumpy-head, grumpy-head, grumpy-head…' over and over until he reached the kitchen and was finally, no doubt to Paul's relief, out of earshot.

'Ssh now, Liam, if you can be as quiet as a mouse, I'll make you some toast and honey. You can take your plate on one of the special trays into the sitting room. Have your cup of milk here in the kitchen whilst I'm sorting out the toast. I'll put the TV on for you so you can watch CBeebies and eat your breakfast.' Ellen desperately wanted Liam settled so that she could return to her bed, alongside grumpy-head.

'Yippee, thanks, Mummy.' Liam was delighted.

'Ssh... and remember, no crumbs on the sofa and be a good boy until Mummy comes back downstairs. I'm going to get showered and dressed.' With the boy settled, Ellen dragged herself upstairs and flopped back onto the bed.

She closed her eyes for a while and tried to doze, but without success. What she really needed was a cup of tea and paracetamol, but the effort of getting back to the kitchen seemed enormous. Ellen prodded Paul, hoping to encourage him to fetch some liquid refreshments, but as he didn't budge, she prised her body away from the sheets for a second time and made her way back to the kitchen, taking a peep into the sitting room as she passed by. Liam was still in the same spot she had left him, transfixed by Shaun the Sheep; the toast hovered in front of his mouth. On any other day, Ellen would have reminded him to eat, as well as watch, but today she didn't really care, so long as he was contented to allow her to emerge from her torpor.

The tea was finally prepared and with painkillers in hand, Ellen made her way back upstairs. She put one mug on Paul's beside table and then propped herself upright in bed so that she could drink her brew and get the painkillers down her as quickly as possible. She swallowed and groaned. It had been billed as a celebration but why had they all drunk so much last night –

except, of course, sensible Ian who still looked sober at the end of the evening.

'I've made tea, Paul,' she said, kicking his leg gently, 'and there are tablets, too.'

Paul grunted and it took him a further five minutes before he was able to get himself to a slumped sitting position. 'Dear God… way too much alcohol, Ellen. Why do we overdo it every time we get together with your friends?'

'They are your friends too and nobody forces you to drink so much. It was lovely that Laura organised the party, but I do wonder how she is feeling this morning. I'm not sure it turned out quite as she'd planned and what is it with Neil? He always seems to end up saying something tactless.'

'What did he say?' Paul seemed unaware that Neil had disgraced himself.

'Seriously, Paul? You can't have missed him announcing his intention to move in with Laura and move Max out to the yard!'

'Brave man,' Paul joked.

'I'll call Laura later. I'll thank her again and that will give me the opportunity to check how she is.'

'Let's have a quiet day today. You didn't have any plans, did you?' Paul said hopefully.

'Liam's been invited to a birthday party, so he will need to be taken to Patrick's house out in Lyndon for two o'clock and then collected at five.'

'I was hoping to watch the football this afternoon. Can't you do the ferrying?'

'I was hoping you would do it, actually.' Ellen didn't see why Paul assumed that she was always free to drive their son around the countryside. She had plans to do some updating on her Facebook page and website, as well as uploading a selection of new photographs; internet enquiries were proving to be a very useful source of business, and Ellen wanted to keep the site interesting and fresh.

'But I was hoping to relax. It's been such a hectic week at work. All the commuting on top of the extra stress at the office is getting a bit much,' Paul said.

'I appreciate that you work hard, Paul, but you are also part of this family.' Ellen was determined not to give in and wasn't inclined to enquire about any 'extra stress'. Men could be so feeble. Wasn't she coping with all sorts on a daily basis in order to fulfil her roles of mother and entrepreneur?

'I've been thinking,' Paul continued, 'I should probably start looking at apartments. A crashpad in the city is becoming less of a luxury and more of a necessity these days.'

'I agree. It does seem to have reached that point. We can probably afford to make that investment now. Things are going really well with The Kernel Kitchen and I'm determined to keep up the momentum and make the business a huge success.'

'Okay, as long as you are in agreement, I might as well start having a look at what's available next week,' Paul said.

*

'Morning, Laura.'

'Oh, hi.' Laura's response was as lacklustre as Ellen feared it might be.

'Thank you so much for last evening. Both Paul and I thoroughly enjoyed your kind hospitality. I'm sorry to say we both felt a little bit jaded when we woke up. Perhaps a drop more wine than was prudent, but nevertheless, it was lovely to get together for a celebration.'

'Oh, you're very welcome and I always enjoy the chance for us to be together, with partners in tow. However, I do think we might all have benefitted from leaving off the alcohol sooner…

but whatever.' Ellen was concerned that Laura didn't sound herself at all.

'Is everything okay? I hope entertaining a bunch of inebriates in your home wasn't too much of a burden?' Ellen felt embarrassed.

'Not at all. A bit of revelry is what's needed every now and then. To be honest I'm annoyed with Neil. Could you believe the cheek of the man, assuming that he could make his new home here with me without so much as a discussion, let alone an invitation?' Laura's voice rose in accordance with her anger.

'Well, yes, I did think that was most presumptuous of him. I hope he's apologised and you've managed to make up?' Ellen asked, curious for an update on the ill-matched union.

'I haven't spoken to him this morning. I called for a taxi to take him home last night. Frankly, I couldn't bear the thought of spending another moment with him.'

'I am sorry, Laura, just when things were starting to look more promising for the two of you. You must feel very let down.'

'Yes, I do. Thank goodness for adorable Max. It seems he's the only male in my life that I can rely on,' Laura said as she reached down to make a fuss of the heavy, hairy beast currently lying across her feet. 'Anyhow, thanks for calling and let's arrange for a coffee date very soon – with April, too. Must go now, Max and I are due to go out for a walk and then we're off to visit Mother.'

'I'd love to join you for a coffee, just text me.' Ellen suspected that Laura was being guarded with the details and that another rift with Neil was inevitable. But he really was such an oaf and surely Laura could see what was apparent to everyone else?

32

Laura replaced the phone onto its stand and stared out of the kitchen window across to the three maple trees, each looking splendid in its attire of emerald green and buttery yellow foliage – an inspired choice to brighten the corner of her garden and today an uplifting sight for a deflated spirit. Her celebratory evening had not turned out as she'd hoped and Ellen was correct in surmising that this morning she was feeling pretty aggrieved.

The perfectly cooked and beautifully presented meal had been a triumph and she always enjoyed spending time in the company of her closest friends, although Paul's remarks to his wife had irritated her, even if Ellen appeared unaffected by his egotism. The lack of sobriety amongst the attendees could definitely be attributed to Neil, who had continued to top up half-filled wine glasses as their owners tried in vain to place a hand over the rim. Where on earth was his propriety setting for social situations and did he really have a devious plan, the details of which were now unfolding?

But Laura's main concern was whether anyone had registered Neil's remark that he had been married to Catherine when they had known each other previously. Would his throwaway comment be enough to raise a suspicion that it had been an extra-marital affair? With the likelihood of another cheat in their midst – who could create havoc for Ellen – the stark reality of her own treacherous behaviour had been as good as announced to the entire group. The selfishness of her own indiscretion was inexcusable and Laura sincerely hoped that on this occasion the amnesic influences of the alcohol might have saved her from detection of that crime.

She went into the hallway and grabbed the first jacket that came to hand and her walking boots; she was overwhelmed by a desperate need to get outside and feel the fresh breeze on her face.

'Come on, Max, let's you and I go out, I feel the beach calling.' Max wagged his tail enthusiastically. All was well in his world – he had his mistress to himself; the unwelcome visitor had miraculously disappeared.

With Max safely settled inside her car, she dashed back into the hallway to find her hat and was just about to close the front door of her cottage when she heard the telephone ring.

'Blast it,' Laura said to no one in particular, but then decided to ignore the call. She wasn't ready to talk to him and locking the door, she walked away before the answering machine clicked in; she didn't want to hear his sorry voice either.

By the time they reached the coast, the wind had strengthened and as she set off from the car park Laura's fine hair began an unruly dance about her face. Max's large ears flapped madly in the gusts – a phenomenon that he seemed to find wildly exciting – but Laura quickly pulled on the last-minute hat.

The pair made their way down to the water's edge. Max was eager to play and swim, and Laura's pleasure derived from witnessing his utter delight at being on the beach; she threw a ball and watched as he galloped through the salty waves to retrieve it. She walked on across the hard surface of the wet sand and looked out to the wide expanse of retreating ocean.

What to do? she wondered. Laura breathed in deeply, threw the ball again for the seemingly tireless dog, and forced herself to consider the issues. The company of a man was a welcome addition to her life, the togetherness was satisfying, but there were definitely drawbacks, especially it seemed when that man was Neil. She found it hard to condone his lack of ambition and drive. Yes, they were now in their later fifties, but she still thrived on new activities and experiences. He appeared content with a mediocre life of procrastination and indifference. The archetypal Neanderthal – coarse and boorish – his needs were primitive: shelter, food, water (in the form of wine) and a mate.

Laura was tired – tired of being the driving force and tired of believing that things would change. She could now see the futility of attempting to mould Neil into the man she believed would be her best match. And just imagine if she had succeeded in some small way? Who could really respect a person who was weak enough to allow themselves to be remoulded? What she'd hoped for was the man who represented her favourably disposed recollections of some twenty years ago. Time to ditch the rose-tinted specs.

In that moment her decision was made. Laura released her hunched shoulders and unclenched her molars (which had been pressed tightly together in concentration), then she breathed in deeply to savour the restorative, tangy sea air. She wasn't going to be alone; she had Max and for the time being he was the best company, the only company, she needed. Perhaps reading her mind, Max rushed to her side to share his rigorous wet dog shake, his jowls flapping, his fur flying, soaking Laura in slobber and chilly seawater. She laughed and got down on her knees to hug her furry friend, feeling as if a weight had been lifted from her weary body. It was going to be okay – *she* was going to be okay.

Both dog and mistress were now damp and Laura was starting to feel uncomfortably cold. It was time they retraced their steps to the car park and made their way back to Ash Green via her mother's care home. Having coaxed Max back into the boot with a number of treats, she jumped into the driver's seat, started the car engine, adjusted the heaters and was soon enjoying the benefits of being blasted with warm air. She rubbed her chilled hands together, then grabbed her handbag to locate a hairbrush and her make-up bag so that she could rectify the damage done by wind and water. Having checked her appearance in the vanity mirror and been satisfied that she was presentable, Laura glanced at her phone before replacing the bag on the passenger seat.

Three missed calls. Would Neil never leave her in peace? Laura tapped in the passcode to access the details and was horrified to discover that the calls were not from Neil, but from The Laurels. Filled with dread, Laura listened to the voicemail messages and was devastated to learn from the first notification that her mother had been taken to Ipswich General Hospital following a nasty fall whilst out in the gardens.

This message had been swiftly followed by a second with an explanation of the incident, which involved Rose managing to clamber over the picket fence which formed a secure boundary to the pond. This astonishing manoeuvre had been achieved with the assistance of a wheelbarrow left unattended by the gardener. Once at the water's edge she had stretched her hand across the surface, attempting to touch Kenny – the largest of the Koi carp – whose iridescent scales had clearly mesmerised the old lady. The splash as she hit the water had alerted nearby staff and thankfully Rose had not inhaled any of the murky liquid, but she had cut her head on a rock and been rather shocked by the whole wretched experience.

The final message an hour later informed Laura that Rose was doing okay; she'd had five stitches to her wound and doctors felt it wise for her to be kept on the ward overnight, just for some further observation. Would Laura please call or visit the hospital as soon as she picked up her messages?

'Right, Max, hold tight, we're off to the hospital to see Rose,' Laura informed the bemused dog, who was unbalanced from his haunches as Laura made a swift exit from her parking space and accelerated away from the beach car park.

33

Late summer was creeping towards autumn, although no one wanted to start thinking about the arrival of shorter daylight hours – much too depressing – and having not seen each other for a few weeks, the three friends were looking forward to meeting up at Amarellos'. For two weeks April had been Gemma's stand-in at the café and although she'd enjoyed the temporary role, she was now looking forward to being a customer.

The aromatic coffee was wonderfully warming, satisfyingly caffeinated and the bliss was further enhanced by the addition of a packet of chocolate chip biscotti for sharing.

'So how is Rose?' April asked Laura as she dunked her baked treat into her coffee, waiting to achieve a consistency that would be gentler on her teeth.

'She's fine now. After the overnight stay in hospital, she was discharged back to The Laurels. The nurse is due to visit her today to check the stitches are dissolving and the wound is healing properly.'

'That is good news.'

'It was the bruising which was most distressing to see, but it is starting to fade now and Mother is enjoying telling the tale to all who are prepared to listen. The facts are being blurred and embellished and she is in her element.'

'But did you discover whether she succeeded in her endeavour to lay a hand on the mighty carp's shiny scales?'

Laura chuckled. 'To her delight she did. Kenny is so tame. It was the large splash as she finally toppled in that no doubt frightened him away to the depths of the pond.'

'You really do have to admire her determination and her ongoing fascination for all that is going on around her,' Ellen said.

'On her better days, she's the same old Rose, which is such a pleasure for me. Of course, I was worried when I heard about her fall. It signals the reality of her increasing bewilderment and her vulnerability. Although I'm still trying to fathom her ability to scale that perimeter fence to get to the water's edge. The staff at The Laurels were mortified that the accident occurred, but frankly who'd have thought one of their residents was up to such a feat.'

'Well, it's good to hear she survived, and with a great "fishy" tale to tell,' said April. 'I want to be just as inquisitive as Rose when I reach my eighties, and I hope Megan will still be visiting her roguish mother in her dotage.'

'So, April, is there something going on with Megan? You hinted on the phone that you're suffering from long distance motherly anxiety. What has happened?'

April sighed. 'Yes, that's a bit of a sorry tale and I really didn't know what to think when Megan phoned in tears, she was so distraught. The problem was, Ian and I were at the beach and I just couldn't make any sense of what she was trying to tell me – not least because of the noisy bloody seagulls!'

'Nasty, evil scumbags!' Laura agreed.

April continued on with her story. 'I was immediately alarmed, as she rarely calls me on my mobile unless she urgently needs to contact me. The trouble is, you always fear the worst and in a little more than a split second I'd imagined all sorts whilst waiting for her to calm herself sufficiently to speak intelligibly. Pregnancy? An assault? A serious medical condition? A car accident?'

Both Ellen and Laura were now on tenterhooks. 'For goodness' sake, what's happened? Is she alright, April? Tell us.' Laura was exasperated by the delay; she had a huge amount of affection for Megan.

'She's been dumped by Dan!'

'Oh no, poor girl. Is she heartbroken?' Laura released her held breath, relieved that Megan's predicament was thankfully less serious than she had feared. Although you could never underestimate the distress of a broken heart.

'You know, I think it was probably an age difference issue,' continued April. 'Megan still wants to enjoy being a carefree student, and whilst her free spirit was clearly an attraction for Dan, it appears he's had enough of her inability to make a commitment.'

'How old is Dan?' Ellen was intrigued, mindful of the age difference between herself and Paul, which she didn't consider caused any issues.

'He's just turned thirty-two and for some reason has now decided it's time they settled down with a mortgage and the shared responsibility of home ownership.'

'But Megan is a few months away from the run-up to her finals, isn't she?' Laura asked. 'She probably hasn't even had the chance to seriously look for a post-degree job yet.'

'That's just it. Dan's been putting pressure on Megan to make decisions and commit to a long-term relationship. Her reluctance seems to have tipped him over the edge and he's suddenly ended the relationship, saying he's wasted enough time on her already.' His comment had annoyed April immensely, not least because his timing couldn't have been worse.

'Poor Megan. I hope this upset doesn't come between her and gaining the excellent degree she's worked so hard to achieve,' said Laura, equally put out by Dan's untimely break-up shocker.

'You and me both,' said April. 'I don't think she's even decided whether she wants to stay in the Bristol area after university. She's loved living there, but I know she didn't want to be tied to an area and limit any job opportunities. Of course, Dan's family are all in the county and so is his work. He has no desire to move away.'

'Megan is right to keep her options open. She's a beautiful and intelligent twenty-two-year-old, the world is her oyster,' Ellen said, whilst privately wondering whether Dan's decision had anything to do with another girl. Young men did seem to have a predisposition to be easily lured away when a willing new female appeared on their horizon.

'So, is she okay now?' Laura asked. 'Has she moved beyond the tears stage yet? Megan is surely the sort of girl who'll find the strength and determination to succeed in spite of Dan.'

'It took a good deal of listening, sympathising and agonising phone calls every evening for a week, until I couldn't stand it any longer and I told her she deserved better and not to allow him to wreck her chances of success.'

'Absolutely right,' said Laura.

'So has the plain truth done the trick?' asked Ellen.

'I spoke to her again yesterday and she seemed much calmer. Even better, focusing on her dissertation.'

'Which is…?' Laura was keen to be included in Megan's choices.

'Ah yes… It's entitled, "Reducing the symptoms of arthritis in patients aged between fifty to sixty years with the help of yoga".'

'So, despite falling nicely into the nominated age group, you'll be unable to assist her. After all, your own experience of studio-based exercise is somewhat limited!'

'Hilarious, Laura.'

'Sorry, couldn't resist... but good for Megan, sounds like she'll be fine. Bloody useless men!'

April and Ellen looked at each other and then at Laura enquiringly.

'Yes, I'm sure it will come as no surprise. It's been fun and best of all I've rediscovered sex... but Neil is just not for me. End of story.' Laura had clearly concluded the relationship and, it seemed, the matter for further discussion.

'How did he take it?' continued April undeterred and then noticing the look on Laura's face added, 'You have told him, haven't you?'

'No, I haven't. Not because I'm in any way undecided – I couldn't be more certain that it's over – but Mother's fall has meant my time has been taken up. He's called a couple of times and left messages, but we've not spoken this week. I'm planning on calling him this evening.'

April was encouraged to note that Laura seemed unemotional and she got the sense that her friend – like her daughter – was going to be fine. It looked as though Neil was finally old news. Hallelujah!

They'd drained their coffee cups and with still more catching-up to do, it seemed there was every reason to have a second cup. Laura volunteered herself to go to the counter to place their order.

'Well, that is good news, isn't it?' Ellen took the opportunity of Laura's absence to quietly confide in April.

'Absolutely. Laura has definitely made the right decision. Neil really was a total bore the other evening, wasn't he?' said April.

'My sentiments exactly. I also think that this time Laura is really committed to her decision. She looks almost relieved.'

Laura returned to the table with a tray of replenished drinks just as Ellen was discussing the price and availability of good apartments in London with April.

'I'm sure Paul's already discovered that you have to be alert and astute when it comes to buying in the city – ready to view as soon as something promising comes onto the market and with the funds to hand to make the purchase. That's generally the case with London properties.' April was keen to share her homebuyer's knowledge.

'So, you and Paul have finally decided to buy the crashpad, have you?' Laura asked as she settled herself back on the sofa.

'Yes, just last weekend, after yet another particularly tiring week of commuting and late evenings. Paul thinks his job will take him into the city more and more in coming months, so it does seem to make sense,' Ellen explained. 'It will also be an investment for the future – a pension pot, and perhaps even a place for Liam's future, if his working life takes him to London.'

'That is certainly planning ahead,' Laura said, 'but I'm sure it's a wise decision, particularly if it improves Paul's quality of life and means that when he is at home, he's able to enjoy time with you and Liam.'

'That would be a bonus,' said Ellen, frowning. 'He's around so infrequently and he can seem very preoccupied.'

'But how do you feel about him being away a couple of nights a week?' Laura enquired. 'Won't you find it harder to cope with Liam on your own, not to mention missing him?'

'What's to miss? No, honestly, I don't think it will be a problem. I'm mostly looking after Liam on my own during the week anyway – with a little help from Izzy – and I've been so busy with the baking I don't have time to feel lonely at the moment.'

'I did take a peek at your website the other day. I was hoping for some light relief having spent far too long reading the web pages of far-flung community cafés, and I loved the new photos. The top-hat cake was inspirational and I bet the clients were thrilled with it,' said April.

'That one certainly was a triumph and I loved doing it. Things do appear to be going rather well and hopefully I'll be able to help you by supplying cakes and treats for your café. And what's the news on that front?'

'We should get feedback and responses from the funding organisations by next week at the latest,' Laura replied, 'but I guess you'll be organising a café meeting shortly, won't you, April?'

'It's definitely on my to-do list, but I've been rather tied up filling in for Gemma. If nothing else interferes, I'll get my head fully back into community café mode tomorrow,' April promised. 'I do think we need to regroup and catch up with our volunteers. We don't want the project to lose momentum and the ladies to lose interest.'

'Did you manage to find any local businesses willing to offer sponsorship, April?' Laura asked.

'Actually, I have had a couple of offers of donations if the project gets the final go-ahead. I agreed that the café would credit these companies by listing their names on our menus. So far that's the garden centre, the picture framer in Benholm and a relative newcomer, The Kernel Kitchen!'

Ellen smiled. 'It seems a rather obvious and inexpensive marketing tool.'

'You truly are the resourceful entrepreneur, destined to go far,' Laura said, and spontaneously all three clinked their coffee cups to salute one another.

April toasted the group. 'To Ellen and the ongoing success of The Kernel Kitchen; to Laura – her resolve and her rediscovered libido; to me, and… and my adventure with Arabica.'

34

Neil was shaking his head as he paced back and forth across the newspaper-strewn carpet of his sitting room; he was panicking. He'd been foolish; no, it was worse than that, he'd been reckless. The victim of red wine, he'd relaxed his guard and now he had a distinct feeling he might have completely screwed things up with Laura.

'And... for the second time... idiot.' The words fell from his mouth in disbelief at the stupidity.

He'd tried to call her but she was doing that thing women do when they are upset with you: making a point by ignoring your messages. He supposed she'd probably need a couple more days to calm down and then perhaps he'd have his chance to grovel. Neil was totally prepared to grovel. He had to face facts; he currently had little going for him if he didn't have Laura in his life.

Unemployment was the least of his worries; homelessness was now just a matter of weeks away and the last thing he wanted to do was start dipping into his funds for another crappy rental property or to use as a deposit for an affordable (and quite possibly undesirable) flat for sole occupation. He feared that streetlife may be looming – although he could never survive that hardship – so his future had to be with Laura. Unfortunately, that particular route was going to require a lot more effort if he was ever going to be able to turn the situation from desperate to daring to hope.

Short term there seemed only one possible option open to him and that was to contact his brother. Ray might just be prepared to give him a bed for a few weeks. Neil reached under the coffee table and eventually located his mobile phone, which he'd thrown in frustration the previous evening following his last failed attempt to get Laura to pick up and speak to him.

*

'Hi, Neil.' Ray didn't really have time to speak to his brother at that moment, but his sibling had been calling incessantly. 'What's up?'

'You okay, bro?'

Ray hated this familial slang and guessed his brother's chumminess was the prelude to a request for a favour. 'I'm just busy, got an important meeting to prepare for and feeling the pressure,' Ray responded, aware that his unemployed sibling was no doubt at home, watching daytime TV.

'I'll make it quick then. Any chance of a loan of your spare room for a week, two tops. Turns out my landlord wants to reclaim his cottage and put it on the market untenanted as soon as, which leaves me without a roof and, before you ask, Laura's unlikely to oblige at the moment. That's a work in progress.' Neil's laugh sounded slightly embarrassed.

'Christ, Neil, what have you done this time?' Ray recalled the evening a month back when Neil had introduced them to his charming ladyfriend; Suzy had said at the time that his brother was lucky to have secured Laura's attentions.

'It's a long story…'

'Laura's a great woman and if you've fucked things up with her you're a complete idiot.' Ray's response held nothing back; he was under no illusions when it came to his brother and how he operated. Ray was also pretty certain Suzy would not want Neil in their spare room for a night, let alone a week or two. His wife described her brother-in-law as a lazy, self-centred bore and claimed the two boys must have inherited the polar opposite set of genes from their parents – that, or one of them just had to be adopted. It was the only explanation for how they could be so completely unalike.

'Oh, you know, I got a bit drunk, might have said a few things that didn't go down well. But it will work out, I just need a base until I can worm my way back into her affections and, better still, under her roof.' Neil laughed and Ray grimaced.

'Sorry, Neil, but Suzy's unlikely to agree and anyway surely you'll have the cash to sort yourself out any day now. Thanks to Catherine's generosity.'

'Hardly generosity. I practically ran that business single-handed. I deserve the money.'

'Well, I suggest you get on to the rental agency straight away and put some of the money to good use. Sorry, but I've got to go… too much to do… good luck, *bro*!' Ray disconnected the call and sighed. He didn't want to speak ill of his own flesh and blood, but his brother was looking like such a loser. Suzy was right, Laura was way too good for the man that Neil had become.

Ray returned to his paperwork and was quite certain he could predict his brother's next move. This would be to scroll through his address book to find someone – anyone – who he could prevail upon to give him a bed.

*

239

Laura was preparing to make the phone call to Neil which would seal his fate and, happily in the circumstances, her own. She gleaned some wicked pleasure to realise that her timing would interrupt his rigid afternoon routine of lying on the sofa, watching *Escape to the Country*, and she was certain that he would see her call as a signal of hope for a reconciliation – another one.

'Darling, hello, thank you for calling me and before you say anything, I am so sorry.' Neil's usual tactic of trying to ingratiate himself began. 'Too much red wine, my downfall, every time.' He laughed, no doubt hoping to lighten the atmosphere, but Laura was counting on her ice-cold tone to seep through the telephone line with chilling efficiency.

'I don't really want to listen to your apology, or discuss your drinking habits. I'm calling to tell you that although I was prepared to give you that second chance, it's now clear a relationship with you is never going to be right for me.' Laura stopped talking for a moment, primarily to gauge Neil's reaction, but also to check that he was still paying attention.

'But, Laura...'

'Look, Neil, I'll always be pleased that we had the chance to be reunited – rekindle the old flame, as it were – but I'm afraid that the fire has well and truly gone out... and... and... well, all of the matches are spent.' Laura found herself in unplanned poetic dumping mode and almost laughed, but then checked herself for the seriousness of the moment.

'Please, darling...'

'I wish you all the very best, I really do, and hope you find true happiness with the right woman, but that woman is not me.' Laura didn't want to hear any more of his pleading.

'But...'

'Take care of yourself... Oh, and one last thing, Max is a pretty good judge of character – call it dog instinct – and he didn't take to you at all.' Laura smiled to herself, patted the head of the dog at her side and disconnected the call. A tsunami of relief washed over her and she knew with conviction that the ghost of Neil had been well and truly exorcised.

'Good job, Max.' She congratulated herself on her decisiveness and bravery to embrace singledom. And her pet on his indisputable intuition and loyalty.

35

Paul was taking some time out of the office to view two apartments located just a couple of tube stops away. As he was due to meet the agent at two o'clock, he'd classified this break from work as a justifiable absence on a late lunch, rather than bunking off.

With Ellen's input he had the list of ideal specifications and the maximum budget for the purchase. The preferences were that the flat should not be a conversion and not on the ground floor. The other stipulations were easy proximity to Paul's city office – otherwise what was the point, if he had to traipse halfway across town, he might as well commute from home – and a small one-bedroom flat would be preferable to a large studio apartment.

'A balcony to provide even a hint of outside space would be the icing on the cake,' Ellen had laughed and Paul wondered if it might be time for the banal baking gags to cease.

Henry, the suave, youthful agent, sporting shiny suit and pointy shoes, was ten minutes late, claiming the traffic had been 'just awful'. Paul wished Henry had taken the precaution of setting off from his office ten minutes earlier and not wasted his precious time. But as he had finally arrived, the viewings could at least get started.

'So here we are, apartment number one,' Henry said, opening the front door straight into the living space of the flat, revealing a pokey, airless room, with kitchenette to one side. The current incumbent had clearly left in a tearing hurry that morning, judging by the state of total disarray. Henry was unapologetic; he either lived like this himself or had seen it all before and was immune to other people's sordid lifestyles.

'Smaller than I was hoping for,' Paul said, trying to look beyond the trail of sullied underpants and socks which led them through to the equally pokey, airless bedroom. 'Not for me, Henry, can we get straight to apartment number two? Time is pressing.'

'Sure thing,' Henry said, and they departed swiftly, each taking a deep gulp of air as the fresh breeze met them on exiting.

The second apartment was better – not great – but an improvement. No balcony as such, but one of those Juliette thingies, facing onto the main road, with French doors and a window box – although the dead plants within it were an ugly addition and a telling sign of neglect.

'Yes, this one has possibilities,' said Paul, envisaging just a weekend with paint pots and brushes along with a new carpet to bring it up to a reasonable standard.

'Just to give you the heads-up, Paul,' Henry interjected, 'this property came onto the market two days ago and we've already had an offer, but the vendor is holding out until after your viewing before agreeing a sale. If you like it and want to offer, you'll have to act quickly and be prepared for a possible bidding war.'

'I don't like it that much, Henry, and I'm not prepared to rush into a purchase.'

'I'm just warning you that it's a seller's market and anything worth having goes in days, but up to you.' Henry seemed disinterested and so confident that he'd sell the flat without any effort that he wasn't even going to try.

Paul had seen and heard enough and excused himself in order to get back to his office. This wasn't going to be as easy as he'd hoped and, with work commitments, getting to see properties was going to be really tricky. What he needed was help and he thought he knew the perfect person to assist him in his quest.

*

She had been delighted when Paul asked for her assistance with his property search. She didn't see this as an imposition; in fact, she was excited to be involved and flattered that he trusted her with the task. They had agreed that as Rosanna was in the city on a daily basis, it would be easier for her to do the legwork – some first viewings during her lunch breaks and then when she'd found a promising proposition Paul would swiftly arrange to view himself. He told her that he'd already missed out on a couple of 'definite maybes', so Rosanna knew her help could be his only route to a successful purchase.

She wasn't, of course, aware that he was already bored with the running around during every spare moment he was in the city trying to find a property; to him it was just a lot of hassle and not something that he would particularly enjoy. House-hunting was the sort of activity that women for some reason got excited about; perhaps it was the opportunity to get inside the homes of others, an insight into how the rest of humanity were living their lives.

At every quiet moment in her working day, Rosanna scoured the property websites and within a week had already dismissed seventeen apartments and arranged to view three hopefuls. She wanted to do a good job and for Paul to be impressed by her commitment to his cause. It would show him just how important their friendship was to her and how fond she had become of him. The night of the passionate embrace had reinforced Rosanna's notion that further intimacy was inevitable. He hadn't said as much but it was obvious Paul's intentions towards her were changing and to her mind it was just a matter of time before they embarked on an affair. She was buzzing with the anticipation. Her confidence had been knocked after Rob's sudden departure, but now she was feeling attractive, desirable again; it felt good.

The second of the three viewings revealed a nearly new, purpose-built, one-bedroom apartment. It was in excellent order, in a great location, with access to a well-maintained shared garden, away from busy roads but close to transport connections. She could hardly contain her excitement and, as it was a city-based day for Paul, Rosanna called his mobile immediately whilst the agent was still with her.

'When can you get here, Paul? I think this flat is perfect for you.' Rosanna was almost breathless with enthusiasm as she described the accommodation around her and gave him a swift FaceTime tour.

'Text me the link to the details and I'll rearrange my afternoon's appointments so that I can get over there later today. Oh, and don't forget to include the agent's mobile number as well, would you, so I can get back to him once I've cleared my diary,' Paul said.

'It's her, not him. The agent's name is Trudi. And yes, will do. I think you'll be pleased with what you see, it's certainly the best of the bunch as far as I'm concerned and, I can tell you, there's a lot of grim offerings out there.'

'That's great, Rosie, and thanks. Speak soon and don't forget to text me.'

Rosanna relayed the details of her conversation to Trudi and told her to expect to hear from Paul shortly.

'It's very generous of you to assist your friend with his flat search, Rosanna. I hope he appreciates all your efforts on his behalf or perhaps there are plans for you to move in together?' The glamorous redhead had seen the look on Rosanna's face whilst speaking with Paul and it was clear she was smitten.

'Oh no, the apartment is just for Paul.' Rosanna was annoyed to find herself blushing.

Trudi shook her client's hand and prepared to return to her office, two streets away, to await Paul's call. Rosanna rushed off in the opposite direction, aided in her haste by her trainer-clad feet; city girls always kept a pair of casual shoes under their desks; it would be unrealistic to think they could really get about efficiently all day in four-inch heels.

*

The promised text arrived on Paul's mobile phone just a few minutes later with the property details attached and he saw immediately the reason for Rosanna's excitement. She'd done a great job on his behalf; this flat seemed to be exactly what he'd given up searching for. Thank goodness for Rosanna. He managed to postpone one of his afternoon meetings and agreed a conference call in place of the other. He then called Trudi and arranged to meet her at the property at 4.30pm.

'Sorry I'm a bit later than planned, Trudi, I struggled to get away from my desk.' Paul's smile was warm and friendly as he strode toward the agent, finding it impossible not to notice her stunning auburn mane.

'Not a problem, let's get straight inside,' said Trudi, finding it impossible not to notice Paul's wedding band.

The apartment, as Rosanna had claimed, was perfect for him; he didn't even envisage having to put on his overalls and practise his decorating skills. Typically, it was at the top end of the set budget, but Paul was convinced it was worth the stretch and made his offer immediately to Trudi. Paul was in an excellent buyer's position – unencumbered and able to complete the deal without delay, an agent's favourite catch – so just a phone call later the deal was done.

*

Rosanna loitered at the train station that evening in the hope of being able to travel home with Paul. She'd even allowed the earlier train, which she was in time for, to leave without her. Rosanna had heard nothing from Paul since her lunchtime call and she could only assume, and hope, that he had managed to find the time to view the flat at Garden Court. She had texted him a little earlier, but he'd clearly not had the opportunity to put an end to her anxious wait for news.

She went in search of a takeaway coffee. It was starting to get chilly on the station platform and she wasn't attired in her warmest jacket; it didn't co-ordinate well with the skirt she had chosen to wear and in Rosanna's mind style always won over practicality.

She found a free seat on a bench in the station's concourse, which allowed her to rest her aching feet – now clad in heels – and put her in the ideal spot to see Paul when he finally arrived. It was another thirty minutes, and another missed train later, before Rosanna saw him.

He looked tired, but the conspicuous smile creeping onto his face as he approached her only served to heighten the attraction she felt. He was obviously pleased to see her, even so Rosanna was taken by surprise when he swept her up in his arms and hugged her.

'Darling, you waited for me,' he said. 'You are such a sweetheart.'

The 'darling' and 'sweetheart' were unexpected and Rosanna was overjoyed to have scored highly in Paul's affections for her efforts. She could only assume that his exuberant behaviour was as a result of a very successful viewing.

'I just had to see you to find out how it went this afternoon, but judging by your good mood…'

'You are looking at the proud, soon-to-be owner of a rather smart London city apartment.'

'I'm absolutely delighted. It's a great flat, isn't it?'

'It's perfect, as you said. I couldn't have managed without you, thank you so much,' Paul said, 'and you must be my first visitor. Hopefully in no more than four weeks' time, I'll be inviting you to a flat-warming and we can share a bottle of champagne to celebrate.'

36

'Oh no!' April exclaimed.

Laura had called with unwelcome news. 'I'm sorry, April.'

'Bloody hell, where does that leave us now?'

'There's a couple more applications still outstanding, but I'm afraid that even if these are successful, we'll still be a long way off the figure calculated to get the community café up and running.' Laura failed to reveal that she had read the emails much earlier, but had been reluctant to break the news to April, who was bound to be utterly disheartened.

'I was so convinced that they'd be falling over themselves to offer us their cash, especially for such a good cause.'

'It's not a great start, but it needn't be the end of the story.' Laura did her best to alleviate April's angst. 'We'll just have to keep trying. There's got to be other grants going if we dig a bit deeper.' She had decided against telling April that her initial enquiries with the district council regarding some financial assistance had also been met with a lukewarm response. It seemed other causes were considered more worthy just at the moment and it was unlikely that the district council would look favourably on their project during the current financial year.

'Damn and blast,' April blurted.

'Very tame for you, April, your restraint is admirable.' Laura was searching for any levity to lighten the mood. 'But look, it's never a given with these things and I hope I didn't give you the impression that it was.'

'Ignore me, Laura, it's not your fault. Thanks for letting me know and for your help. I need to have a think and if you do have any ideas or fall upon some other potential funders, please call me straightaway.'

'Of course I will. We'll talk again in a few days.' Laura decided it was best to leave April to dwell for a while, although knowing her friend it was more likely she'd seethe for a while. 'Will you join us for Pilates tomorrow evening? The relaxation would certainly ease all that the pent-up frustration you're feeling right now.'

'You must be joking, Laura, I am so not in the mood.'

*

'No sign of April?' Ellen had just arrived at the community centre and was arranging her mat adjacent to the patch of floor already commandeered by Laura.

'I'm not surprised. Apart from the fact that she can't accept the merits of Pilates, she's still feeling down about the stumbling block for her café plans now that the money is not forthcoming.' Laura had tried to call April earlier, but had only been able to leave a voicemail.

'What? There's no grant?' Ellen was not the only one of the potentially interested parties who had been denied the update. Neither Laura nor April had been willing to spread the news, both hoping to find a way forward before dampening spirits. 'Well, that certainly is going to be a problem.'

'We're not giving up just yet, but you're right, it could be curtains for April's project.' Laura lay down on her mat, closed her eyes and prepared her mind and body for the start of the class. She didn't want to talk about the café this evening, she wanted to enjoy the gentle exercise and soak up the soothing tones of the whale music, which would be starting shortly in preparation for the warm-up exercises.

'By the way, Paul and I have bought an apartment,' whispered Ellen, but Laura chose not to respond, being already in the relaxation zone. She left Ellen to assume she either didn't hear this piece of news, or chose not to for the moment.

But Laura had, of course, heeded this snippet of information and couldn't help but imagine the implications now that Paul had acquired a separate space in which to bed down. The word love nest came to mind, but she forced herself to clear her head of all but serene thoughts; she focused on the streamlined beauty of a large, marine mammal, effortlessly moving through the quiet of the deep blue ocean.

*

Ellen was disappointed not to effect any reaction to her good news, but the class had now started and she had to respect Laura's desire to pay homage to Joseph Pilates – the much-revered promoter of physical fitness. She turned her attention to the teacher's instructions and settled into the practised routines with a comfortable sigh.

When Paul had shared his news, Ellen had clapped her hands in delight.

'Well done you. You've worked some magic there, sifting through the dire offerings, and I'm amazed you found the time. Who'd have thought it, in a matter of weeks we'll have secured an ideal property in the ideal location. It's got to be a great investment, too.' Ellen was revelling in a sense of the grandeur she associated with securing a second property to add to their assets list.

'Well, it wasn't an easy task, but certainly it's going to be worth all that effort.'

Clearly the main purpose of this acquisition was to make life easier for Paul, to reduce his commuting and so reduce his levels of fatigue, but there was a small part of her that welcomed the prospect of more time to herself – on her own, but with Liam, of course. The sense of self-sufficiency her venture engendered had given her a new outlook on her life and she didn't want anything, or anyone, to stand in her way.

*

Laura was blissfully engrossed in the Pilates stretches and testing her stamina as the movements became more complex and strenuous. She was squeezing the soft ball between her knees, whilst lying on her back with her legs raised; the exercise was intense, but essential for firing up the inner thighs and abs; quite clearly it had to be done. This was followed by the bridge and squeeze, which would definitely keep her buttocks firm and svelte. Laura was determined to keep in good shape, primarily for herself, but she had decided that, given time, there was no reason why another man couldn't be given a chance in her post-Neil life. Having a man around – albeit a more carefully selected individual – definitely had its benefits and she wanted to ensure she was in tip-top condition for that moment.

*

Ellen was squeezing the soft ball between her knees and planning an online shopping spree once she got home that evening. She thought IKEA would be the answer for fitting out the new apartment. Simple, on trend and stylish, but ultimately very affordable. She was looking forward to searching for items suited to the modern architecture of the apartment block, a different proposition to the choices she had made when furnishing her Suffolk Pink farmhouse.

The Pilates teacher encouraged the class to take up the plank position, the final exercise which signalled an end to the punishing effort and the start of the relaxing, cool-down element of the session. She knew that Laura liked to challenge herself to keep the hold position for the entire minute, but Ellen lasted to the count of thirty-eight seconds and then collapsed, face down onto her mat. Not bad, she congratulated herself on her stamina – better than April could have managed; shame she wasn't there. It was rather satisfying to have an exercise companion with depleted fitness levels; it made you feel so much better about your own achievements. The class ended and Laura jumped to her feet and rolled her mat in readiness to leave; Ellen struggled to her feet and crammed her mat in her leisure bag.

'Time for a quick drink at the pub?' Laura asked.

'Just a quick one. I've plans to check out IKEA's website for furniture in preparation for Paul taking possession of the flat.'

'So, yes… you've found an apartment.' Laura had known that it would be impossible to avoid all conversation on this topic.

'After we made the decision to go ahead, he obviously put his mind to it and – bless him – despite being frantically busy at the office, he's managed to secure a property in record time,' Ellen said.

'That's great, Ellen. I hope it answers Paul's needs,' Laura said, although secretly wondered what exactly his needs might be.

*

April had heard the phone ringing but chose not to stir herself from her position on the sofa. She was warm and comfortable with the TV remote and a glass of rather excellent Merlot all within easy reach. Ian was at his carpentry evening class and not due back until at least ten o'clock.

She lounged for a further ten minutes and then decided she'd run herself a bath. With the taps turned on fully, April planned to fill the bathtub nearly to the top – no Ian around to chastise her for wasting water – and as a treat she added a good splash of her favourite relaxing bath oil. The perfect evening and not a second thought for the Pilates session she was missing.

April was still trying to come to terms with the likely stalling of the community café project. She hadn't the heart to tell Laura when they last spoke that responses to the village survey had also sadly been few. After the initial trickle of replies, over the last couple of weeks even the trickle had dried up. She'd been out and about speaking to residents, but was surprised by their lack of enthusiasm. It seemed that getting a great cup of coffee on their doorstep was not as high on their priority list as April had imagined. Great cake – now that was different, they were all praising Ellen's skills, but they could buy her delicacies in the village shop now that Ellen had an agreement to supply John and Sue's convenience store.

Dropping her clothes onto the bathroom floor, April stepped into the bath and eased her body into the comfort of the bath water. The water was hot, but she chose not to turn on the cold tap, deciding her muscles would relax in the heat and her facial pores would get a good steaming while she soaked. She closed her eyes and cleared her mind of all but the most tranquil of thoughts. She envisaged a delicious chocolate and raspberry torte – her most favourite dessert – and promised herself she'd ask Ellen to make her one next time she saw her.

'Hi, honey, I'm home. The class finished early this evening.' She heard Ian's call as the front door slammed shut.

April's heart sank, as did her whole body as she slipped below the water level, eyes and ears firmly closed to the outside world. Needing the peace and serenity to be maintained for as long as possible, she held her breath, endeavouring to delay the moment she'd have to emerge from the bottom of the tub and face her current dilemma once more.

37

Rosanna's eyes glanced to the bottom right-hand corner of her computer screen at the digital time display; she couldn't resist checking for an update on the minutes, willing it to be 5.30pm, signalling the end of her working day. Concentrating on her tasks was impossible; she was just too distracted by thoughts of meeting Paul at the apartment. The day had arrived when his conveyancing solicitor was due to confirm that the purchase was completed and Rosanna couldn't bear to think there might be any delays which could interfere with their plans. Paul was due to pick up the keys from the agent following his afternoon meeting and had arranged that he would only text her if there was a problem with this agreed timeframe. Thankfully, he had not sent a follow-up message, so their much-anticipated get-together was definitely on.

Having been away in Dublin visiting her mother for two weeks, she hadn't seen him for a while, not even as a train buddy, and she had missed him. It was always good to return to her family home; Rosanna enjoyed seeing relatives and visiting friends from university days, many of whom had not ventured outside of Ireland. She enjoyed the status she held as the most-travelled one of her peers; having worked in France and now London she was considered quite the most cosmopolitan amongst her group of friends.

The instant she could, Rosanna would get out of the office and make her way to the newly acquired apartment. She was looking forward to celebrating the occasion with Paul and with any luck there would be champagne to share – maybe more, she was open to suggestions. She grinned, thrilled by the thought of being alone with him – alone in his flat. Their relationship must now be on a one-way trajectory to the longed-for intimacy. The prospect of Paul staying in London at least one, maybe two, nights a week opened up interesting possibilities for them as a couple and she hoped she was correct in assuming that this evening's rendezvous would be the first of many. Thoughts of Paul invaded her consciousness on a daily basis and securing opportunities to be in his company were high on her list of priorities. She didn't doubt that he felt the same.

The clock now informed her that it was 5.27pm and Rosanna began a search through her handbag. Having found her indispensables kit, she pulled the brush through her hair, and applied lipstick and a generous spritz of perfume to her cleavage and wrists. She switched off her computer, grabbed her jacket and left the building. A new phase in her friendship with Paul awaited and an invitation for a sleepover was surely imminent.

*

Paul swiftly made his way up the open stairwell to the front door of the apartment. Once inside his new part-time abode he was relieved to see that all had been left clean and tidy, all fixtures were in place and the fittings negotiated in the sale – curtains, fridge-freezer, corner sofa – had been left by the previous owner, as agreed. He switched on the fridge, deposited the bottle of Prosecco inside and left the supermarket bag containing nuts and crisps on the kitchen work surface.

He shivered – the flat felt chilly – and having been unable to locate the instruction booklet for the boiler during his quick search of the kitchen drawers, he reverted to uneducated fiddling with the control knobs until the apparatus fired up. Paul wanted the environment to be cosy, warm and welcoming for his first guest. He was looking forward to seeing Rosanna; he had missed her company whilst she'd been away on holiday. It had been a particularly gruelling time at work during her absence and not having her around to soothe him, act as a sounding board and lighten his mood had made it a great deal tougher. This knowledge was testament to just how much he relied on her to make his life seem more bearable.

Paul had not shared his concerns about the current issues at his company with Ellen; she was really busy now that she was supplying the village shop with cakes and biscuits as well as private clients. Rosanna always had the time to listen and, being a fellow city worker, she understood his world and the stress of commercial business. Home baking was a whole other realm – a sort of paid hobby.

He heard the doorbell ring and a broad smile appeared on his face, replacing the work-weary frown. She had arrived and was just the other side of his new front door, right there on his doorstep.

'Welcome to my city pad,' he said, opening the door and ushering her in quickly. Before she had a chance to return the greeting his arms were around her, his mouth on hers and she happily responded to his urgent embrace.

'Mmm, that's a lovely welcome. I may visit again if your greetings are always so exuberant,' Rosanna laughed.

'How was your holiday? God, I've so missed you, Rosie.'

'I enjoyed the break away from work – and from living with Caitlin and Stephen – but it's good to be back. It's especially good to see you. The weather was foul – rained nearly the whole time I was away.' Rosanna wandered into the sitting room.

'Take your jacket off and make yourself comfortable on the sofa, I'll open the fizz.'

'Well done for negotiating such a good price for the sofa. It fits this corner perfectly. It would have been difficult to replace and it's such a gorgeous shade of French grey, which works really well with the soft dove grey colour on the walls,' said Rosanna as she flopped down and ran her hand over the texture of the upholstery.

Paul smiled in agreement as he returned from the kitchen with sparkly refreshments in plastic beakers and two bags of crisps. The reality was that he hadn't actually noticed what colour the walls were painted; he just knew that no DIY was necessary and that this flat was going to be the answer to his prayers.

'Here's to crashing in the city. Thank you so much for all your help, Rosie, I really couldn't have done this without you.'

They drank quickly and Rosanna chatted about her trip and news of the antics of her Irish friends.

'Sheelagh's finally met a man – an utterly gorgeous Italian guy called Michele. Seems she's fallen madly in love, but can't let her family know anything about the relationship, especially her father.'

'Why?'

'Because Michele is a Catholic and her pa's a bigoted old bugger. She's cleverly manufactured a new work colleague, called Michelle, who she's been spending lots of her free time with, helping her care for her poorly mother, which of course entails the odd overnight stay to help with the nightshifts! The length we girls have to go to have some fun,' Rosanna laughed.

'Sounds as though, despite the foul weather, you've probably had a better time than I have.' Paul updated Rosanna about his work worries. He didn't really want to think too much about it, especially not this evening, but he was starting to have serious concerns that the whispers and rumours amongst his colleagues about impending job losses were well-founded.

'Oh dear, that doesn't sound great... but let's not spoil our evening...'

'You're right, it's good to have a catch-up, but if you wouldn't mind putting your drink down and refraining from talking for just a moment, I'd very much like to kiss you again.' Paul didn't wait for her agreement to this proposition; their embrace was passionate, leaving Rosanna slightly breathless.

Without another word passing between them, Paul slowly unbuttoned her blouse, revealing the purple lace of her bra and the smooth, pale flesh of her youthful body.

'You are a beautiful, sensual woman, Rosie,' he murmured. 'I can't tell you how I have longed for this moment...'

His temptress smiled encouragement and Paul set about removing the individual items of her clothing, kissing each part of her unacquainted nakedness as it was exposed to him. His desire for her was intense, irrepressible, overriding all other thought.

38

April was finally emerging from her doldrums and had been spurred into action by news just received from Laura of an organisation which might be able to assist with the café funding. She was preparing to scrutinise the information on their website, but before attempting to explore the charitable offerings decided that fortification for the task would be essential. With a cup of coffee and a slice of Ellen's lemon and pistachio loaf on the desk beside her, she settled down to read the details and, with the old – and painfully slow to function – laptop now fired up, decided she'd check her emails once more in the hope of finding responses to the village survey.

Sadly, she was discouraged on both counts: on close inspection the grant provider didn't seem that promising – they had little previous involvement when it came to community cafés – and her inbox revealed an absence of survey results. A newsletter from The Kernel Kitchen caught her eye; clearly Ellen was using every marketing tool available to get her message out, keeping in touch with all her contacts and sharing details of new recipes and special offers.

April sighed. Of course, she was pleased for Ellen, but she dearly wished she could see a way to emulate her friend's achievements. How satisfying to experience the gratification of accomplishment. But she resolved not to sulk and returned instead to unread emails, starting with the job alerts messages. She'd signed herself up to receive vacancy notices after leaving the dental practice; however, the weekly communications she'd received had been given no more than a cursory glance once she'd got the barista bug. Now that the café plan appeared to be thwarted, April was inquisitive and opened the pages of the *Jobs Today* website to find out what she might have been missing.

April trawled through the list; there were plenty of opportunities for care assistants and she wondered whether she had the right disposition for this role – it would certainly address her need to do something worthwhile. But clearly, this was a complete departure from her area of expertise and experience and now was not the time to be unrealistic; she must suppress all urges to have a go at every career on offer.

It was then that she saw it – an advertisement that caused her such excitement that she gasped and leapt to her feet, dropping her reading glasses with a clatter onto the wooden floor.

'Ian! Ian, where are you?' She felt sure he was somewhere in the house, but with no response forthcoming she could only assume he was hiding out at the bottom of the garden in his recently installed man cave.

*

Ian's new hobby had required a dedicated workroom. The space for practising his carpentry skills had to be sited away from the house so that he could make as much mess as he deemed necessary and be able to concentrate without distractions. April wasn't sure she liked the idea of being classed as a distraction, but she couldn't imagine this reference was meant for anyone (or anything) other than herself. It was just two weeks after he had commenced his evening classes that he proclaimed he felt passionate about this new pastime and was sure he had a talent for woodworking. The following weekend he had visited Mega Spec Sheds, placed his order and arranged for Derek to sort out the installation of a concrete base on which his acquisition would rest at the far end of the garden.

The monster shed arrived shortly after Derek had prepared the site and was just in time for Ian's latest project: the construction of a footstool. Ian had kitted out the workshop with all the tools he needed to get started and couldn't wait to begin crafting. However, in April's view, mega shed was a mega eyesore and directly in her line of sight when at the kitchen sink.

'Before you get started on any project, I'd really like you to paint that shed. It urgently needs toning down from ugly orange to, say, a beautifully muted grey-green. This Willow shade looks perfect,' April mused as she checked out the colour chart she'd picked up from the hardware store whilst in town. She was adamant, so Ian bought the paint and the following evening gave Derek another call.

*

'Ian... Ian, are you in the shed?' April yelled from the back door, but as he didn't answer she knew the only thing for it was to put on her wellies and wander the length of the lawn to his male retreat.

She pulled open the workshop door to find that her husband was indeed ensconced with his tools, bits of wood and sawdust. However, he also appeared to be preoccupied with the effort of attempting to wrap a large handkerchief around his hand.

'What's happened?' April shrieked, seeing the blood soaking quickly into the inadequate bandage.

'It was the marking knife, it slipped,' Ian explained, whilst holding his arm high in the air and pressing on the wound, remembering his basic first aid training.

'Let me have a quick look.' April wasn't great with blood, but wanted to check how serious the injury was and whether stitches might be required. She peeped cautiously at Ian's palm and concluded that a sticky plaster would not suffice and a trip to A&E now loomed for them both.

'It will be okay, April, just find me a better bandage in the first aid box, we must have something suitable, or some of those butterfly strips to close wounds.'

They made their way back to the kitchen and April handed Ian a clean tea towel to replace the sodden handkerchief.

'Try not to drip blood on the floor, Ian,' she said as she rummaged amongst plasters, antiseptic cream, packets of paracetamol (goodness, she'd be in trouble if the pharmacist knew she was in possession of so many) and throat lozenges. 'No, there's nothing in here that's up to the job.'

April got her jacket and bag and found Ian's fleece; she took the car keys off the designated hook and grabbing one more tea towel – just in case – they left for the hospital.

Located on the outskirts of the town, the general hospital was only a twenty-minute drive away, but April groaned inwardly at the prospect of who knew how many hours in the emergency department. She just hoped it was early enough in the evening for the department to be free of drunks throwing up in the waiting room.

'I told you to be careful with those carpentry tools.' April was alarmed by the amount of blood, but equally annoyed that Ian hadn't taken adequate care.

'It was a freak mishap. I was distracted by the sound of a woman's voice yelling at me.' Ian was smiling and from this she gathered not suffering as much distress as she was.

'Just tell me if you start to feel faint.'

'Anyway, why were you yelling?'

'Oh yes, I'd almost forgotten.' April was reminded that she had been trying to locate her husband to tell him something important. 'You'll never guess who's advertising a part-time job.'

'Okay, I'll never guess, so tell me.' Ian didn't feel like playing a guessing game, even if they still had another five miles of travel time before they reached the hospital.

'Richmond & Marsh!' she exclaimed excitedly.

'Richmond & Marsh, your old estate agency firm?'

'Of course, there's no other Richmond & Marsh, is there?'

'And this has got you all excited because…?' Ian was lost.

'Because since I left they've appointed a new manager. Annabel has gone solo, which is the first sensible thing she's done, considering how impossible it is for anyone to work alongside her, and the advertised role is for part-time hours. You know the reasons why I left were firstly, Annabel, and secondly, the painfully long hours. I really am too old for a fifty-hour working week,' April patiently explained.

'Dear God! You're not seriously thinking of returning to estate agency, are you? After everything you said about it being a thankless task, with nothing but stress, targets and ungrateful clients!'

'Well… yes, I did say that… but this is part-time and I was a bloody good estate agent.'

'I know you were, but we've moved on and now you are a bloody good barista, aren't you?'

39

Laura retrieved the soft cloth from the pocket of the car door. She wiped the glass but the effect was limited; the moisture was unlikely to disappear completely until the warming air from the vent under the windscreen finally worked its magic. The weather was atrocious as she set out on her way to visit Rose. Large droplets of rain, falling in torrents, left her struggling to see the road ahead, despite switching the wipers to their fastest setting.

Soon patches of light fog replaced the rain as she continued on her journey and relaxing slightly she allowed her thoughts to feature her life without Neil. Laura had heard nothing further from him and, despite the reprieve, she was inexplicably miffed that he hadn't made any more attempts to contact her. April had mentioned that she'd seen the rental cottage was already being advertised for sale on *Rightmove* – '*a two-bedroom cottage with immense charm and potential, in need of some modernisation*'. Clearly Neil had moved on. But where to? Laura wasn't sure she really needed to know his whereabouts, but was naturally curious.

Nearing her destination, Laura glimpsed the entrance to The Laurels through the mist and murk; she was thankful to finally get off the main road away from the spray and onto the long private driveway. The care home, with its Gothic Revival architecture, was just visible in the distance through the trees and the hazy veil. She was reminded of her first visit to the property when the weather had been equally dreadful, and her initial impression had been of an eerie and uninviting building, with its turrets, steeply pitched roofs and the mythical beasts glaring down on her. Thankfully on entering, Laura had found the interior striking in its contrast to the exterior; the imposing building had been skilfully converted to offer modern comfort and the truly welcoming atmosphere one hopes to find in a nursing home. An exemplary case of repurposing, had been April's comment when she'd accompanied Laura on one of her visits to see Rose.

Having reached the grounds of The Laurels, Laura released her tight grip on the steering wheel and rolled her shoulders to ease the tension that had built up during the journey. She was just rounding the final bend of the driveway, when she noticed that her prized Burberry raincoat had slipped off the passenger seat and onto the sandy floor of the foot well. It was as she stretched across to reach it, momentarily glancing away from the road, that her peripheral vision detected a vehicle coming straight towards her.

Laura instinctively stamped hard on the brakes and turned the steering wheel sharply to the left away from the obstruction. The roadway was wet and her car slid off the tarmac and down the slight incline, coming to an abrupt halt as it bumped against an earth mound at the base of a tree.

The next thing she knew there was a face peering through the window and a tapping on the glass.

'Hello, hello, are you alright?'

Taking a few seconds to focus clearly, Laura saw the man. He repeated his concerned questioning and knocked persistently on the glass to attract her attention. She was grateful to determine that she did indeed seem to be 'alright', although shaken by the trauma of having lost control of the car and being pinned to her seat by the safety belt snapping into action to restrain her.

Before she had a chance to answer, the man started to pull open the car door and was soon leaning in to inspect the state of her for himself.

'Are you hurt?' Again, the man was pressing her to communicate. Although unable to find her voice, her sense of smell had been unaffected by the shock as she breathed in his very recognisable scent – Calvin Klein's *Escape for Men*.

'I didn't see you coming around the corner until I was almost upon you. This weather is appalling. Do you think you can move?'

Mr Klein's relief was palpable as Laura finally spoke. 'I'm okay, I think… just a bit shaken,' she said as she released the seat belt and got herself out of the car with the scented man's reassuring hand under her elbow.

'Thank goodness, you do appear to be unscathed, although I think your car is going to need some repairs. Looking at the crunched front end there's a chance that the radiator has been damaged. You probably shouldn't try and drive it, even if we could manage to get it back up the bank and onto the roadway, which I doubt.'

He seemed terribly concerned for her wellbeing, and she felt compelled to try and explain the rather drastic avoiding actions she had taken. 'It was my fault. I'd taken my eye off the roadway to retrieve my coat. I rarely meet a vehicle coming in the opposite direction on my visits to The Laurels. I stupidly wasn't paying enough attention. Although I think you may have been travelling slightly too fast for the weather conditions.' Laura did feel responsible, but didn't believe Mr Klein was entirely blameless.

'Let me drive you back to the house and we'll get you a cup of their restorative, if rather insipid, tea. Your car isn't causing an obstruction, so we'll sort that out later.'

As she tried to walk, Laura felt her legs trembling and was relieved to be assisted into the passenger seat beside her rescuer. She sat silently as her driver made his way to the next passing place on the roadway, where he performed an impressive three-point turn, enabling them to drive back in the direction of the care home.

'You must tell me straight away if you start to feel faint, or get double vision, or a headache, or feel sick, or dizzy.' Mr Klein ran off the list. 'And you'll have to watch out for any symptoms of whiplash over the next day or two.'

'What are you, a doctor or something?' Laura was impressed by his apparent knowledge of all things medical.

'Sorry, where are my manners? I'm Dr Dunlop, James Dunlop.'

'And I'm Laura, Laura Thompson. I'm on my way to visit my mother, Rose.'

James parked his Volvo by the front door, in the bay marked 'Doctor', then made his way around to Laura's side of the car to help her out.

Once inside, James took control of organising tea and calling the recovery company to get the car transferred to a garage. Laura sat in the staff room, away from the residents, so as not to alert her mother to her current predicament and alarm her unnecessarily.

'Here is your tea, Laura,' James said, placing a tray on the table in front of her. She was pleased to see there were two cups, indicating that he was going to join her, and he was soon sitting in the armchair opposite. 'Are you still feeling okay?'

'Thank you for the tea and yes, I'm sure I'm fine.' Laura smiled and was now able to fully appreciate the man behind the aroma. Dr Dunlop had Romanesque facial features, he was very tall, slightly too slender, with dark brown eyes and silver hair. She decided she liked his appearance as much as the smell of him.

James and Laura chatted companionably as they drank their tea – not insipid, as he had clearly overseen the teabag count in the kitchen. James explained that he had been visiting one of his long-term patients, a new resident at The Laurels, who was just recovering from a chest infection. Today was his first visit and he apologised again for not being familiar with the narrow roadway and taking the corner a little too fast, particularly with the poor visibility.

'Lesson learnt,' he said.

'How was your patient?' Laura asked, feeling his wellbeing probably deserved more recognition than her own.

'Definitely on the mend and, better still, quite settled here at The Laurels.' James explained that the elderly gentleman, Ernest, had lost his wife two years ago and had not been coping well at home. Recently he'd also started to shows signs of dementia and it was clear then that he shouldn't be living on his own. 'And what about your mother? Has she been living here for a long time?'

'Rose has been in residence for nearly a year and seems very happy here. I can't praise enough the commitment of the staff and the care that they show to all their residents.'

James had finished his tea and looked at his watch. 'I should be getting back, afternoon surgery starts shortly. Will you be okay, Laura?'

'I'll be fine. I've texted my friend April, and she's going to come and pick me up in an hour. That will give me time to seek out my mother and spend some time with her. Thank you so much for your kindness.' Laura had been touched by his gentle and caring demeanour; more than bedside manner, he seemed a really genuinely nice man.

James was smiling at Laura. 'Well, I'm sorry that you ended up in the ditch, but it has been a pleasure to make your acquaintance. I'll give you my card so you can contact me, just in case there are any issues regarding the car.'

'Thanks, but really I do feel I was the one at fault,' Laura said, but took the card from James anyway – just in case.

40

Looking up to her kitchen clock, Ellen saw that she had a further forty minutes before she was due to collect Liam from school. She was busy putting the finishing touches to the design of another wedding cake and was desperate to get the task completed before her son was home and wanting her attention. Ellen was beginning to realise how popular alternative choices were when it came to weddings – in every sense. This particular special request wasn't exactly difficult to execute, requiring only the dextrous placement of a selection of Warhammer warriors on the top tier of an otherwise ordinary sponge cake, covered with fondant icing – black icing, of course.

With ten minutes to spare another commission was completed, and Ellen boxed the cake and hid it safely away. The figures would no doubt fascinate Liam if he caught sight of the cake and she'd have a terrible job keeping his hands off them.

As she started out on her walk towards the village school, Ellen's mind began to consider the ways in which she could further develop her business. The cake orders were steadily arriving, but the private catering was still an option she wanted to investigate. She thought it might be worth some specific advertising to highlight this service and it was probably also time to come up with an updated replacement poster to display in the village shop window. She had taken some photos of the prepared dishes for the one dinner party she had been asked to cater, which might be suitable for the website; uploading the images would be a job for later once Liam was in bed.

The dining event had taken place a few days previously and was arranged to mark the impending moving out and on of Caitlin's sister.

'Stephen is so pleased that we'll shortly see the back of her, he wants us to have a celebratory farewell meal,' Caitlin had explained.

'Please tell me she's not actually aware that's the sentiment behind his suggestion, is she?'

'Oh, I think she knows that it's probably time she left us in peace, but it is great news that she's finally getting her life back together. She's really excited about living in London, and I'm so happy she's found a flatshare,' said Caitlin.

Ellen had enjoyed preparing the three-course meal for Caitlin, but it had proved to be quite a challenge to get the dishes prepared at home and then transported. She had allowed additional time to spend in Caitlin's kitchen to ensure the venison casserole was properly reheated, with a last-minute seasoning check, as well as overseeing the cooking of the hasselback potatoes – Stephen's particular request. Starter and dessert courses had been selected to ensure there would be no need for any last-minute preparation or cooking.

As this was her first attempt at offering private cuisine, Ellen had been anxious that everything would go smoothly, perhaps even more so as she didn't want to disappointment a friend. She had made a special effort by also helping Caitlin to put the place settings on the dining table, and agreed to stay on until the family were ready to be seated and commence the meal. Stephen's brother and sister-in-law had been invited to join the party, it being a good excuse for a family get-together.

'Stephen's sister-in-law is such a good cook, I would have panicked if left to cater for this evening by myself. I'm so pleased you're here to help,' Caitlin said.

The guests had arrived a little after the appointed hour and drinks were served in the sitting room whilst Ellen remained in the kitchen to garnish the starters. She was stacking up the empty plastic containers as a young woman entered the cook's domain. This was Ellen's first encounter with the guest of honour who Caitlin had said was upstairs in her room getting ready, a process which apparently involved a longwinded hair-straightening exercise. And it had definitely been worth the time taken as far as Ellen was concerned; Rosanna had the most glorious mahogany mane. Although the sisters shared some familial resemblance, there was no denying that Rosanna was the easy winner when it came to facial symmetry and associated good looks. She was stunning, immaculately dressed, despite the informality of the family gathering at home, and she had on the most stylish court shoes – two tone, snakeskin print and black patent, pointy and high.

'Oh... err, sorry... I didn't realise that you were still here.' Rosanna seemed terribly flustered at discovering Ellen at her workstation.

'I'm Ellen.'

'Yes... I guessed you might be.' Rosanna was inexplicably blushing and excused herself in a hurry and strangely without accomplishing whatever the task was that had brought her into the room.

Well, that was rather rude; Ellen had been taken aback at the time and could only assume that the young woman was either shy or sadly lacking when it came to social graces.

41

The damage to Laura's car was not extensive, but as it was still awaiting repair at the garage she was temporarily reliant on friends and the meagre bus service in order to venture outside of the village. April had offered to pick her up on the way to Great Aldebridge where they planned to visit Amarellos' before doing a supermarket shop together on the way back home.

'Thanks for this, April,' said Laura as she closed the car door behind her.

'Not a problem, it will be lovely to have your company for a coffee and definitely good to have you along as a distraction whilst I'm doing the dreaded weekly shop.' April loathed supermarket shopping. It was an errand she preferred to delegate to Ian, but as she was out of work just now she really didn't feel there was any excuse not to undertake the chore herself. Actually, she believed she did a better job of it than Ian, who, even with the benefit of a detailed list, was always tempted to go off-piste and return home with unusual groceries. If the supermarket had an item labelled '*superfood*' Ian just couldn't resist. Dealing with the cherimoya had required a Google search to discover whether this light green, heart-shaped item should be added to a fruit salad, or steamed and served as part of a main course.

'I don't understand why you dread it so much,' Laura said. 'I love a browse along the food aisles, checking out what's in season and taking the time to plan meals.'

'Any news on when your car might be ready?' April asked, choosing to ignore Laura's notion that there was joy to be had in a supermarket.

'They're hoping to return it to me tomorrow afternoon. But I've not found it too difficult without transport courtesy of my lovely friends. The village shop is also pretty good at providing for most of my needs, and Max and I have enjoyed exploring the local walks for a change. Having said that, a walk at the beach is the first thing we'll do once we are mobile again.'

'So, have you found an excuse to call the number on the card given to you by the delectable doctor?' April had been excited on the journey back from The Laurels the day of the accident to hear all about Laura's chance encounter.

'Not yet, but I do think it would be polite to call him to let him know when I get the car back, don't you agree? After all, he was very concerned to make sure I was alright and offered to assist in any way he could.'

'Definitely, you must give him the update. It's something he will clearly be very grateful to hear.' April laughed. 'And don't forget to find out his current marital status.'

'Ah yes, that I have already ascertained with the help of the lovely care assistant at The Laurels.' Laura explained that she had taken the opportunity to discreetly quiz Sarah on her recent visit to see Rose. 'It appears that James is a widower. He sadly lost his wife to cancer eighteen months or so ago. How terribly sad is that.'

'Poor dear man,' April responded. 'It must have been awful for him. I don't suppose being a member of the medical profession is necessarily any help in those circumstances.'

They both silently reflected on the random nature in life of sheer good luck, or, more poignantly, dreadful misfortune which could in a moment tear apart a contented existence.

'Incredibly fragile, isn't it... life, I mean,' April finally whispered.

Laura nodded and sighed.

The town centre car park still offered plenty of free spaces and April manoeuvred her car into a convenient spot close to the pathway leading to the shops. They were soon on their way to the café, but had planned a slight diversion via the gorgeous gift shop where they hoped to seek out a suitable birthday present for Ellen. Having browsed all the wonderful art and craft items and the homewares on display – useless trinkets and trifles in Ian's world, April mused – they eventually decided upon a set of enamel cake tins and a recipe card box in a gorgeous pigeon grey colour, which would complement Ellen's kitchen perfectly. Happy with their purchases and ready for refreshment, they made their way to Amarellos'.

'My treat to thank you for your taxi services,' said Laura as she went to the counter having instructed April to find a free table.

With coats removed and cups of coffee on the table in front of them, they relaxed, but April wasn't surprised that Laura swiftly raised the matter of the community café – a topic of conversation which until this moment had been avoided.

'There's a parish council meeting next week and I will be asked for an update on the café project.'

'Yes, I suppose we can't continue to steer clear of the subject,' said April. 'I'm just so disappointed that the residents of our village haven't embraced the idea more enthusiastically. Ian says it must be to do with the average age being in the region of fifty-five and the fact that we live in a wealthy community of car owners.'

'Ian is clearly exaggerating. The average age could only be forty, surely! It's possible that we could scrape together the sums needed to get the project going. A couple of our grant requests have been successful, but they are for small amounts. If I remember correctly, one is for about £300, the other £500. With those sums and some business sponsorship, we could perhaps open a café…' Laura was clearly trying to give a succinct round-up of the scenario, but April couldn't fail to be aware of the unspoken 'but' coming her way.

April sighed. 'Okay, so what should we do?'

'I'm sorry, April, but I'm not sure there is anything else we can do just now. If there really isn't the support from parishioners, nor a clear intimation that the café will be viable, the council won't want to endorse the venture.' Laura looked relieved to have finally expressed her opinion. 'I really hate being the bearer of bad tidings.'

April frowned. 'I did see Derek the other day and we chatted about whether the older folk could be encouraged to make use of such a facility, but his view, having spoken to some of them, is that they just don't seem to get it. The older ladies are happy attending the Women's Institute meetings and the luncheon club – especially now these are being held in the smart community centre – and the older gents are very contented either at the pub with a pint, or at their allotments with a flask of tea.'

'What about the young mums?' Laura asked.

'I spoke to Caitlin and she said that most of the mums go off to work after dropping their kids at school – there really aren't that many at-home mums – and those with pre-school kids already go to baby gym and messy play sessions held in the centre, where free refreshments are served to the parents.' April sighed.

'The negative responses to our funding requests and the lack of enthusiasm for the project are a serious setback. So, be honest, April, do you really think there is anything to be gained by continuing to pursue this plan?'

'Truly, I'm not sure anymore.'

'I tell you what I think: how about we park the idea for the time being and revisit it again in the future? It's never been a bad idea, in fact, it could have been an excellent amenity for lots of our villagers, but if there isn't the appetite for it just now, is it really worth us expending any more time or energy?'

'I know you are right. I'm just finding it difficult to accept. Councillor Graham is going to be so smug, which is annoying. I'll let the would-be volunteers and any other potential supporters know. I really don't want to have to acknowledge that this is another of my brilliant ideas that hasn't worked out. Ian probably won't be surprised. I suspect he was dubious about a successful outcome.'

'That's way too harsh and I'm sure it's not even true in the case of the café. Ian is always supportive of your ideas, but unlike you he does tend to have an aversion to anything with an element of risk.'

'He's not averse to a risk when it comes to playing with carpentry tools!'

'Poor Ian, is his knife wound healing?'

'He's on the mend and desperate to get back to work in his shed.'

'So, did you have any more thoughts about returning to estate agency or was that just momentarily tempting because it's Richmond & Marsh – "the devil you know"?' Unbeknown to April, Laura was secretly hoping her friend might have this diversion to take her mind off the café project stalemate.

'Having considered the prospect for a couple of hours – whilst sitting in the waiting room of the A&E department with Ian – I don't think it would be the right thing to do,' said April. 'Anyway, I'm not quite ready to give up on my dream of being a barista.'

Laura was alarmed, having been certain that April now realised that particular dream was presently unfeasible. 'Really, I thought we'd…?'

'I found a website the other day with details of mobile coffee franchises. It was really interesting and I have to say I quite fancy the idea of driving around in one of their *"fully equipped, state-of-the-art vehicle conversions, serving outstanding coffee at corporate and outdoor events, music festivals and exhibitions"*.' April reeled off the marketing spiel. 'Just imagine the freedom, not to mention the opportunity to meet people from all walks of life. The festivals really got me excited, I've always wanted to go to Glastonbury and the Isle of Wight.'

'Okay. Not sure I share your enthusiasm for this, but… whatever you think.' Laura was lost for words.

'Oh, I know what we should do. Let's text Ellen and see if we can call in with her present tomorrow morning. We should definitely try to see her on her actual birthday.' April was distracted by the possibility of cake, for there would surely be cake.

'Yes, let's, it would be nice to see her.'

42

Paul's apartment had been adorned with the modern selection of furniture and household items that Ellen had delighted in spending time selecting from IKEA's inspirational webpages. She would remain unaware that it was the ever-obliging Rosanna who had taken time off work to sit in and wait for the delivery to reach No.5 Garden Court. Once received, the young woman had used her feminine skills to tastefully arrange the additional pieces of furniture and had found a home in the kitchen cupboards and drawers for the basic set of utensils and crockery, which Ellen had assumed her husband would require in his part-time residence.

'This all looks fabulous, Rosie. You've made it look like a proper little home,' Paul had complimented his helpful assistant following her efforts.

'I just wanted it to look nice for you, and comfortable for the nights you are here. There's not much in the way of pots and pans, but I don't suppose you'll be doing much cooking.'

'I doubt it. I'll probably rely on the microwave, takeaway meals and dining out.'

'Yes, of course. You want to be able to relax and switch off after a hard day in the office. This flat will be your sanctuary,' Rosanna said.

'You are so right, darling, and how about making this a much more appealing domestic haven by moving your lovely self in as well?'

'Really, you'd be okay with that?'

'I most certainly would be, and then I'll have the pleasure of your delectable company whenever I'm in residence.'

He had known he wouldn't need to insist; Rosanna had been only too happy to pack up her possessions at Caitlin's house and make the apartment her home. And so, their lunchtime liaisons – the sequel to the initiation of the affair – had been replaced by long evenings and nights together whenever Paul required No.5 as his city quarters.

Paul was very happy with this new arrangement – complacent even. He was quite certain he was going to continue to have it all.

'Victoria sandwich cake and the opportunity to devour every delicious morsel,' Paul said to himself, whilst seated behind his office desk one morning as he reflected on recent events. Then he grinned, noting that the cake puns continued to prevail, but the unconscious baking reference reminded him of Ellen. He was gratified to discover that he wasn't finding it in any way difficult to compartmentalise this duplicitous lifestyle. After all, Ellen was unlikely to ever have the need, or the desire, to come into the city, so there was barely a fear that she would discover the set-up at her husband's *pied-à-terre*.

This was the ideal scenario: an amenable wife at home, happily pursuing her new career and untroubled that he would be spending a few nights a week away *and* a tantalising young woman in residence, to keep him entertained, at his London bolthole.

*

Rosanna pressed the snooze button of the IKEA alarm clock for the second time; she was far too comfortable to even consider becoming properly awake just yet. It had been late when she returned to the flat the previous evening, having been out for cocktails with a friend at a nearby bar. There were so many great places to eat and drink in the vicinity, she was spoilt for choice, and her rent-free accommodation allowed her the funds to be sociable and enjoy herself without having to consider the budget.

Living in London was working out just fine for Rosanna, who had been delighted to vacate the cramped single room in her sister's house to take up occupation of Paul's trendy apartment. It was a young woman's dream situation: a smart city property, close to her office, with Paul's company on just two or three nights a week. The rest of the week and, more importantly, the weekend, Rosanna had the flat to herself, which meant she was able to invite her work friends to hang out with her in her new home. This did, however, have the disadvantage that she became rather popular with her colleagues, who anticipated her provision of a place to stay following a heavy evening out at a nightclub.

Rosanna was barely conscious and jumped at the sound of another early morning alert, this time an incoming text to her mobile phone, which vibrated against the steel top of the bedside table. Paul's message was to let her know that he'd be coming to the flat after work and had booked tickets for them to go to the theatre. An evening out to see a show was a lovely surprise, but as she looked around her she panicked at the state of disarray.

'Shit! What a mess,' Rosanna swore, realising that she was now definitely going to be late for work. There was no choice but to take some time to have a tidy-up before leaving the flat.

Rosanna had discovered that her flatmate was annoyingly fussy when it came to her keeping the apartment clean and uncluttered.

'Rosie, darling, you really should ensure the bin is emptied before it starts to overflow, otherwise we'll have food debris sticking to the inside of the lid. That's just disgusting,' he'd said.

Apparently, the bath also needed a wipe around after every use in order to prevent the appearance of a tidemark. Paul had even provided her with a bathroom cloth for this purpose, which he said was to be kept in the vanity unit under the basin for ultimate ease of use. However, she reminded herself that this current arrangement suited her very well, so Paul's previously hidden OCD would just have to be tolerated.

She was enjoying the buzz of having an older, experienced man as her lover, but she was especially pleased to find herself rehomed as a result of their liaison. Paul's maturity and professional standing were an absolute turn-on for Rosanna; Rob had been so unworldly in comparison. Their lovemaking was exciting and at times exhilaratingly experimental. Paul appeared to be a man with long-suppressed needs and desires, who was now trying to make up for lost time, hoping his young girlfriend would indulge him by exploring some of his fantasies. It still caused her to grin when she recalled him asking her to wear nothing but her stilettos – a vagary she was happy to oblige – but this particular fancy had instantly become less appealing when she'd accidentally impaled his shin with the point of her heel.

Rosanna had just zipped up her navy shift dress when her mobile phone rang. 'Morning, darling, just checking that you got my text. You haven't replied. I thought you'd be excited about a night at the theatre.'

'Oh yes, of course… that does sound lovely. I'm sorry, I'm just running a bit late this morning.' Rosanna was rushing around, with the phone pressed firmly to her ear with the aid of her shoulder, as she deposited the final items of discarded clothing in the laundry basket and crammed a takeaway carton into the bin as she spoke. Damn it, that bin would need emptying again!

'Okay, I'll let you get on, but I'm so looking forward to seeing you. I've had a hellish weekend. Liam had an ear infection and was miserable. Ellen was behind with her orders for the village shop, so I ended up childminding for two days,' Paul whined, and Rosanna murmured her sympathies, whilst plumping up the cushions on the sofa. Once satisfied with the uniform arrangement of the soft furnishings, she went in search of her shoes.

'Poor darling,' she said, 'never mind, you'll have a change of scene this evening, and we can immerse ourselves in culture and intellectual stimulation before allowing ourselves some naked fun at bedtime.'

Rosanna had intended her response to excite him; she loved the sense of power she was able to exert over him – it was intoxicating. He was clearly beguiled by her. She had come to believe that time spent with his lover was a most welcome contrast to Paul's life at home. Distance from the village had enabled Rosanna to banish all thoughts of his other existence, with Ellen, into the dark recesses of her mind where conveniently they became imperceptible.

Paul had introduced her to an interesting and highbrow, grown-up world. She considered him erudite and well-informed and was convinced these qualities were becoming her own, loving the idea that she now appeared so much more sophisticated.

'I must dash.' At last she'd located her shoes under the sofa.

'See you at six.' But she didn't hear Paul's sign-off, having already disconnected the call.

*

Paul put his phone back into his pocket – time to make some progress with the pile of paperwork on his desk. There was so much of the stuff in need of shuffling about and offloading before he could contemplate a relaxed evening. That morning, as he arrived at his office, he'd discovered an unannounced board meeting was to take place, with a number of the non-executive directors attending alongside the other executive management. These meetings always caused an unsettled atmosphere around the workplace, with staff unnerved by the presence of the company gods.

Imperative on these occasions to look super busy and efficient, Paul knew he was going to struggle, feeling utterly jaded following the family weekend. A good strong coffee might just fix this and he took the chance to slip out of the office to get himself a takeaway.

'Your usual, Paul?' The agreeable and ever cheerful Chinese barista lady smiled and welcomed him into her tiny coffee bar pop-up located on the corner of the street.

'Lovely, thanks, Lin, but could you add an extra shot today and I'll take a croissant as well, please.'

'Sure thing, but you should watch out for caffeine overload,' Lin laughed as she added the extra measure.

The sunlight was bright as Paul stepped back out onto the pavement. The warmth of the sun rays on his skin was soothing and, dreading the prospect of an immediate return to his desk, he decided he had time for a restorative stroll in the nearby park. He was soon happily perched on a bench enjoying the warm, gentle breeze. Taking a gulp from the beaker, he winced and immediately jerked back his head, removing his lips from the rim to avoid a further scalding. The coffee was hot-as-hell and with limited time available to drink the pick-me-up he took the plastic lid off the cup in order to hasten the cooling process.

Paul closed his eyes and lifted his face towards the sky to bask awhile and dream of his lover's silky-smooth skin and irresistible pert breasts. Closing his eyes was his second mistake, his first having been to imagine that he had time to chill out in the park.

If his eyes had been open, he would have noticed the large dog as it careered towards him. The searingly hot liquid meeting the skin on his chest made him painfully aware of the collision and he jumped to his feet, holding the sodden shirt away from his body. The dog galloped off with the croissant between his teeth; no owner appeared to be present, but Paul suspected an owner of such a crazy dog would now be in hiding.

'What's happened to you?' Paul's secretary looked aghast as he re-entered his office with a large brown stain down the front of his previously pure white shirt. 'And where have you been? Mr Cunningham was asking to see you.'

'Fuck… how long ago?' Paul knew his boss would be irritated that an underling could not be located on demand.

'He called down just after you nipped out. I told him you'd popped out briefly for some paracetamol.' Suzy seemed pleased with her inventiveness to handle the potentially tricky situation.

'Shit, now he'll think I've come into work with a hangover.'

'Sorry, Paul.' Suzy was crestfallen, having been convinced her boss (who she was only slightly in love with) would be thankful for her quick thinking. 'Anyhow, I assured him I'd ask you to go up to his office as soon as you returned.'

Paul put on his jacket, which sadly only partly managed to disguise his predicament, and then ran up the stairs, two at a time – in defiance of office H&S policy – to the top floor.

'Ah, there you are, Paul.' Cunningham's PA picked up her phone to alert her boss of his arrival.

The interview was excruciating. Paul couldn't have felt more uncomfortable – still horribly fatigued, looking uncharacteristically unkempt and feeling increasingly hot and bothered – hearing the news he had been dreading.

43

Sitting with the phone resting on the arm of her chair, Laura turned the small rectangular card over and over between her fingers. James had been eager for her to take his contact details and it had seemed a genuine gesture; however, would it be a wise decision to call the mobile number printed in the right-hand corner just underneath his impressive list of medical qualifications? He had been kind to her when they met and, despite the unfortunate circumstances, she did sense that they had made a subliminal connection – if that terminology was the correct one to fit the vibe.

'So, Max,' Laura turned to the dog for his endorsement, 'do you think I'm mistaken? Is it possible that I might have misread the signals?' She was afraid of embarrassing herself and imagined the horrible discomfort if she were to make contact with him only to discover he couldn't recall who she was.

A few moments later, she chastised herself; what on earth had happened to her philosophy of living life to the full and not allowing fear to dictate her choices? If she didn't call then she would never know whether he had hoped she would get in touch. Laura took a deep breath and pressed the digits of his number into her phone, hoping her timing was well-considered and that it would be late enough for James to have finished his day at the surgery.

'Hello, Dr Dunlop speaking.' He sounded tired and slightly hesitant.

'Hi, James, it's me, Laura.' She prepared to wait a moment to allow him sufficient time to place her, but she need not have worried.

'Hello there, how lovely to hear from you. I trust all is well and that you have your car back on the road.' James appeared to perk up.

'Yes, all okay. The insurance company has paid up and I'm mobile again.' Laura was enjoying hearing his soothing tones, especially as he sounded appreciative of her call. 'And that's what I called to say… and also to thank you for your kindness.'

'That's great news and I have to admit that I do feel some of the responsibility for the mishap lies with me. But I'm pleased all is now well and thank you for letting me know.'

Laura didn't want to allow the conversation to end so quickly and her prayer was answered when something else to chat about popped into her mind. 'How is Ernest getting on? I asked Rose last time I was at The Laurels, but she was having a bad day and thought I was referring to one of the Koi carp.'

James laughed. 'Yes, the names those fish acquire do seem to present opportunities for confusion. Ernest is doing well, thank you for asking. He's over his chest infection and seems to be settling in really well.'

'I'm very pleased to hear it. So as long as he continues to improve I guess you may not be required to attend to him again at the care home.' April didn't want Ernest to be unwell, but she did like to think there might be another opportunity to bump into James again – so long as it didn't cost her another £200 in excess charge.

'Probably not,' said James. 'I don't currently have any other registered patients in that particular residential home.'

Laura was searching again for something further to add, when James saved her the trouble.

'I hope you won't consider this presumptuous, Laura, but I did wonder whether you would allow me to take you out for dinner one evening. My way of apologising for the accident, but also I'd really like to see you again.'

'That sounds absolutely lovely, thank you, James.'

'May I call you, perhaps next week when I am less busy and we can arrange a suitable date?'

'Please do.' Laura felt herself flush with pleasure at the thought. As she disconnected the call, she turned to Max and added, 'Well, haven't we both been very clever? That could be the best ever decision we've made.' He wagged his tail, excited to hear her happy tones.

*

April prodded her fleshy belly as she stood in front of the bedroom mirror wearing just her underwear. It really was time to make a proper effort; she'd had enough of feeling podgy and out of condition.

The verdict, which she had reached that morning, saw her quite uncharacteristically arriving at the community centre for the Tuesday evening Pilates class, to the astonishment of both her friends. Laura and Ellen appeared to be pleasantly surprised, although she suspected they would think her incentive was the post-session pub visit rather than the exercise class. They couldn't be more wrong; new April had emerged.

Having managed to park – as Laura had so nicely described it

– the community café idea for the time being, April felt recovered from her despondency and ready to ditch the associated comfort eating. She felt a fresh determination to get back to a fitness regime and a healthier eating programme. It had also occurred to her that she must ensure her stamina levels would be up to attending music festivals and that she'd be looking great alongside the youths when she set up her mobile coffee shop.

'You'll have to take it easy this evening and not overdo it, April,' Laura urged. 'It's been a while since you've done a class – you don't want to pull a muscle.'

'Have you ever known me overdo it?' April didn't feel this was something she personally needed to guard against. She would do her best, but could revert to her fall-back pose of lying on her mat with her eyes closed if it all got too much and give the impression that she was mindfully reflecting with a spiritual purpose. The Pilates teacher was easy-going and never seemed too bothered; she was used to April doing her own thing whenever she did get around to attending a class.

The hour-long lesson drew to a close, Laura and Ellen got to their feet and April opened her eyes. They all made their way to The White Swan and once inside ordered a bottle of wine to share; abstention from alcohol would be a step too far, but April vowed to keep to one glass only.

'So, Laura, have you and James managed to arrange an evening out together yet?' April was keen to know of any progress and jumped in with her question as soon as the three were seated at a table.

'He did call, as promised, and we're going out together next Wednesday evening,' Laura said.

'How exciting, you'll have to tell us how it goes.' April was hopeful that James would swiftly usurp any remaining yearnings for Neil should Laura be harbouring any – which on reflection thankfully seemed no longer to be the case.

'And what about you? Any more thoughts on getting yourself one of those franchises?' asked Laura. 'I can see you at the Burghley Horse Trials with your coffee cart. I'd come too, I hear they have fabulous stalls and that the shopping experience is exceptional.'

'Never mind the shopping, what about all those handsome horses and their talented riders?' April said. 'Actually, I am still mulling it over, but in the meantime, Gino has asked me to help out at Amarellos' for a couple of weeks. Gemma's got a holiday booked.'

'That's brilliant, you're fast becoming his favourite stand-in barista,' said Ellen.

'Well, whilst it's lovely to feel needed, I do find it tiring working there, I can't lie. I'm sure I'd manage better if when I got home, after a day on my feet serving the public, Ian had made some efforts towards preparing our evening meal. Now there's an idea… I don't suppose you could oblige by preparing me a week's worth of delicious suppers, could you, Ellen?' April said in jest (although secretly wondering if this might be an option worth exploring further). 'Which reminds me, how's it going with the dinner party orders? Any more requests?'

'Sadly not, which is disappointing. I've been giving some thought to further publicity. Having said that, it was quite hard work preparing the three courses for Caitlin. Oh, that reminds me, I didn't tell you both that I finally got to meet her sister, Rosanna. It was my last chance, of course, because she's now moved out and is flat-sharing in London.'

'So I heard. What's she like?' April asked.

'There's a similarity between the sisters, but honestly she's the beauty of the two. A stunner – long, dark, shiny hair and an apparent penchant for gorgeous shoes. I don't really know much about Jimmy Choo's fine products, but I'm guessing she does. But I was surprised by how shy she seemed to be – quite ill-at-ease, in fact.' Ellen remained curious about Rosanna's behaviour that evening.

April barely registered Ellen's comment as she continued on, but she did catch sight of a puzzled expression on Laura face.

'Oh, she's just young... but, I meant to ask, Ellen, did you eventually get a birthday present from Paul?'

April had felt indignant on discovering, when she and Laura had called in with their gifts, that Paul would not be returning home on the evening of his wife's birthday. Ellen had seemed totally unconcerned, claiming she was busy anyway and happy to prepare a special meal for them both at the weekend by way of celebrating.

'He's been spending most Monday and Tuesday nights – sometimes also Wednesday – in London, so I knew he'd not be home on my actual birthday. But he did arrive with a huge bunch of flowers and a bottle of Prosecco on Friday evening and we had a really lovely supper together – all my favourites: goat's cheese tartlets and salted caramel chocolate pots.'

'Prepared by your own fair hands, no doubt?' April's mouth watered at the thought.

'Oh, you know I love cooking. It was no hardship and anyway Izzy wasn't available to babysit.'

'You know what would be a fabulous treat to make up for the lack of a proper birthday celebration? You and Paul could go to a show in London. When was the last time you had a night out in the city? And you've got the apartment there now.' April thought her suggestion was inspired. 'If Izzy's not available you can always leave Liam with me.'

'Actually, Paul and I do love a musical and we haven't been to the theatre since well before Liam was born.'

'Then it is definitely time you went,' said Laura.

'Poor Paul, he had such an embarrassing day in the office last week.' Ellen relayed the tale of Paul's coffee catastrophe and having to bluff professionalism in front of the director, whilst sporting a very visible, large brown stain down the front of his shirt. 'He scalded himself too, the coffee was boiling hot.'

'Nasty... but is everything okay... with Paul, I mean... and work?' Laura asked.

'Funny you should mention that because I don't think things are great at his company at the moment. But he assured me that he'll be fine.'

'So, the theatre, what do you think, Ellen?' April wasn't letting it go.

Ellen thought for a moment. 'Actually, you've given me an idea. It's Paul's birthday next month, I could get some tickets and surprise him. He loves my surprises.'

44

'Night, April,' Gino called as she pulled her coat on and threw her bag over her shoulder in the staff room. 'Thank you so much for your help this week and I look forward to seeing you on Monday morning.'

It was the end of her first week back on duty behind the counter of Amarellos' and the days had felt exhaustingly long, her feet and calves were incredibly achy and she continued to need up to three apron changes a day due to regularly splattering herself with coffee sediment. The customers had been annoyingly rude, so it was a blessing that the staff were so patient and helpful. April rolled her shoulders and then flexed each ankle to ease the tension in her lower legs.

'Night,' April said, concealing her sigh. Just at that moment, she wasn't relishing the prospect of another week's work, but at least there was the weekend in between to give her time to recuperate.

As she drove towards Ash Green, April's thoughts turned to her husband, who she suspected would be at the bottom of the garden in his man cave working on his secret project. As soon as Ian's hand had healed sufficiently, he was back in his shed every evening straight after supper, staying there until bedtime. Between the frequent polishing of the mid-life-crisis sports car and the woodworking, April felt ever so slightly left out and lacking an absorbing pastime of her own.

'I need a hobby... expand my horizons... Oh, who am I kidding? I'm too knackered to get started with a hobby, whatever it might be,' April argued with herself as she made her way home.

It was close to seven by the time April drove onto her driveway and her heart sank at the sight of the unwelcoming darkness of her home. As she got out of the car, she could hear the grating sound of a saw blade being drawn back and forth through a piece of timber, pinpointing her husband's whereabouts. Once inside, she trailed from room to room, drawing the curtains and switching on table lamps in an attempt to create a cosy atmosphere for the evening. Then with a sudden burst of energy, she ran upstairs propelled by the urge to cast off her workday clothes. April deposited her black skirt and mottled white shirt in the laundry basket, pulled on her joggers and sweatshirt, and then ran back to the kitchen to find herself a glass of wine.

After the first gulp of the ruby red liquid, April went to the freezer for a rummage. It was a huge relief to find two leftover portions of lasagne, which the oven could reheat for her without requiring supervision. With dinner sorted, she took her wine into the sitting room, slumped on the sofa and turned on the television. Having flicked through the channels a couple of times, she settled for a rerun of *Prime Suspect*.

'You're home then,' Ian called as he opened the back door and entered the kitchen.

'I hope you're not covered in sawdust,' April called back.

'No, I had my overalls on and I've left them hanging in the back porch. How was your day?'

'Very long, and I think I may have inhaled too much ground coffee dust, because my windpipe hurts.'

'So, you enjoyed it then. I wondered how long it would be before you found a few more reasons not to love working at Amarellos', you know, to add to the throbbing, sore feet at the end of the day,' Ian said, laughing.

April chose to ignore his comments. 'Dinner's in the oven, but you'll need to put some peas in a saucepan.'

'Could you put the peas in the saucepan? I'm going up to get showered, I have sawdust in my hair,' Ian replied.

'Whatever... it can wait until you're done. The lasagne is cooking from frozen so it will be a while. Just pop the peas on once you get back downstairs.' April didn't think she had the energy to get off the sofa and anyway Ian could do it, after all, he had been home from work for ages.

It was close to eight o'clock by the time the lasagne and peas were plated and, having got through a second glass of wine, April was almost past caring about solid food. But they sat at the kitchen table together to eat the meal.

'When are you going to let me take a peep at your handiwork? It must be close to completion by now,' April asked her shower-shiny husband.

'It's a surprise and I don't want you to peep until I'm totally happy with the result.'

'And when do you think that might be?'

'Err... another day or two. I'd like to get it done this weekend. Had you any plans?' Ian said, clearly hoping she didn't have any that involved him.

'My feet really do hurt, and I think I've a varicose vein appearing on my calf,' April said as she twisted her head around, straining to see if there was a purple thread appearing under the skin, or was it just a smudge of dirt. 'So, my plan is to lie on the sofa for two days, if that's okay with you!'

*

Laura had finally found what she hoped was a suitable card from the shop's spinning stand, which had now completed three rickety rotations. She was in the newsagent's, on an errand for Rose who wanted to mark the occasion of Ernest's birthday at the end of the week. The two octogenarians had become friends and partners in crime; their latest antic was to steal biscuits from the plates of dozing fellow residents at afternoon tea and hide them away for their own midnight feasts.

With her chores in town completed, Laura made her way to Amarellos' for a coffee, hoping she had timed her visit perfectly and April could take a break and join her.

'Hi, April,' Laura called out cheerfully as she strode into the coffee shop.

Startled by the call, April swivelled around, failing to upturn the bag of coffee beans she was pouring into the grinding hopper. A few hundred escapees tumbled onto the counter and the floor, resulting in a round of applause from her colleagues who continued to marvel at the frequency of April's party piece.

'Oops, sorry. I didn't mean to make you jump.'

With the brown beans swept away and deposited in the waste bin, April's young colleague, Jason, took over and made coffee for the two ladies, urging April to take a break.

'How's it been going? If you don't mind me saying you are looking a little flushed and tired.'

'Thanks for that... Could you keep your observations to yourself, please.'

'Sorry... I know it's hard work.'

'I have a new respect for all employees who spend their entire working day on their feet. I think I'm developing varicose veins, too.'

Laura was doubtful, but went along with the inspection of April's right calf.

'Only two more days and then you're done. I bet you've enjoyed keeping up your barista skills and earning a bit of cash.'

'I'm not sure I care about either anymore. Honestly, Laura, you have no idea how picky and demanding some of the clientele are. One woman complained that there was too much milk foam on the top of her cappuccino and insisted I make her a fresh cup. I so wanted to shout at her that a cappuccino is a frothy coffee!'

'It's your break, let it go. I have exciting news!' Laura felt her face spontaneously beaming.

'Oh my goodness, yes, the date with Dr James. How was it?'

'It couldn't really have gone any better. He's such a gentleman – intelligent, so beautifully spoken and so interesting. Apparently, he'd also checked with Sarah at The Laurels to find out about me after our first meeting. It's just as well Sarah is so inquisitive when it comes to the personal lives of others.'

'And…?'

'And we had a lovely meal together in a pub not far from The Laurels, and we talked and we laughed… He has a great sense of humour.'

'And…?'

'And he's lovely and brilliant company and we've arranged to see each other again next week.' Laura was enthralled by James. 'And he loves dogs. He has a crazy Jack Russell, called Darcy!'

'Perfect… he sounds just perfect, Laura.'

They finished their coffees and it was clear from the expression on Jason's face as he passed by their table that April's break was over.

'Is there something wrong with Jason today?' Laura asked.

'He received a text this morning from Gemma. She's having a whale of a time in Greece, met someone fabulous apparently.'

'Poor Jason,' Laura said. 'No one could have failed to notice how infatuated the young man is with his absent work colleague.'

As Laura got up from her seat, ready to leave, she turned to April with a parting question. 'Any plans for next week to celebrate being a free woman again, or are you going to use the time to do some further research into the coffee cart franchises?'

'The franchise idea will have to wait a few days, I'm afraid. I have another mission in mind. It involves a trip to London, a bit of shopping, lunch out and some surveillance of Paul's apartment.'

'Pardon?'

'It took a while for my brain to join together all the pieces of information and reach the logical conclusion – I blame the menopause – and I suppose there is a chance that the conclusion is flawed.'

'April, what are you talking about?'

'Caitlin's sister, Rosanna – she of the long, luscious locks and the Jimmy Choo shoes – seems to bear a remarkable likeness to Paul's mystery travelling companion, don't you think?' April said. 'Interesting too that she has recently moved into a flat-share in London.'

Despite the fact that April's conclusion mirrored her own, Laura feigned disinterest. 'How do you even know where the flat is?'

'Well, when Ellen was telling me about the proposed purchase, I made a point of noting the details so I could do the usual estate agent thing and get onto the property websites to check it out. Couldn't help myself!'

'*Rightmove* is still on your favourites list then?'

'Absolutely. So, I've earmarked Wednesday for my outing. I don't suppose you fancy joining me?'

45

Ellen had booked tickets for *The Jersey Boys* having scrutinised the reviews for numerous shows in the London theatres. A musical was the perfect choice; both she and Paul loved a musical. Feeling very pleased with herself, she had presented her surprise birthday gift to Paul earlier, anticipating his delighted response. However, it was perhaps time to give up on her surprises, as his reaction mirrored that which she received for their anniversary treat: he showed no sign of being thrilled by her thoughtful gesture. This time he categorically refused to go to the theatre with her, claiming that he couldn't think of anything he'd like to do less to celebrate his birthday.

'But you love a musical, we both do, and it's been ages...'

'No!'

Completely aghast at his kickback, Ellen felt there was nothing further she could say, so she left him to his inexplicable strop, slumped in his armchair in front of the fire.

Having taken refuge in the sanctuary of her kitchen, Ellen set about wiping surfaces and loading the dishwasher. A good tidy-up usually helped when she needed to calm herself. She wasn't exactly angry, more flummoxed by his uncharacteristic outburst. Searching for an explanation, Ellen wondered if there was more to his problems at work than he had already shared with her. Perhaps his job was, after all, in jeopardy. Having recently taken on the city flat, the timing wasn't great, but the flat would always be a good investment. And hadn't he reassured her just the other evening that his job was secure? It seemed she would remain baffled. She really couldn't be bothered to try and discuss this, or anything else, with him at the moment.

After a further rub-down with a dry cloth, the granite surfaces had a satisfying shine and Ellen looked for something else to take her mind away from thoughts of her grumpy husband. Liam was on a play date at George's house and not due to be collected until later. With no wish to rejoin Paul in the sitting room, despite the enticement of a comfy chair in front of the fire and a new edition of *Good Food* magazine to flick through, Ellen preferred to keep some distance between herself and her spouse.

The laptop was on the kitchen table and with nothing else requiring her attention, she fired up the technology to check her emails and the contents of the website. Updating dinner party menu suggestions to keep the ingredients seasonal and adding new photographs was a task that Ellen prioritised, although she remained undecided about including the images of the Warhammer wedding cake. That one really did push the boundaries to the astonishingly unusual and surely had to be a one-off request.

There were three unread emails: two were junk mail and she immediately pressed the delete button, but the third looked interesting and Ellen was thrilled to see that she had finally received an enquiry for catering a dinner party.

Dear Ellen, I really hope you will be able to provide a catering service for me and perhaps I should explain how I have come to hear of your wonderful business. I own a ski chalet in the Alps, which is let for holidays via the company where a delightful young lady called Rosanna works. By virtue of all the contact we have I consider Rosanna a friend and it was whilst we were chatting a few weeks ago that she told me all about the fabulous farewell dinner party her sister arranged for her before she moved. I was intrigued to learn that the wonderful spread was not provided by Caitlin, but by yourself. Knowing from our conversations of Rosanna's penchant for fine dining, I was keen to find out the name of this special caterer. I did have to press Rosanna to reveal your identity – she said that you worked only locally, so would be unlikely to want to help me. I live in Hempstead and I truly hope that you will not consider this distance a problem. Now that I have also discovered your website, I am sure you are just the person I am looking for to cater the dinner party, planned for next month, with ten guests. I would be so pleased to hear from you. Kind regards, Wendy

Immensely cheered, Ellen put aside the petty irritation caused by Paul, which twenty minutes earlier had threatened to ruin the rest of her day. She read and reread the email to minutely absorb its encouraging content. New business by word of mouth – she smiled at the metaphor – and from outside the village; this was exactly what she had been hoping for. It was just a shame that Rosanna hadn't been more immediately helpful – such an odd and self-obsessed young woman.

Liam would need collecting in an hour, but Ellen decided she would acknowledge the email and perhaps do some baking for Caitlin by way of a thank you for getting the dinner party enquiries rolling. Looking up at the wall clock – a fun design with the hands depicted as a fork and knife, which always made her smile – she calculated there was just enough time to make some of the cheese scones that Caitlin loved so much.

*

The sound of cooking utensils being pulled noisily from cupboards and deposited on the granite surface in the kitchen alerted Paul to the fact that Ellen was baking again. He knew this activity was a favourite stress-buster and Ellen had clearly started her task in order to shut him completely out of her mind. And who could blame her? He felt bad; his behaviour was disingenuous, but surely by now she must realise how he hated surprises. What she didn't know, however, was that he had so very recently been to see *The Jersey Boys* with Rosanna and was not in the least bit interested in repeating the experience.

More alarming for Paul had been Ellen's suggestion that they would be able to make a night of it and stay at the apartment. The sheer panic at this most appalling predicament had been all-consuming in that moment and it had been fear that had led to his outburst. In hindsight he realised just how thoughtless and hurtful his reaction had been. It had been naive to imagine that Ellen would never want to go to the London flat and the presence of a female incumbent would be more than tricky to explain, particularly in a one-bedroom apartment.

It was only a month or so ago that Paul had thought he had everything going his way: a great job, paying him plenty of money; a city crashpad with resident lustful young woman; and a much-admired idyllic property in a great Suffolk location, providing the perfect home for his talented wife and child.

Just a few weeks later things were looking decidedly less than ideal: his director had now given him his notice of impending redundancy; his flatmate was treating his smart apartment like a student let; and his wife was blanking him. Paul feared his perfect world was about to combust and he wasn't sure what to do next to stop that happening. Falling out with Ellen or giving her any cause to be suspicious of the crises he was facing must be avoided.

He sat staring into the flames of the open fire for a while longer contemplating his plight. He then stirred, attempting to prise himself from the armchair with the aim of seeking out his wife, hoping to blag his way out of the mess he'd created. As he got to his feet he heard the front door being opened.

'I'm off to fetch Liam, see you in half an hour or so,' Ellen called out as she departed.

Paul watched his wife walking down the driveway, through the sitting room window, with her basket of just-baked goodies over her arm and began to question his sanity. 'You idiot,' he said out loud, whilst punching his forehead with a clenched fist. Could he really continue leading this double life? Was a relationship with Rosanna really worth risking everything and could he actually envisage a life with her (would she even want that) and not Ellen, if the worst happened and their affair was discovered?

It was clear that none of his recent decisions had been given sufficient thought or clarity; he'd been totally selfish, led entirely by lust and blindly flattered to receive the attentions of a younger woman.

He slumped further down in his armchair and before he could dispel them, thoughts of Rosanna took control of his reasoning; ensconced in his flat, always welcoming and willing, Paul wasn't entirely convinced he was ready to give her up. So long as Ellen never discovered the truth, perhaps the answer was to be a lot more cautious and prevent any possible chance of his secret being revealed.

46

As April crossed the Millennium Bridge, she watched the murky waters of the Thames flowing either side of the structure. The façade of her favourite art gallery was in sight and she was looking forward to spending time perusing the exhibits on display at the impressive Tate Modern. She was also planning to treat herself to a light lunch in the café overlooking the river.

Ian didn't share her love of contemporary art and would have found the visit dull and uninspiring. This had provided an excellent excuse to exclude him, as he would never have approved of April's ulterior motive for her day in the city. His ignorance of her hidden agenda was for the best; in fact she considered that she was doing him a favour to keep him in the dark. However, she had hoped that Laura would be willing to accompany her. Her friend had declined, claiming she didn't want to leave Max for the entire day, but April suspected Laura wasn't convinced about the wisdom of her intention to stalk Paul.

April had been feeling burdened by the need to know whether her suspicions were correct. She truly hoped she had got it all wrong, that Paul was not being unfaithful to Ellen and that Rosanna had a double. There were lots of attractive twenty-something women, with long dark hair and fabulous footwear; she'd already seen quite a few of these lovely specimens since arriving in London.

The artwork at the gallery had left her feeling energised and enlightened, and the tasty lunch had restored her in readiness for a stroll along the Thames on her way to the underground station at London Bridge. Her schedule for the afternoon was to take the tube to Bond Street, browse the shops on Oxford Street and then make her way to the Barbican area where Paul's apartment was located.

The start of a light rain shower caused April to tug at the zip of her jacket as she arrived at the main shopping streets. Having already been in the city for close to four hours, fatigue and the drizzle threatened to curb her enthusiasm, and with the prospect of a further four hours before she'd be back in the comfort of her home, she began to wonder if she had taken on an impossible mission. Employment in the field of surveillance might not, after all, be a realistic prospect; it involved too much hanging about and in every kind of inclement weather. April knew for certain that a career as a private eye could have no place on her '*possible jobs*' list.

Her weariness was partly dispelled by the anticipation of all the lovely goods on display in the huge department stores. A shopping trip to Ipswich was always a nice outing, but it was such a treat to explore the stores of the capital and a Selfridges experience was obligatory when in London. Mindful of the journey home and her unemployed status, April felt obliged to restrict her purchases to small, lightweight and relatively inexpensive items, but she just couldn't resist a tiny decorated box containing handmade truffles. It would make a lovely gift – perhaps a nice present for Laura's mother, who shared April's passion for chocolate.

By four o'clock April's energy was sapped and she made her way up to the fourth-floor café. She checked her street guide and underground map as she sipped her tea, working out her route and timings, with the aim of reaching her destination at Garden Court just before 5.30pm, which she had determined to be the appropriate hour.

Her careful pre-planning for this final and vital stage of her city excursion had allowed her plenty of time to successfully locate Paul's apartment. On arrival, she cast her estate agent's eye over the building; it was modern and well-maintained, clearly of the 80s era and therefore a more pleasing design than the square box residential structures of the 60s and 70s. The location was also ideal for easy access to transport link, and with independent shopping and bistros in the street nearby – April's rating was a nine out of ten – it looked as though Paul and Ellen had invested well. She scoured the immediate area for a suitable spot which would conceal her presence, whilst maintaining an uninterrupted view of the entrance to the block of flats. From her vantage in the communal gardens, behind a bed of mature shrubs, she hunkered down, thankful that the rain had ceased and the wind had dropped a little. April folded her umbrella away with some relief; having an aversion to the cold and wet she knew she would struggle to keep up her vigil if she became shivery and sodden.

But fortune was kind and her persistence was rewarded when, after a further fifteen minutes, she heard the sound of shoe heels click-clacking along the pavement. April's initial delight at recognising the young woman as the very same goddess she'd seen accompanying Paul on his train journey was swiftly replaced by utter sadness. She felt physically sick; her suspicions had been correct. This harlot was spending time with Ellen's husband, actually living at the apartment. He was having an affair, but armed with the evidence, what on earth should April do now?

The burden of her suspicions was now replaced by the burden of a dreadful knowledge.

*

Ian had spent a thoroughly enjoyable evening in his shed. April was not due back from London until late; she'd been vague about timings, but that was fine by him. He'd also enjoyed a microwaved meal on a tray in front of the television, with an old episode of *Top Gear* for company. April hated all motoring shows – well, it was actually Jeremy Clarkson that she was averse to.

The woodworking was turning out to be a most satisfying pastime and even April had been full of praise when he had finally revealed his secret project. The pair of garden planters, with perfectly crafted dovetail joints, was painted the subtle Willow shade – using the remainder of the tin after Derek had finished painting the shed.

'Ian, they really are fantastic. Who'd have thought you'd turn out to be so adept at crafting,' she'd complimented him.

April was clearly impressed by his successful execution of the task and Ian was prouder of this achievement than he had ever been about any of the software programs he developed at work. This knowledge caused him to wonder whether April was not the only one to have missed out on a true vocation. However, he was soon distracted from this thought by the sound of a car turning into the driveway.

'Hi, I'm back, at last!' April gasped as she practically fell through the front door, clearly exhausted by her outing. 'Could you get me a glass of wine, darling, I'm shattered.'

'Did you have a good time?' Ian asked as he made his way to the fridge to fulfil her order, whilst she struggled out of her shoes and coat in the hallway.

'I've had a most successful day. It was a mission accomplished.'

'I didn't realise there was a mission involved?' Ian wondered if he hadn't been listening properly when she was telling him about the proposed trip to London. If that was the case, he was likely to be in trouble.

'Let me sit down with my drink and then we must talk.'

Having a moment to recall whilst April glugged her wine, he was pleased with himself to be able to confidently ask, 'So how was the gallery?'

'I really enjoyed it. It must be a couple of years since my last trip to the Tate. I'm never disappointed, it's an exciting gallery.' April was now relaxing, with her feet up on the sofa, but Ian felt a sense of foreboding.

'Have you eaten? I could make you some beans on toast, or scrambled eggs?' Ian wanted to show willing, especially as April had mentioned the need for them to talk, but his culinary repertoire was limited.

'It's fine, darling, I had a toasted sandwich with a coffee at the station whilst I was waiting for the train.'

'That's good, but shall I fetch the box of cheese straws for you to nibble on with your wine?' Ian was stalling any tricky conversation.'

'No thanks. I just need you to listen, Ian. I did something in London which you may not approve of – no, I must be honest that's not quite right – something that you most definitely will not approve of.'

'Dear God, April, what have you been up to?' Ian was alarmed; knowing his wife anything was possible. His mind cavorted between her forgetting to pay for items before walking out of a shop, poking someone with the pointy end of her umbrella or, horror of horrors, crashing into an art installation whilst at the gallery. What sort of an impact were her actions going to have on his wallet?

'I held a sort of stake-out.'

'Go on!' Ian said, trepidation causing him to fail to recall April's loathing of that particular expression.

'I went to Paul's apartment to see for myself whether Rosanna is also residing at the property,' April blurted, 'and I saw her – the very same person – going into the building.'

'Sorry, who is Rosanna?' Ian was clearly struggling to keep up with the story.

'The attractive young woman, with luscious locks, the one I saw with Paul on the train a few months back, and who, it turns out, is also Caitlin's sister!'

'Who's Caitlin?'

'Never mind that. I've outed the bastard.'

'Hold on… you are saying that what you saw today is clear proof of Paul's infidelity?'

'Finally, you've got it,' April said as she got to her feet to make the journey to the kitchen for a refill.

'And now what?'

'Exactly, what shall I do now? Clearly Ellen must be told.'

47

It was a big day for Ellen, she was going to a proper business meeting; it had been a while since she had done anything quite so grown up. Her excitement was at level pegging to her apprehension, but she felt ready to step out of the comfort zone of her daily territory and take on the challenge. The first pleasurable activity of the morning had been choosing an appropriate outfit, and she had opted for smart trousers, a blouse and jacket. As she checked the impact of her choices in the mirror, she felt assured, her courage suitably buoyed.

Wendy had invited Ellen to her home to discuss the proposed dinner party, which was clearly intended to be an elaborate event, requiring a top-notch menu. Armed with her laptop and a notebook, Ellen got into her car and set the satellite navigation to the postcode Wendy had provided. The device informed her that her proposed journey time would be thirty-five minutes with an ETA of 11.20am – ideal for the 11.30 appointment, allowing for any delays she might encounter due to agricultural machinery, or sheep escapees on her route along the country lanes.

Ellen was eager to meet with Wendy and hoped this would be her first step towards extending her current services into catering for upmarket dinner parties. She also saw this as another stride forward to achieving her goal of returning to the world inhabited by professional adults, where she would be making executive decisions and earning a decent income to enhance her independence. Who knew, if this all went well, she might even need to consider employing someone to assist her in her catering business.

At 11.26, she turned off the lane and followed the track on the approach to Wendy's home. As she spied the property in the distance, Ellen could imagine the description April would bestow upon this vision: '*a fabulous Dutch barn, retaining its pastoral origins, now stylishly converted to provide a substantial country house, in a tucked-away, idyllic rural location with uninterrupted views over open farmland*'.

The thought brought a smile to Ellen's face – lovely April, such a good friend, always supportive, fun to be with and unremitting in her enthusiasm for all things allied to real estate. Could being a coffeehouse employee really satisfy her?

Ellen parked close to the timber house alongside a mud-splattered Land Rover. She walked towards the oversized doorway – allusive to its origins as the haycart gateway – which was now transformed into an impressive glazed entrance, revealing a cavernous vestibule within.

A fair-skinned, attractive woman of tall stature and with broad shoulders, distinctly Germanic in appearance, came to the glass door.

'Hello, you must be Ellen,' Wendy welcomed her visitor. 'So lovely to meet you. Come through to the kitchen and we can have a cup of coffee and some cake. I'd love you to try some of my *Torten*.'

'Thank you, that sounds wonderful.' Ellen followed her host through the open-plan accommodation and into the spacious, ultra-modern kitchen. She sat on a stool at the polished concrete countertop and took in her first impressions of Wendy as she prepared two cups of espresso and cut two large slices from the multi-layered cake, topped with nuts and fruit. She had the air of a woman who appreciated the finer things in life.

'Thank you for making the journey over to see me. I thought it would be useful for you to visualise the setting and get some idea of my kitchen layout and dining arrangements, etcetera.' Wendy pulled out a stool and sat beside Ellen, presenting her with the traditional fare of her motherland.

'Absolutely, it's good to be here,' said Ellen. 'Thank you for the invitation and for giving me the opportunity to offer my assistance.'

'Rosanna told me that the food you prepared for her party was absolutely delicious, on a parallel to the meals she has enjoyed at Brasserie Amélie in Ipswich,' said Wendy. 'I have to say that was an impressive accolade – who doesn't love that restaurant.'

Ellen smiled, this was praise indeed, but she was slightly taken aback to learn that a woman of Rosanna's tender years had had the opportunity and the finances for fine dining experiences at such an exclusive venue.

'Delicious cake, Wendy, and a treat to try something different.' Ellen loaded her cake fork with a second mouthful, noting the exquisite silver cutlery, whilst Wendy explained her German ancestry and how she came to be living in East Anglia. Another woman relocating to satisfy her husband's career progression, Ellen noted.

'So, my dinner party… As I explained on the telephone, I'm expecting ten guests and I would like to offer a four-course meal. We will be celebrating my husband's fiftieth birthday and the guests will include a few close friends, but the others will be his business partners,' Wendy said. 'I'd really like it to be a special occasion, but I don't want the stress of doing the preparation myself. And that's where you fit in… I'm so delighted that you will be able to help me.'

'It will be my pleasure and I have had some thoughts on menu ideas for all the courses.' Ellen opened her laptop and the two women talked through the proposals.

'I like the idea of a seafood ceviche as a light first course, followed by the green herb salad and champagne vinaigrette, and your freshly baked parmesan and black pepper bread as an accompaniment sounds wonderful.' Wendy appeared to be impressed by Ellen's menu creation. 'And chicken as a main course would go down well with my husband – he's not a great fan of red meat. But I'd like to finish off with a rich, delicious chocolatey dessert – as an indulgence for me.' Wendy confessed to her sweet tooth.

They discussed a few more of the details and Wendy explained that she had a florist friend who would be providing table decorations and an excellent local wine merchant, who would assist her with wine to complement the dishes.

'Let me prepare you another coffee,' said Wendy, 'and I'll show you the dining area… oh, and the crockery and cutlery I plan to use for the party. If there are any other kitchen appliances or utensils that you will need, then we can check those out too.'

With the tour over, the pair sat down again to enjoy a second cup of coffee. Despite some anxiety about discussing the costs for her services, Ellen had boldly approached this matter determined not to undersell herself; she wanted to leave this client in no doubt that she was the talented proprietor of a high-quality service and deserved to be paid accordingly.

'If you would like me to, I'd be happy to prepare a portion of the chicken dish and the chocolate pot dessert for you to taste-test before the evening.' Ellen was keen to ensure that Wendy had a preview of her choices, just in case she wished to amend any of the dishes. She was clearly a woman who moved in the sort of circles where people hosted lavish dining occasions and getting this one right could be hugely advantageous.

'That would be marvellous. Goodness, I am so grateful to Rosanna for leading me to you.'

'Well, yes, I am too.'

'A genuinely lovely girl. She's been so helpful in our dealings with her for letting our chalet. So conscientious, always cheerful and chatty.'

'Err... actually, I've only met her once and that was very briefly, but she is indeed rather lovely and I suspect has plenty of admirers.'

'One particular admirer, it would appear.' Wendy smiled. 'I was so pleased to hear that she is finally over Rob. She was so upset when he went off without her.' Wendy was obviously very friendly with Rosanna, privy to her confidences and seemingly familiar with all her personal life events.

'Yes, Caitlin did mention that. She was lucky to have a big sister to look out for her when she became homeless, although I think perhaps the hospitality was wearing a bit thin by the time Rosanna moved on. I'm sure Caitlin hopes that the flatshare works out.'

'I hear he's an older man, which might be just what Rosanna needs right now. Such a romantic story – how they met on the train and how their friendship turned into a love affair,' Wendy continued on, sharing the details of Rosanna new relationship.

'I'm not sure Caitlin has been introduced to Rosanna's new friend,' Ellen said. 'To be honest, I wasn't aware that she'd moved in with a man. I thought Caitlin said it was a work colleague she would be living with.'

'No, not a work colleague, she's living with Paul. She was so delighted when he bought the city apartment. Apparently, he also owns a home in Suffolk but needs to be in his London office frequently so has invested in a second property to ease the strain of commuting. It has all worked out very well for Rosanna.'

Ellen felt her cheeks instantly burning. Wendy's innocent words had unleashed a havoc with the power of a tornado, causing jumbled and confused thoughts to whip around her mind. Her heart started to pound and her stomach to twist and clench as she struggled to suppress the retching. She desperately needed to get out of Wendy's home and away from the horror of this shocking revelation. Doing her utmost to avoid embarrassing herself – and Wendy – Ellen brought the meeting to a swift end.

'Oh goodness, is that the time?' Ellen managed to get the strangled words out of her mouth, despite the nausea, as she looked up to the stylish Thomas Kent clock face on the kitchen wall, whilst also grabbing her bag and laptop. She jumped to her feet, made off to the hallway and yanked open the front door, almost falling down the steps in her haste to depart. 'I've another appointment I must get to, please excuse me, Wendy, I really must dash. I'll email you to confirm all the details.'

'Excellent and thank you,' Wendy called out as Ellen ran to her parked car. 'Apologies if I have delayed you.' Ellen was concerned that Wendy might be perplexed by her sudden exit; had she noticed the florid flush which had flooded her complexion?

She saw Wendy's wave in her rear-view mirror and weakly held up her hand to acknowledge the friendly gesture. Pushing her foot down hard on the accelerator, Ellen sped off down the narrow lane and as soon as a suitable spot to pull off the road presented, she halted the car.

Smashing her hands down hard onto the steering wheel, she screamed out, 'You bastard, you utter bastard!'

The bawled verbal abuse was the outlet for the anger welling up inside her and the horror that was currently consuming her. The roared outburst also caused the rapid departure of a group of roosting crows from the hedgerow beside the car; their hideous squawking and the crescendo of flapping wings startled Ellen. She gasped and then started to weep, the tears rolling down her hot cheeks leaving ugly streaks through the foundation so carefully applied that morning. The crying was followed by sobbing and a frantic search through her bag for a hanky. With a wad of tissue paper in her hand, she blew her dripping nose noisily and thoroughly.

After a further ten minutes of mopping and deep breathing, Ellen had calmed herself enough to drive again; her mind was now fixed on returning home safely to her son and to consider their options in the light of Paul's treachery.

48

Ian had finally managed to convince April not to make any approach to Ellen with her findings. It hadn't been easy; April's view was that Ellen should not be left in the dark. Ian's view was that Ellen might be happy to remain in the dark – well, perhaps not, but he was concerned that she might disbelieve or even take out her wretchedness on the bearer of the bad news. The friendship was so important to his wife that he thought his solution was preferable.

'All I ask is that you don't say anything just yet. I'll speak to Paul. I'll invite him out for a beer and find out what, if anything, he's up to,' Ian said.

'Really, you'd do that?'

'Yes, I'm prepared to try.'

'If you could, I think that might be the best first step. But if he's not forthcoming, I'm going to find it really difficult to remain quiet.'

'Fine, just give me a chance to do it my way.' April could be impetuous and Ian was keen to save her from herself, although he wasn't exactly looking forward to broaching this personal subject with Paul. 'I'll give him a call, or perhaps text him. Hopefully he'll be home one evening this week and not in London the entire time, shacked up with his hot girlfriend.'

'Don't even jest, Ian, there's nothing amusing about the destructive predicament Paul appears to have created.'

'Sorry, perhaps that was in bad taste, but let's not make this our problem, please, April,' Ian begged her.

'Okay. And I won't say anything to Laura either for the time being. Right, I'm off to the village shop to get some more milk, we seem to have run out again. Are you taking supplies down to your shed now that you've got a kettle in there?'

'It's thirsty work and the sawdust gets in my throat. I need copious amounts of hot liquids while I'm in my crafting zone.' Ian smiled. He was loving his new hobby and had begun to seek out other carpentry courses he could sign up for once the current classes came to an end. He had lots still to learn and his enthusiasm was running high.

*

April took a detour on her way to the store; she wanted to check out the gardens at the community centre and see how the plants were looking. She'd heard from Laura that the new benches had been installed – with Derek's assistance – and as necessity required had been bolted to the ground to prevent them disappearing from their positions. Sadly, petty villains lurked even in this rural idyll. As she approached, she was pleased to see that a couple of elderly ladies were taking advantage of the new seats, chatting whilst resting their legs. But April did a double take as she noticed that they had a thermos flask between them and were sharing a cup of coffee from a plastic cup. April looked on incredulously; how could they think that arrangement was better than enjoying a freshly made coffee in a china cup provided by a welcoming community café?

She sighed and forced these annoying thoughts out of her mind. Instead, April admired the shrubs she had planted with Laura all those months before, which were clearly thriving and providing an attractive display alongside the seasonal bedding plants. The disarray caused by the building work to the old village hall was now undetectable; the splendid gardens and the smart community centre in the village green setting were a credit to this lovely Suffolk location. April smiled with satisfaction and a sense of pride in her rural locality. It was such a shame about the halting of the community café; she still maintained that this was an opportunity missed.

'April, hello! Admiring our beautiful garden, are you?'

April turned and saw Laura and Max striding across the green on their way over to see her. 'Hi, how are you?' April called back and then panicked; Laura was bound to ask about her recent trip to London.

'We are well and just back from a mammoth walk across the fields. So many rabbits. Max has been having a simply marvellous time.' He certainly looked very pleased with himself, as did Laura.

'How are things with you, Laura? Have you seen Dr James again?' April's quick thinking told her to keep Laura focused on her new, favourite topic in the hope of distracting her.

'Actually, we're getting together this evening. I'm so looking forward to seeing him again, but determined to take things slowly this time.'

'When you're ready and not before, of course, do let us meet him. Not for a cross-examination but because he does sound rather perfect for you.' April was curious, as always.

'Absolutely, we'll all go out somewhere really nice – the six of us – perhaps Brasserie Amélie.' As soon as it was said, Laura appeared desperate to take back the suggestion. 'No perhaps not there. But that reminds me, how did you get on in London the other day?'

'The Tate was wonderful, I really enjoyed it, thanks.' April averted her eyes and started to walk off in the direction of the store. 'Sorry, Laura, must go. Ian's waiting for me to return with the milk for his cup of tea. He'll be wondering where I am. We'll talk soon.'

April departed at a pace, leaving Laura with the impression that Ian must indeed be really desperate for that tea.

*

Ian was halfway through his pint of beer by the time Paul arrived at the pub.

'Sorry I'm late,' Paul said as he approached the table Ian had found for them in the busy bar area, 'the train was delayed and the traffic leaving Ipswich this evening was crazy. I was running so late I've come straight here to meet you. I did text Ellen, but I haven't even had a chance to call in at home. Anyhow, I'm here now so I'll just get a drink and some crisps, I'm famished. Can I get you another?'

'Yes please, another pint would be most welcome.' Ian felt the need for alcohol, but looking at the strained expression on Paul's face it appeared he was the one most in need.

It took a while to get served, but Paul finally made his way back from the bar, sat down with Ian and began to gulp down his beer.

'It's good to get out. We haven't done this for a while and your timing couldn't have been more perfect. I've got so much going on at the moment, a drink with a mate is just what I need,' said Paul.

'I thought things were destined to be easier now you've got

the flat and don't have to do so much commuting?' Ian asked, happy to allow Paul to take the lead.

'Ian, you just don't know the half of it. I seem to have well and truly fucked up on all fronts and God knows how I'm going to make things right again.' Paul clearly did have lots on his mind and seemed desperate to talk. Ian had never seen his friend so on edge and began to feel anxious; his counselling and empathy skills were lacking (as April was always telling him).

'What's up?'

Paul shared his troubles and Ian learnt the true depth of the mire that was currently consuming his friend – and it was so much deeper than either he or April could have envisaged. When April had told him about Rosanna, Ian couldn't help a nanosecond of envy; charming a beautiful young woman into bed sounded exciting and invigorating. But it was just that very brief moment, as Ian acknowledged the disastrous repercussions of such a digression; he valued his relationship with April far too much to risk it on a flirtation, or a casual romp. She was hard work at times, but he knew for sure that she was the only woman for him.

'What should I do, Ian?'

'Stop hiding the truth.' Ian thought it was obvious that Paul must now face the consequences of his actions. 'There's really no other option.'

'I can't believe I've allowed things to get so out of control. Losing my job is a nightmare, but I'll be entitled to a redundancy package and I guess there's no reason why I can't get work somewhere else. But none of that matters if I've truly messed things up with Ellen.'

'You have to talk to her and you have to make sure she knows that this – for want of a better word – dalliance means nothing to you.' Ian felt sorry for Paul, but was also annoyed that he could have been such a prat. 'Don't wait for someone else to tell her what's been going on.'

'No one else knows and you won't say anything, will you?'

'I won't, but it's not impossible that someone else will tell her if they get wind of what's happening.'

'Surely no one will… I just hope Ellen can forgive me. I've been such a prat.'

*

Paul made his way home, feeling better for having shared his worries with Ian, which was inexplicable really because the problems remained huge and still needed dealing with. The alcohol had helped though; he felt calmer and much more relaxed than he had done two hours and four beers earlier. The whisky chasers with the last two had definitely washed away the stress.

Ian had seemed certain that the only choice was to confess all and pray the relationship with Ellen could be salvaged. Paul thought about this as he walked home – having agreed with Ian that his car should be left in the pub car park – and wondered if perhaps he could avoid telling Ellen everything. Yes, that was surely an option. He could explain about the redundancy; she would be upset and concerned about their finances, but he'd be able to reassure her on that front. Was there really any need to tell her about his affair as well? That would be cruel; she'd have enough to deal with knowing he was losing his job. It had been a fling, nothing more – a dalliance, as Ian had so delicately described the episode. He liked Rosanna, but he had never actually intended it to be a serious relationship. It was a time-limited arrangement; Rosanna doubtless felt the same way.

Paul reached Walnut Tree Farmhouse without being conscious of the process of getting himself there. After a second search through his jacket pockets he located his key and let himself in, gently closing the front door after him. The house was in darkness. Ellen must already be in bed and with any luck was now asleep. He was relieved. His news would have to wait until tomorrow morning. As he removed his shoes on the doormat, he dragged his tongue away from the roof of his mouth, disgusted by the rancid, parched palate. A large glass of water would ease his thirst and dilute the alcohol, which would hopefully also prevent a full-blown hangover. Having come to this sensible conclusion he tiptoed off towards the kitchen; being noisy and waking Ellen would be reckless, so he made his way as quietly as he could across the hallway, relying on the glimmer of moonlight through the small window to illuminate his way.

With limited visibility, Paul failed to notice the large holdall blocking his passage; when he tripped the fall was dramatic and the stone flagged flooring was utterly unyielding.

49

The doctor came through the swing doors into the waiting room and sat on the seat beside Ellen and Liam.

'Well, it's good news, it seems your husband has been incredibly lucky,' he reassured her, adding that scalp injuries often appear very serious because of the numerous blood vessels located close to the skin surface. Lots of blood, but not necessarily irreversible trauma to the vital organ, it seemed. 'He is suffering from concussion as a result of his injury, but the MRI scan has confirmed that thankfully there is no damage to the structure of his brain.'

The muscles in her shoulders that had been held so tensely were instantly relaxed and a surge of relief swept through Ellen's body. She kissed her son's head. 'Daddy is going to be okay, just like I said.'

The doctor informed her that Paul was likely to have double vision for a time, as well as an almighty headache. *That just serves him right*, thought Ellen, her anger immediately returning now his recovery had been assured.

'He'll probably feel groggy and a little confused. It may take a while before he is able to recall events.'

How very convenient, thought Ellen.

'We'll just sort out the discharge paperwork and then he's all yours. He'll be fine at home as long as he's in your care. You'll need to keep a close eye on him and contact us if there are any signs of his condition deteriorating – any concerns at all,' said the medic.

How absolutely bloody perfect, thought Ellen.

She was instructed that someone must stay with her husband for at least the first forty-eight hours. He must get plenty of rest and be protected from any stressful situations. As a safeguard, he must refrain from drinking alcohol and not return to work until completely recovered, which could take two weeks, perhaps more.

Ellen decoded the doctor's orders to her incredulous self: *So, I'm expected to take care of my unfaithful husband, ensuring his wellbeing and restoring him to health when what I'd actually like to do is kill him.* But reason swiftly told her to keep such thoughts concealed – never to be spoken – for fear that others might question her culpability. Because she was guilty, wasn't she? Guilty of setting the scene, where in the darkness the strategically placed obstacle would indisputably have the potential to fell him.

Ellen cast her mind back over the previous fourteen hours which had seen her floundering through the waking nightmare that followed her conversation with Wendy. It hadn't required a degree in genius to work out that Rosanna was staying at the flat in London with Paul. Her Paul. They had obviously been having an affair – and conceivably for a considerable time – which unquestionably explained Rosanna's discomfort when she had met the wife of her lover, in her sister's kitchen. But how could Paul have behaved so despicably, and why would he?

Reliving the immediate agony of the discovery caused her whole body to tremble. She feared what her future might look like now that the perfect world she had so recently created for herself had been exposed as being based on a big fat lie. As she stared blankly at the indecipherable posters taped to the waiting room wall facing her, Liam got to his feet and went to fetch her coat from the seat nearby, where she had discarded it earlier.

'You need to put your coat on, Mummy, to stop you being shivery cold.'

'Thank you, darling,' Ellen said, whilst being woefully aware that a warm jacket could never alleviate the chill currently coursing through her veins.

With the doctor departed on an urgent errand and Liam preoccupied with the mismatched selection of toys in the plastic box, Ellen experienced a series of nauseating flashbacks. The pictures evoked memories of the minutes just prior to Paul's 'accident': she saw herself busy in the utility room, at the back of the cottage, icing a cake in an attempt to lessen her rage; she recalled vividly the venom and the revulsion she felt towards her errant husband at that moment. Then she had heard it – the awful thud.

Ellen knew she had stalled, taken a moment to wipe the icing sugar off her hands, before steeling herself and making her way towards the origin of the noise. Her wrath had been so overwhelming she was quite bewildered by the level of her upset on discovering Paul in a pool of blood, lying motionless on the flagstone floor. But she was certain it had been instinct rather than genuine concern that had led her to seek medical assistance. He deserved to suffer, didn't he, in equal measure to her own suffering?

So here they were, at the hospital, with the assurances of the medical team that he was going to be okay. Paul had survived virtually unscathed; her son would still have his idiotic father in his life and she would still have an adulterous husband.

*

Once back at home, Ellen swiftly despatched Paul into the guest bedroom at their farmhouse. He'd uttered just a few words on their journey from the hospital; these were a mumbled declaration of his love for her and something nonsensical about a new job. Apparently unable to offer anything in the way of meaningful conversation, Ellen knew that any confrontation would have to be put on hold until later.

The two hours remaining before dawn only allowed her to catnap before the clock hands raced to 8.30am when Liam had to be delivered to school. In making this journey she was immediately breaking the doctor's first directive about not leaving the recuperating man alone. But she figured that little harm would come to the snoring human lump she'd left under the duvet. Her concern was more for her son and the worry that he may be too tired for the classroom following his disturbed night. However, she needed the boy safely out of the way. Paul had some explaining to do and Ellen had a huge amount of anger to vent.

Returned from the school run, she made herself a cup of coffee and took it upstairs. Paul stirred as she sat down in the chair next to the bed. He groaned and put a hand to his aching head; she passed him a couple of painkillers and a glass of water.

'It's time you started talking.' Ellen was determined to remain calm and in control, despite the anger she felt at her husband's deception.

'Could we talk later, darling, I'm not feeling at all great just now,' Paul whined.

'No, we are going to talk now.' Ellen smarted at his ill-advised use of a term of endearment.

*

Paul didn't feel up to discussing anything at the moment, but his wife seemed insistent. Wishing only to be allowed to lie down and left alone, he pulled himself to a seated position and adjusted his boxer shorts, which he was surprised to find himself wearing. He usually slept in the nude, but then again he usually slept in the master bedroom, not the guest room. Ellen must have thought he'd be more comfortable alone in a large bed, whilst he was recovering from his cracked head. He put the much-needed painkillers onto his tongue and glugged at the water in the glass, proffered by Ellen. He winced, his eyes narrowing; the pain in his skull was excruciating.

His wife continued to look at him expectantly, so he concentrated really hard and searched the cognitive process of his scrambled brain for an inkling of the subject requiring his input.

Paul recalled going to the pub with Ian and remembered that they'd talked a lot, which was unusual because Ian wasn't much of a talker. Further thought brought the realisation that Ian had listened whilst he had regurgitated details – more detail than perhaps had been prudent. Not only the loss of his job, but also his relationship with Rosanna. It came to him that the bit about having lost his job was the morsel of information to be shared with his wife. Yes, that was it; something he'd said to her last evening had obviously alluded to the loss of his job. She must be waiting for him to explain about the impending redundancy and his expectations for finding employment; it was not the time to confess to an affair, not now, nor anytime soon if he could avoid it.

*

'Stop looking so vague, I'm not going to let a severe concussion excuse you from explaining to me what the hell is going on.' Ellen was incensed by Paul's delay in answering.

'I'm sorry, Ellen, you are obviously very cross with me, but I need you to know that I really do love you.'

'And?'

'Err… yes, and I don't want you to worry, but I'm going to be made redundant. I've got another month or so before I'm out… and I'll be entitled to a lump sum… severance and holiday pay, so we'll be fine and I'm sure I can soon find another job.'

'And?'

'I'm really sorry, Ellen, but I'm not sure what you're getting at and I'm feeling rather dizzy just now. Would you mind if I laid back down and closed my eyes for a while, please?'

'Before you do that, perhaps just think a little harder about what else you might need to tell me.'

Paul struggled for a few moments. 'Oh yes, the apartment… well, don't worry about that, I might still need it if my next job is London-based and anyway it will continue to appreciate in value,' said Paul, 'which could be very advantageous right now.'

'Or perhaps we could get a tenant for it.' Ellen raised her eyebrows in disbelief; the redundancy news wasn't great, in fact it was potentially another huge complication to their lives, but putting that hurdle aside for a moment, she wondered how much longer he'd continue to skirt around the matter of his secret lover.

'Do you think that would be a good idea? Having tenants can cause all sorts of problems.' Paul seemed confused by her comment.

Ellen looked at her husband with absolute incredulity. 'Well, you obviously did think it was a good idea – a bloody marvellous idea – and from what I hear Rosanna is also very happy with the arrangement!'

50

April had risen earlier than usual, after a restless night, and was sitting at the kitchen table with the last few mouthfuls of muesli left at the bottom of the bowl. Despite the label describing a luxury cereal (with oodles of nuts, fruit and seeds), she hadn't enjoyed the mixture; it could never be a substitute for eggs, bacon and plenty of buttered toast. The only pleasure the oats and dried fruit offered was the virtuous feeling of having commenced a planned healthy eating day in the appropriate manner.

Once the crockery and cutlery had been deposited in the dishwasher, April made her way to the sitting room to ensure the sofa cushions were sufficiently plumped and that there were no cups lurking on the coffee tables. The next job to tackle was the laundry and once she heard the water beginning to trickle into the washing machine drum, she walked aimlessly away from the appliance, having nothing pressing arranged for the rest of her day.

Undertaking a bit more research on mobile coffee franchises would probably be worthwhile, but she also needed to give some consideration to her recent conversation with Gino.

'Gemma is returning to Greece.'

'Another holiday?' April was surprised that a barista's salary could offer a lifestyle which included numerous trips abroad.

'No, this is to be with her man,' Gino explained. 'It was more than a holiday romance with Christofer, he's the son of a café owner and Gemma is preparing to move on and take up a position in his family's bar on the beach.'

'I see... how very reminiscent of *Shirley Valentine*,' April had said, although Gino didn't seem to understand her meaning.

Gino was under the impression that April would be eager to fill his vacancy, but she wasn't at all sure. She'd begged him for a couple of days to think over his offer, but her thoughts were currently distracted whilst she pestered in anticipation of a call from Ellen.

Ian's efforts had exceeded her expectations; he had evidently managed to get Paul to divulge the details about his affair with Rosanna. It can't have been easy to get him to talk about something so personal and potentially explosive, but Ian had promised to try, and had succeeded. She was very proud of her husband.

When Ian returned home from the pub, he'd clearly had quite a few drinks and was uncharacteristically merry. Whilst initially disconcerted that the two men had merely got drunk together, April was prepared to accept that the alcohol had eased the process of frank discussion for the two men. The slightly inebriated Ian explained that he had told Paul he must go home and talk to Ellen, tell her everything, and then pray for forgiveness. The additional news about the redundancy was a shock, but frankly for April this paled into insignificance alongside Paul's infidelity.

So, if Paul had kept his promise, why on earth had she heard nothing? Ellen would surely know that she was here for her, prepared to listen and sympathise, offer support and advice. It was driving her crazy not knowing what was going on at Walnut Tree Farmhouse but, unnatural though it was to her, April knew she must not wade in with her commiserations. Ellen would guess that Ian had shared Paul's revelations with his wife. April imagined Ellen needing time to recover from the shock of finding out that her husband was a cheat and a liar; as a couple they would no doubt want some privacy whilst finding the strength to confront the maelstrom of hurt, anger and pain. Poor, dear Ellen; this was going to be devastating for her friend.

April had hoped to get the chance for a further chat with Ian before he left for work that morning, but he had been more uncommunicative than usual – a situation not helped by a hangover. It appeared that he was starting to lose enthusiasm for his current job; this lack of motivation was worrying and completely out of character. He'd been with the same company for ten years, but since becoming enthralled by carpentry he didn't seem to have as much of an interest in the world of IT. She'd noticed he had subscribed to *Good Woodworking* magazine, which he read avidly each month; an unsightly tower of these publications was now growing beside his armchair and threatening to topple.

The telephone rang. 'At last,' April said as she ran to answer.

'Hi, it's only Laura.'

'Oh, hi.' April tried to hide her disappointment.

'Listen, I don't want to worry you, but I've just seen Caitlin in the village shop and it seemed Paul was taken by ambulance to A&E last night with a head injury.'

'Oh my goodness, what on earth happened?' April's mind was racing and she didn't want to imagine what might have occurred when Paul had told Ellen about his affair. Would Ellen really be so enraged that she would attack her husband? Dear God, perhaps Ian had been right all along and they should not have interfered.

Laura continued with her report. 'Ellen did the school run this morning and apparently Liam couldn't wait to tell George at the school gate about going to the hospital in the night because, and I quote, "Daddy's head got bashed in and the ambulance lady said I was very brave because I didn't cry".' Laura relayed the tale as told by Caitlin, as told by Liam.

'Is he okay? What on earth has happened? Did Caitlin find out anything else?'

'Caitlin said that Ellen didn't stop long as she had to rush back home to keep an eye on Paul, but apparently when he got back from the pub last evening – slightly inebriated it appears – he lost his footing in the hallway and cracked his head on the stone floor.'

'Yikes, that sounds very nasty. Do you know how serious it is?' April panicked at the thought of Ian's influence on the amount of alcohol Paul had consumed, which could have contributed to his fall.

'Apparently he's got concussion but hopefully nothing more. Do you think we should call in? She may need our help,' Laura asked.

'Err, I'm not sure.' April was struggling to know what to say next. 'I think what I'll do is send her a text, let her know that we've heard about Paul's accident and ask if she needs anything.'

'Okay, if you think that's best... but send it from us both. I think Liam is going to George's house after school, so he's taken care of.'

'Laura...' April hesitated and then asked, 'are you free this morning?'

'I can be, I've nothing specific planned, just walking the dog, as usual.'

'How about walking the dog in my direction after your romp across the fields? Call in for a coffee, would you?'

'Will do... see you in about an hour.' Laura disconnected the call and went in search of wellies.

*

'Why didn't you tell me what you'd been up to?' Laura was sitting at the kitchen table, her hands wrapped around her mug of strong black coffee, warming through nicely as she listened to April's story.

'I knew you didn't approve of me stalking Paul. But, Laura, I just had to do something. Then Ian was adamant that he should talk to Paul before I revealed my findings to Ellen. I felt she had a right to know. I'm sure I'd want my friends to tell me if I was in her shoes, wouldn't you?' April explained.

'Honestly, April, I really don't know... but I appreciate that you were faced with an impossible decision.'

'In hindsight, Ian was probably right and his approach to Paul was a better way. I'm just hoping that despite his accident, Paul has now told Ellen everything – the affair and the redundancy. What an appalling mess and what a hideous quandary for Ellen,' April said. 'You know for one awful moment when you told me about Paul's injury, I was fearful that Ellen had been driven to assault the cheating bastard!'

'Ellen...? No way, I can't imagine Ellen being physically violent, even if she had felt Paul deserved it.'

'So, what do you think she should do?' April asked. 'I'm sure I'd want to confront Rosanna, wouldn't you?'

*

Would she? Laura felt desperately troubled as she contemplated Ellen's current situation.

She had been that mistress, unconcerned about the effect her relationship with a married man might have on his family. Laura shuddered at the thought of her insensitivity and total lack of consideration for those affected by her indiscretion. Rosanna was young, naive and impressionable, but what had her own excuse been?

Laura was relieved that the incriminating detail – alluded to by Neil – which implied their previous relationship took place during his marriage to Catherine, had never been questioned by April or Ellen. Despite Neil's drunken admission at the dinner party, her friends did not appear to have grasped the significance of his garbled words. Laura continued to sincerely hope this was the case; she couldn't imagine April not trying to get to the bottom of something intriguing, especially if she got any hint of a scandalous liaison.

But could she be wrong? April did love to solve a mystery and perhaps she had registered the irrefutable truth. Nothing had been said, but perhaps she would now see her opportunity to test Laura, even imagine her friend would be grateful for the chance to purge her guilt in the light of witnessing Ellen's plight.

Laura didn't want to confess, she wanted to forget. What had occurred in the past could never be changed and discussing her actions would not acquit her of the thoughtless deception that she'd been party to at that time. She had to live with the knowledge that her behaviour had been unforgivably selfish; the shame would serve as her punishment.

Laura made a pledge to herself and to God – although not religious, including Him was a mark of her absolute sincerity – she was going to do anything and everything in her power to support Ellen through this trauma. Her friend must survive and thrive beyond her husband's infidelity and Laura would ensure she did.

The beep of April's mobile phone halted any further self-judgement.

*

April grasped the mobile phone the second she heard the alert of an incoming text. Quickly opening the message, she read Ellen's words aloud for Laura.

Thanks, both. Guess Ian has told you all. Need time to think. Will call soon. E x.

'Good, at least she has responded and she knows we're here for her if she needs us. But for the time being we should wait until she's ready to contact us again. Agreed?' said Laura.

'Guess so… but what…' April was stopped mid-utterance by what she might guess was Laura's deliberate butting-in tactic.

'So… Megan… she's completed her uni course. What plans now?'

Always delighted to have an opportunity to talk about her clever daughter, April was diverted.

'I'm so proud of her. Seems she's likely to get the 2:1 result. We'll hear soon. So, with any luck there will be a graduation ceremony to attend, which will of course require a shopping trip for a suitably smart outfit… I wonder if a hat would be too much?'

'Probably. I bet those ceremonies are very moving occasions for the parents – and an exciting rite of passage for the students.'

'I'm sure there will be tears… of joy, of course.'

'And has she made any decisions yet?'

'She's applied for a couple of jobs in London. As we suspected she's ready to move away from Bristol now. Dan is history – she survived the heartache. Another rite of passage!'

And here we are, full circle, back to the same old subject, Laura mused. The trials and tribulations of relationships and the yearning to find a soulmate, because you should never stop believing that such a being exists. Time and unfailing persistence are all that's required.

There was indeed an inkling of hope for Laura because the liaison with James seemed to be going extremely well and was showing all the signs of becoming a very special relationship. She was trying to keep her optimism in check but dared to imagine this man featuring as her partner well into the future. To add to her pleasure, Max was happy. He was besotted with Darcy the Jack Russell; she was his new best friend and playmate. Max never minded James coming to stay; he never snarled at Laura's latest manfriend.

51

SOME TIME LATER

With his hands on his hips, Paul stretched backwards, arching his spine in an attempt to relieve the tension in his lower vertebrae. Cobwebs and specks of dust were glued to the perspiration on his bare flesh. Every muscle in his body ached, but it was satisfying to attribute the discomfort to hard, physical work and to know that his efforts were contributing to the progress of a significant task.

Paul was clearing the rundown barns at Walnut Tree Farmhouse, removing decades of rubbish, decay and plant growth which had all but destroyed the structures. Having spent his working life cooped up in offices, confined in his company car, or shoulder to shoulder with fellow commuters, this work, by contrast, was agreeably rugged and manly. No suits or Prada Derby shoes for him these days – just a pair of ancient jeans and steel-toe work boots.

His recovery from the concussion and the subsequent blinding headaches had kept Paul away from his office for three weeks, but as soon as he'd been deemed fit for work, he had returned to complete the notice period, which ensured he received his full entitlement of redundancy pay.

The dismissal was a massive blow to his self-esteem, but Paul was soon appreciating how the enforced change of lifestyle might be fortuitous. He had little desire to return to a commute of two hours followed by ten more stuck inside an airless office. Recuperated from his injury, he had emerged minus work-weary eye bags and belly paunch – the maturing man's accoutrement to accompany a sedentary job. Clearly the long hours had hindered his wellbeing and proved disastrous for the health of his marriage. He was still desperate for a second chance with Ellen and hoped his willingness when it came to volunteering his labour for the mucky work at the farmhouse would warrant her respect.

Unsurprisingly, the aftermath of the revelations had been horrendous. Ellen's fury, well, frankly that had been quite terrifying; Paul was shocked by the vociferous response from his naturally placid wife. Her anger seemed barely to diminish, despite his ongoing attempts to assure her that he was utterly contrite. Rosanna meant nothing to him; it had been purely about male ego and feeling virile again. But in the end, it had been an atrocious mistake.

Thankfully, ridding himself of Rosanna's presence in his life and his flat hadn't been difficult to achieve. He'd not seen her again after his accident; instead he'd sent a text saying that the relationship had to end. Her text in response had amounted to just two words – 'au revoir' – lacking any feeling or sentimentality, which reaffirmed his understanding that it had been nothing but a fling for them both. He'd heard recently that she was relocating to Europe, which was excellent news; he didn't want to imagine the impact of his furious wife and his former mistress bumping into each other anytime soon. Succeeding in his quest to make amends with Ellen was likely to depend upon Rosanna being well out of sight.

*

Rosanna had packed up her meagre belongings, the total filling just two suitcases, and moved out of the London flat a week after receiving Paul's message. Typically male, he'd dumped her by text – such a coward – but she didn't make a fuss. Firstly, because a fuss might lead to her sister discovering the identity of her flatmate, and secondly because… well, the truth was that the relationship had just about run its course.

It had been fun, the courting and the dating had been thrilling – a welcome distraction at a time when she'd felt abandoned, with her self-esteem knocked by Rob's abrupt departure. She'd loved the flirting and the exciting anticipation of spending time with Paul once the friendship had been established. It hadn't really come as a huge surprise when Paul made it clear he saw no barrier to upgrading their friendship and embarking on an extra-marital affair. Looking back to the early days of their liaison, she'd been content to enjoy toying with the man-woman friendship, but in her limited experience men always wanted more and it had suited her to allow this to happen.

The convenience of the accommodation that the affair ultimately provided was – in her view – just reward for her amenable availability. But a girl should never overlook the importance of self-respect and there was no doubt that their relationship was devoid of any true connection of shared loving. He was obsessed with her, not in love with her, despite his post-coital claims.

During the post-dumping analysis, Rosanna acknowledged that her friend-with-benefits was also too old for her, too needy, too pernickety and crucially too spoken for. Although she had never really wanted him just for herself. He was better off with Ellen, but whether Ellen was better off with Paul… well, that Rosanna thought was debatable.

Time was up on this unorthodox episode in her life and she didn't consider it required further scrutiny; any consideration for Ellen was not her concern now and never had been. She was Paul's wife and he had made his choices.

The offer of a short-term contract at her company's office in France had been most serendipitous. She was soon on a plane and dismissing any further thought of Paul from her mind; the Mediterranean beckoned.

*

Pangs of conscience continued to invade Ellen's mind for many weeks, despite her endeavours to resist the intrusions. On that fateful night, it had been many more minutes than admissible before she'd been sufficiently composed to make the call to the emergency services. Her overriding concern had been to shield her son from the terror of witnessing his father lying prone in a bloodbath. Guilt-ridden, she had swiftly wiped the floor and stowed away the holdall in the understairs cupboard before the paramedics arrived at the farmhouse. In retrospect, she validated her actions as those of a woman in a state of shock – a wife devastated by the discovery of a husband's betrayal.

The anger and tears that followed eventually gave way to a bitter acknowledgement that her husband had done the unforgivable: he had destroyed her trust. Having had too much time to dwell, her disordered mind settled on a simile to reflect her contemplations. She compared the breaking of this trust to the broken pieces of her once treasured Clarice Cliff vase – an anniversary gift from Paul which had been hurled in a fit of rage: gluing the pieces together might give the appearance that the precious item was whole again, but the repair would forever be perceptible.

Ellen's grief had been compounded by learning that Paul had

felt unable to share the worries of his impending unemployment; this was not only demoralising, it was unfathomable. It seemed to say so much about how their relationship had failed. Their bond of marriage had been fragmented by Paul's deceptions, as had her respect for the man.

Propriety forced her to commit herself to a modicum of soul-searching as she attempted to explore her own part in the rift which had appeared in her partnership with her husband. Could she really consider herself blameless? Perhaps not entirely. Her determination to fulfil her dream had seen their relationship relegated to second place, with first place being reserved jointly for Liam and her business aspirations. But a temporary relegation could never justify Paul's actions. And so, the impossible question remained: should she allow his duplicity to seal the fate of their family unit?

Paul was clearly repentant; he was attempting to be obliging in any and every way, with gestures of support and affection, and with a constantly cheery disposition. But she struggled not to be irritated by his overly jovial demeanour, which seemed to border on insincerity, or perhaps just insanity. He must have been slightly insane if he thought his affair with Rosanna would remain undiscovered.

As far as her friendship with Rosanna's sister was concerned, Ellen had revealed none of the sordid details to Caitlin. Why compromise a lovely friendship? Caitlin didn't need to know of her sister's actions; it would be of help to no one – least of all Ellen. Contemptible village gossips might endeavour to seek out titbits of information, but Ellen was determined to maintain her silence and, therefore, her self-respect.

Mercifully, there was, amongst all the hideous upheaval, a hint of a silver lining and a way forward was presented in the form of the financial windfall. Ellen was adamant that a portion of the income from her husband's redundancy package should be put towards the renovation of the outbuildings at Walnut Tree Farmhouse.

With the local council's approval sought and permission for a change of use received, she planned to convert the barns to provide bed and breakfast accommodation for paying guests. Ellen's previously acquired knowledge of the best local craftsmen guided her to the talented architect whose inspiring drawings would see the realisation of three stylish annexe rooms. The guests would enjoy smart, contemporary comfort, within the barn-style accommodation, overlooking the glorious open countryside surrounding the farmhouse. They would be served splendid breakfasts, just across the courtyard from the annexes, in the currently under-used dining room of the main house.

Her growing business prowess was now bolstered by a determination to take control and never again allow others to presume she was guileless.

*

Megan had been living back at her parents' home for seven weeks. She had returned to Ash Green for much of June and July, prior to taking up her new position at King's College Hospital. April, in particular, had appreciated having the company of a much-matured, self-assured and contented daughter. They had shopped, visited galleries, been to see movies and enjoyed meals out together.

Ian hadn't felt excluded, he was happy to see mother and daughter revelling in some girl time. He'd been occupied with another carpentry project and was spending most of his free time in the shed making a dresser which was destined for a prominent position in the new breakfast room of Ellen's guesthouse.

Whenever he took a look outside his workroom and

discovered a spell of glorious weather – which the confines of the shed kept hidden from him – he'd allow himself some time off from his labours to go cruising in his beloved open-top car. He would venture out and spend a blissful hour or so navigating the country lanes, without having to listen to April moaning about her hair getting messed up.

He did join the two ladies in his life on moving day, when his taxi service and physical strength were required to transfer Megan and her belongings to her new home in London. The arrangements were ideal from everyone's point of view. Megan was to rent Ellen and Paul's flat, and she was thrilled at the prospect of living in a swanky apartment and working in the capital.

*

April felt the tears stinging her eyes as she hugged her daughter when it was time to depart. She was leaving her beloved only child alone at the flat, and she urged Megan to keep herself safe and call home at least twice a week.

'Mum, I've been living away from home in Bristol for the last three years. I'll be fine.'

'I know, but I've got used to having you around again. I'm going to miss you. Take care and always be streetwise, this is the metropolis!'

'Come on, April, we should get back to the car before a traffic warden comes along.' Ian hugged the beautiful young woman, who only yesterday had been his little girl.

Calls home confirmed that, as anticipated, Megan quickly settled into her new environment; she was enjoying her new job, had made friends and was loving living in the city. April made a couple of visits to see her daughter, but soon realised that she was surplus to requirements. She told herself firmly that this was how it was meant to be, she had done her job and, on this occasion, it was a job that she had done well – very well indeed. Megan was happy, confident and very capable, living her own life. But April knew that she would never stop worrying about her daughter; it wouldn't be normal not to, would it?

Worrying aside, April knew that with Megan now forging her own career path and Ian deep in sawdust, she must stop flitting and get focused. There had been a period of distraction whilst she endeavoured to be the best possible, supportive friend during Ellen's initial trauma. Both she and Laura had offered to assist Ellen in any way they could, and despite their moderate skills had been keen to shop, chop and stir to enable the proposed dinner party for Wendy to go ahead. It had been the first predicament requiring the friends to work together to find a solution.

'I'm sure we'd make a good team and it's lovely of you both to offer, but catering for Wendy is now inconceivable,' Ellen had insisted. 'Even with you two alongside, I just don't have the energy – or the mental capacity – so, April, I'm afraid I'm not going to be complicit in sneaking you a look around Wendy's gorgeous barn.'

'Now that is a shame. I've been past the entrance a couple of times on my way to the farm shop. I bet it's utterly fabulous inside.'

'It is indeed a rather splendid home,' Ellen confirmed.

'Just as I'd suspected. An example of repurposing done particularly well,' April said, unable to suppress her inner estate agent who was desperately trying to put a market value on Wendy's cavernous timber-framed home.

'That particular problem has now gone away,' Ellen had told them. 'I sent Wendy an email saying that unfortunately I was now indisposed, taking care of my husband following an injury he had sustained.'

'She's going to be so disappointed... and who will cater for her now?' Laura had said.

'It will be fine. In the message I included a link to Waitrose Party Food to Order, assuring Wendy that they provide an exceptional service.'

'Do they?' April asked.

'I've absolutely no idea. I've read positive reviews but never tried their entertaining range. All I do know is that I never want to see Wendy again. It is such a shame, I did like her, and she was not to blame, but no... no...' Ellen clearly was unable to say another word on the matter.

So, with the vacancy for a sous chef having been withdrawn, April had time for further deliberation and came to the conclusion that being available to cover staff absences at Amarellos' for Gino was one thing, but filling Gemma's shoes permanently was really not for her. A full-time barista she was never destined to be. And as for the mobile coffee franchise, well, it had been a pretty good idea, but another one to park, especially as there now appeared to be a more exciting opportunity in the offing.

*

With Liam at school, Ellen was busy baking and with the last of the bowls washed and the cake tin containing the coffee and walnut mix in the oven, she filled the kettle in preparation for a tea break. As she sat at the kitchen table, she flicked through a magazine, stopping at a feature entitled 'Being Married'. The author wrote: *Marriage makes you vulnerable, but it also makes you strong. It brings out the best of you, as well as the worst, and ultimately it changes you in ways you could have never imagined.* As Ellen read the two sentences, she thought the author must have been looking over her shoulder as she wrote. She had been young and vulnerable when she'd married Paul; there had been moments when she'd felt she was doing an exemplary job of being a wife and mother and moments when these roles had made her feel crazy and exasperated, even a failure. Now she was determined to focus on being strong and leaving behind vulnerability. Ellen bristled with a sense of empowerment that was utterly satisfying.

Whilst the fine dining element of her business would now by necessity be sidelined, Ellen firmly believed that her home baking and the B&B would run well in tandem, with one complementing the other perfectly. In fact, she'd been busy honing her bread and pastry-making skills with a view to offering freshly baked fare as part of the Walnut Tree Farmhouse breakfast experience. She had never been a great lover of the traditional English cooked breakfast, but she had a friend who was a connoisseur when it came to synchronising frying pans for a morning feast.

The celebration cake orders continued to arrive in her email in-box and Ellen knew she wanted to concentrate her efforts on forging ahead with this side of the business. Over the past year, she had acquired a reputation as the go-to person in the area for speciality party gateaux. She had a significant following of clientele and this aspect of her baking empire gave her the most pleasure; the opportunity to get creative was incredibly fulfilling.

With the anticipated growth of the two ventures, Ellen knew that the skills of an accountant could be beneficial to ensure that all financial matters were kept in order. This again was not her forte; she'd always left the sums to Paul who had a better head for keeping on top of their accounts and any tax implications of commercial activities. Thankfully she had a clever friend who'd have no trouble at all putting in place a system to maintain and audit the books, and more importantly to analyse the financial status and operating results of the two businesses. She would definitely want to have an accurate record of profitability for the business partners.

*

April was certain she had finally found her niche, as destiny, or more specifically Ellen, appointed her the partner-in-charge of housekeeping and breakfast preparation at the Walnut Tree Farmhouse Bed & Breakfast establishment. She definitely had the wherewithal and the talents to successfully undertake both of these duties; what thrilled her the most was the opportunity to join Ellen and Laura in a new venture.

The repurposed outbuildings were being completely transformed from shabby, falling-down barns into simply stunning annexe rooms. All April needed to do to complete the perfect stay for their guests was to provide spotlessly clean, welcoming rooms and delicious, hearty breakfasts.

This wasn't quite the same as being the proprietor of her own coffee shop, but not such a huge departure from that previous good idea. After all, she was going to be responsible for the entire proceedings in the B&B's dining room, offering their clients the best-quality full English breakfasts, including deftly prepared hot beverages. She wondered whether Ellen might consider investing in a scaled-down version of a barista-style coffee machine, and she included a note on her spreadsheet entitled WTF B&B to remind her to raise this excellent idea at the next partners' meeting.

April was convinced that this was it – the dream career – and she was proud of her new job title. Business partner – a status so much more prestigious than that of a mere employee.

*

Laura had quietly endured the discomfort of self-scrutiny, whilst her blameless and dear friend coped with the fallout of her husband's betrayal. By chance, the proposed new partnership gave her the ideal opportunity to atone in some small way, and Laura was determined to be on hand to help and support Ellen. She considered the plan inspired and was keen to utilise her expertise by acting as the financial consultant to ensure money matters were kept in order. She was also keen to assist with housekeeping duties at busy times; after all, she wanted to play a significant role in ensuring the success of the B&B. The time had come for her to hand over her parish councillor duties; she had found it an intriguing and rather quaint, villagey role, and it had enabled her to become firmly established in the Ash Green neighbourhood, but she was done with volunteering – a business opportunity awaited.

She had always been convinced that retirement in Ash Green would be her nirvana for the latter years of her life. Having survived the Neil-blip and emerged sagacious and reinvigorated, everything appeared to be moving forward wonderfully well. Laura had a flourishing relationship with a charming and genuinely lovely man and quite unexpectedly an interesting new pastime; she was looking forward to joining April and Ellen's adventure.

*

Their plans moved on apace and a month later the three friends joked that it was essential for the B&B to swiftly provide them with financial reward. They needed to recoup their increased expenditure at their favourite coffee shop and the local hostelry where they had taken to holding business discussions.

'Ian has suggested a chalky teal finish for the breakfast room dresser,' said April whilst they were holding one of their meetings at Amarellos.' He thinks this will work well alongside the Emma Bridgewater pottery we've decided upon.'

'So, as well as his creativity, your IT boffin is revealing his aesthetic sense when it comes to interior décor. Hidden depths indeed,' Laura added.

'Yes, who'd have thought it… and it's a bit of a diversion from his sports car and the ritual of the weekly bodywork polish.'

'As long as he has some time and energy left for your bodywork, April!' Laura laughed.

'Now, Laura, that's a subject not suited for sharing,' April responded, in line with her recent pledge of discretion and propriety.

'That reminds me, James said something rather interesting the other evening,' Laura said, smiling at the thought of her gorgeous man.

'Do share…' April was eager to learn more.

'Dog-friendly establishments… It's not that easy to find overnight accommodation which caters for dogs and humans together. I think we should consider providing lodgings for canine companions. I'd be happy to come up with some suggestions of what we'd need to make this work.'

'Would our housekeeper mind an open invitation to moulting, hairy dogs and all the additional vacuuming that would require?' Ellen enquired.

'If it makes our B&B more appealing to a wider audience, why not?' April said.

The three women agreed there was little doubt that they were an excellent team, open to each other's suggestions and prepared to engage in amicable debate over such matters as the pros and cons of Egyptian cotton bed linen – should it be 400- or 1000-thread count? – and the optimum size for bath towels.

One particular detail on which they did all concur was the absolute necessity for their breakfasts to include local produce – free-range eggs, bacon and sausages – alongside Ellen's delicious home-baked bread and pastries.

'I'm planning on increasing my brood of girls. I'll need at

least seven more hens if we're going to be self-sufficient in eggs for breakfasts, as well as for my baking.'

'You'll also need a more substantial henhouse for the ladies. That's sure to be another project that Ian would be delighted to take on.'

'Marvellous, and I've recently seen just the thing in *Practical Poultry* magazine. He could use it for his design. So cute, with pink and blue painted slats, a ladder to access the inside and proper perches. Perfect for keeping the birds safe from foxes.'

'That reminds me,' said April, 'I had a peep in your hen run the other day. Aren't those three original old birds getting rather portly? Has Liam been heavy-handed with the pellets?'

'They may not be youngsters, but they are not yet past it. And yes, tending to the hens has become Liam's favourite job, so I suspect he could well be overfeeding them.'

'How do you think Liam's doing?'

'Well, you can never be sure, but on the surface I'd say he's doing okay. He doesn't seem to miss Paul not being around. Well, let's face it, Paul used to be at work more often than at Walnut Tree Farmhouse.'

'And the fortnightly visits... is that arrangement working out?'

'Seems to be. Paul picks him up at the appointed time on a Saturday and returns him the following evening as we've agreed. Liam always says he's had a good time with his dad. He especially appreciates having someone who's willing to go to the park for a kick-about with a football.'

'It's ideal that Paul isn't going to be located too far away. How's he getting on with the renovations at the cottage?' April asked.

'From what I hear he's making progress. Since he's finished helping the builder with the annexe rooms here, he's had more time to work there, along with the benefit of the extra skills he acquired. He did mention that Neil had left the place in a bit of a state...'

'Which reminds me, did you ever discover where Neil had sloped off to, Laura?'

'Weirdly, yes I did. I bumped into his brother, Ray, in Ipswich. He told me that Neil and Catherine are back together. Ray was incredulous... as was I.'

'No way... but they got divorced, didn't they?'

'Apparently not. Neil was clearly being less than candid. He called a halt to the proceedings as soon as we split up, so just days before the decree absolute. Somehow – God knows how – he managed to endear himself to Catherine once again.'

'What a rogue,' said April.

*

Ray's revelation had confirmed Laura's suspicions that Neil had ultimately been looking for a lifestyle provider. What she didn't share with her partners was that she'd been so incensed on hearing this news that, as soon as she was back at her home, she had written a missive intended for Catherine. The woman surely had to know the whole truth. But once penned, the letter was secreted away in Laura's bureau drawer for further consideration. The writing of it had assuaged her anger but, although well-intended, the sending of it might not be beneficial. Taking a course of action which involves meddling in other people's lives really ought to be avoided and perhaps Catherine should be left to make up her own mind – just as Laura had.

*

Preferring to banish the topic of miscreant men, Ellen swiftly continued with her update on Paul's recently acquired abode.

'Paul negotiated an excellent deal. It certainly needed a lot of work, but, of course, my only concern is that he can provide a nice, comfy home for Liam's visits.' She was sure her friends knew her priority was her son's wellbeing, but it satisfied a need to say the words out loud.

April sighed. 'I'm afraid I still find it all very sad, but I suppose it is... well, what it is...'

'I really did try, April, but I realised that I could never completely forgive Paul or trust him again. So, it's better that we part. Of course, for Liam's sake I'm determined that we should remain on good terms. I want to ensure he still has his dad in his life.'

'It's such a courageous decision, and you certainly seem to have surfaced from the wreckage miraculously unscathed. All credit to you, Ellen, you have transformed not only your fabulous pink farmhouse, but also your very self. The warm-hearted, but timorous, young woman I met just a few years ago has emerged as an awesome beauty. You have unfolded your wings and look set to fly.' April was pleased with her poetic analogy. 'And I'm determined to hang on and take the flight with you.'

'Thank you.' Ellen flushed with a sense of pride in her undisputed resilience. 'I am feeling good about myself. I have a confidence that I haven't known in a long time, if ever... and with you two beside me, goodness, I think I might be verging on invincible.'

'We certainly are an indomitable trio,' said Laura.

Seated on their favourite sofa by the window at Amarellos, Laura and Ellen realised that April's attention was waning; she appeared to be staring blankly into the distance through the glass.

'What's up, April?' Ellen asked.

'Something has just occurred to me,' April said after a further moment of quiet thought. 'How much spare land is there beyond the immediate garden at Walnut Tree Farmhouse?'

'Some… I'm not exactly sure.' Ellen was intrigued. 'What's on your mind now?'

'Yes, April,' Laura quizzed, 'what scheme are you conjuring up this time?'

'Well, it's all this talk of eggs, bacon and sausages. Apart from making me feel hungry, which is really bad news on a calorie-counting day, I was just visualising the setting – a larger henhouse and spacious run for our flock of free-range hens. Won't that be hugely appealing to guests – fresh eggs, straight from the backyard to their breakfast plate?'

'Undoubtedly, and an excellent marketing opportunity.'

'Actually, I wonder… is there a spare acre of land at the farmhouse?' April was now visibly excited and clearly concocting one of her great ideas.

'Where are you going with this, April?' Ellen and Laura said in unison.

'We could ask Derek to put up some fencing and perhaps Ian could construct an ark… after he's finished the henhouse, of course.'

Laura and Ellen were bemused.

'And then we could acquire a few Saddleback pigs, too!' April squealed in delight. 'Have I never mentioned my yearning to be a pig farmer?'

For exclusive discounts on Matador titles,
sign up to our occasional newsletter at
troubador.co.uk/bookshop